Vladimir K. Dobrev
Invariant Differential Operators

De Gruyter Studies in Mathematical Physics

Volume 49

Vladimir K. Dobrev

Invariant Differential Operators

Volume 3: Supersymmetry

DE GRUYTER

Mathematics Subject Classification 2010
17A70, 17BXX, 17B20, 17B35, 17B62, 17B60, 17B81, 20CXX, 20G42, 33D80, 58B32, 81R50, 81Q60, 81T30, 81T60, 83EXX, 16S30, 22E47, 81R20, 47A15, 47A46, 53A55, 70H33, 32C11, 46S60, 58A50, 59C50

Author
Prof. Vladimir K. Dobrev
Bulgarian Academy of Sciences
Institute for Nuclear Research and Nuclear Energy
Tsarigradsko Chaussee 72
1784 Sofia
Bulgaria
dobrev@inrne.bas.bg;
http://theo.inrne.bas.bg/~dobrev/

ISBN 978-3-11-052663-9
e-ISBN (PDF) 978-3-11-052749-0
e-ISBN (EPUB) 978-3-11-052669-1
ISSN 2194-3532

Library of Congress Control Number: 2018952171

Bibliographic information published by the Deutsche Nationalbibliothek
The Deutsche Nationalbibliothek lists this publication in the Deutsche Nationalbibliografie; detailed bibliographic data are available on the Internet at http://dnb.dnb.de.

Typesetting: VTeX UAB, Lithuania
Printing and binding: CPI books GmbH, Leck

www.degruyter.com

Preface

This is Volume 3 of our monograph series on invariant differential operators. In Volume 1 we presented our canonical procedure for the construction of invariant differential operators and showed its application to the objects of the initial domain—noncompact semisimple Lie algebras and groups. In Volume 2 we gave detailed exposition with many concrete examples of the application of our procedure to quantum groups.

Chapter 1 of Volume 3 has an introductory character. It contains standard material on Lie superalgebras. After exposing the generalities of Lie superalgebras we present the classification of finite-dimensional Lie superalgebras, mostly by using the root systems. Most attention is given to the basic classical Lie superalgebras as they are mostly used. The affine case is also briefly introduced. Then we give the fundamentals of the representations of simple Lie superalgebras. The classification of the real forms of the basic classical Lie superalgebras is also presented.

Chapter 2 treats in detail the conformal supersymmetry in the 4D case (the superalgebra su(2, 2/N)). This case has wide applications in string theory and conformal field theory. We stress the algebraic treatment using Verma modules and their singular vectors. We recall from Volumes 1 and 2 that the duality between singular vectors and invariant differential operators is the corner stone of our approach. Here most emphasis is put on the singular vectors and invariant submodules in the case of positive energy UIRs at the unitary reduction points since these are very important in the applications. We also study in detail the character formulas of the positive energy UIRs with many explicit special examples, e. g., the BPS states. Chapter 3 takes up two examples of conformal supersymmetry for $D > 4$, which are treated in great detail. One example is the positive energy UIRs for $D = 6$ using the superalgebra so(8^*/2N). The other example is the superalgebras osp(1/2n, $\mathbb{R}$), which are suitable for conformal supersymmetry for $D = 9, 10, 11$ for $n = 16, 16, 32$, respectively. For both examples we present the classification of the positive energy UIRs. Character formulas are also discussed.

Chapter 4 is in a field which is the intersection of two major developments in physics starting in the seventies and in the eighties: supersymmetry and quantum groups, respectively. We present the general definition of quantum superalgebras and some examples. The case of multiparameter deformation of the supergroup GL(m/N) and its dual quantum superalgebra $U_{u\mathbf{q}}$(gl(m/n)) is treated in great detail. We present also the induced representations of the quantum superalgebras $U_{u\mathbf{q}}$(gl(m/n)) and $U_{u\mathbf{q}}$(sl(m/n)).

Each chapter has a summary, which explains briefly the contents and the most relevant literature. Besides this, we present a Bibliography, an author index, and a subject index.

Note that this volume is only half of what was announced, since later it turned out that the intended material would need many more pages than customary. Thus,

https://doi.org/10.1515/9783110527490-201

the material was split into Volume 3 (the present one) and Volume 4. The fourth volume will cover applications of our approach to the AdS/CFT correspondence, to infinite-dimensional (super-) algebras, including (super-) Virasoro algebras, and to (q-) Schrödinger algebras.

Sofia, April 2018 *Vladimir Dobrev*

Contents

1 Lie superalgebras

Summary

Supersymmetry was discovered independently by three groups of researchers: Goldman and Lichtman [205], Volkov and Akulov [380], Wess and Zumino [382]. Other important early contributions were made in [178, 340, 218, 100, 192, 66, 163, 383, 315, 384, 306, 333, 109, 170, 248, 343, 237, 359, 77, 164, 93, 284, 396, 45, 47].

This introductory chapter contains standard material on Lie superalgebras based mainly on the papers of Kac [241, 243] (see also Cornwell [86]).

1.1 Generalities on Lie superalgebras

Let F be a field of characteristic 0. A *superspace* is a $\mathbb{Z}_2$-graded linear space M over F, that is, M is decomposed in the direct sum of two linear spaces over F: $M = M_{\bar{0}} \oplus M_{\bar{1}}$. The elements of M_i, $i \in \mathbb{Z}_2$, $i = 0, 1 \mod 2$ are called a *homogeneous elements*; $m \in M_{\bar{0}}$ are called *even elements* and those from $M_{\bar{1}}$ are called a *odd elements*. For a homogeneous element $m \in M$ we define the *parity* of m, denoted $p(m)$; $p(m)$ is 0 if m is even and 1 if m is odd. If we write $p(m)$ for $m \in M$ without explanation this means that we assume that m is homogeneous.

M is called finite dimensional if $M_{\bar{0}}$, $M_{\bar{1}}$ are finite dimensional, and in that case the *superdimension* of M is the pair $n_{\bar{0}} \mid n_{\bar{1}}$, where $n_{\bar{k}} = \dim M_{\bar{k}}$.

A *subsuperspace* is a $\mathbb{Z}_2$-graded subspace $N \subset M$ so that $N_{\bar{k}} \subset M_{\bar{k}}$.

Let M, N be two superspaces. The direct sum $M \oplus N$ is a superspace with $(M \oplus N)_{\bar{0}} = M_{\bar{0}} \oplus N_{\bar{0}}$, $(M \oplus N)_{\bar{1}} = M_{\bar{1}} \oplus N_{\bar{1}}$. The tensor product $M \otimes N$ is a superspace with $(M \otimes N)_{\bar{0}} = M_{\bar{0}} \otimes N_{\bar{0}} + M_{\bar{1}} \otimes N_{\bar{1}}$, $(M \otimes N)_{\bar{1}} = M_{\bar{0}} \otimes N_{\bar{1}} + M_{\bar{1}} \otimes N_{\bar{0}}$. Analogously the *homomorphisms* from M to N, i. e., $\mathrm{Hom}(M, N)$, form a superspace with $\mathrm{Hom}(M, N)_{\bar{0}} = \mathrm{Hom}(M_{\bar{0}}, N_{\bar{0}}) \oplus \mathrm{Hom}(M_{\bar{1}}, N_{\bar{1}})$; $\mathrm{Hom}(M, N)_{\bar{1}} = \mathrm{Hom}(M_{\bar{0}}, N_{\bar{1}}) \oplus \mathrm{Hom}(M_{\bar{1}}, N_{\bar{0}})$.

A *superalgebra* is a superspace $\mathcal{A}$ which is also an algebra. If $X \in \mathcal{A}_i$, $Y \in \mathcal{A}_j$, then $X \cdot Y \in \mathcal{A}_{i+j}$. An *ideal* $\mathcal{I}$ in a superalgebra $\mathcal{A}$ is an ideal of the algebra $\mathcal{A}$, which is also a subsuperspace of $\mathcal{A}$. A *subsuperalgebra* of $\mathcal{A}$ is a subalgebra $\mathcal{B}$ of $\mathcal{A}$ which is also a subsuperspace of $\mathcal{A}$. Clearly $\mathcal{A}_{\bar{0}}$ is a subsuperalgebra of $\mathcal{A}$.

An endomorphism D of a superalgebra $\mathcal{A}$ is called a *derivation* of degree s, $s \in \mathbb{Z}_2$, if

$$D(X \cdot Y) = (DX) \cdot Y + (-1)^{sp}\, X \cdot (DY),$$

where $X, Y \in \mathcal{A}$, $p = p(X)$. Note that if D, D' are derivations of degrees s, s', respectively, then $D \circ D' - (-1)^{ss'} D' \circ D$ is also a derivation.

An agreement. All definitions which carry over verbatim from the nonsupersymmetric case will be used without further notice. All formulas given for homogeneous elements are extended by linearity for arbitrary elements. In generalizing formulas from

https://doi.org/10.1515/9783110527490-001

the nonsupersymmetric case we use the signs rule: if something of parity $p \in \mathbb{Z}_2$ passes through something of parity $q \in \mathbb{Z}_2$, then the sign $(-1)^{pq}$ appears.

The tensor product of two superalgebras $\mathcal{A}$, $\mathcal{B}$ is a superalgebra with product $(X \otimes Y) \cdot (X' \otimes Y') = (-1)^{pp'} X \cdot X' \otimes Y \cdot Y'$, where $X, X' \in \mathcal{A}$, $Y, Y' \in \mathcal{B}$, $p = p(Y)$, $p' = p(X')$.

Let $\mathcal{A}$ be superalgebra with unity, M be a superspace. The left action of $\mathcal{A}$ on M is a morphism of superspaces $\mathcal{A} \otimes M \to M$ so that $X \otimes (Y \otimes m) = (X \cdot Y) \otimes m$, $1_{\mathcal{A}} \otimes m = m$, where $X, Y, 1_{\mathcal{A}} \in \mathcal{A}$, $m \in M$; then M is called a left $\mathcal{A}$-module.

Schur's lemma. Let M be a superspace, let $\mathcal{O}$ be an irreducible family of linear operators on M, $\mathcal{O} \subset \mathrm{gl}(M)$. The centralizer $C(\mathcal{O})$ of $\mathcal{O}$ in $\mathrm{gl}(M)$, i. e., $C(\mathcal{O}) = \{B \in \mathrm{gl}(M) : [B, D] = 0, \forall D \in \mathcal{O}\}$, is given by either $C(\mathcal{O}) = \{\mathrm{id}_M\}$ or $\dim M_{\bar{0}} = \dim M_{\bar{1}}$ and $C(\mathcal{O}) = \{\mathrm{id}_M, B\}$, where B is a nondegenerate operator permuting $M_{\bar{0}}$ and $M_{\bar{1}}$, and $B^2 = \mathrm{id}_M$.

A *Lie superalgebra* is a superalgebra $\mathcal{G}$ in which the product of $X, Y \in \mathcal{G}$, which we denote by $[X, Y]$, satisfies

$$[X, Y] = -(-1)^{pp'}[Y, X], \tag{1.1a}$$

$$[X, [Y, Z]] - (-1)^{pp'}[Y, [X, Z]] - [[X, Y], Z] = 0, \tag{1.1b}$$

where $X, Y, Z \in \mathcal{G}$, $p = p(X)$, $p' = p(Y)$. Note that (1.1) represents an application of the signs rule. (Indeed, in the nonsupersymmetric case p=p'=0 and we obtain the defining relations of a Lie algebra; note only that the Jacobi identity is usually written as $[X, [Y, Z]] + [Y, [Z, X]] + [Z, [X, Y]] = 0$.) Analogously it is clear that $\mathcal{G}_{\bar{0}}$ is a Lie algebra.

Let $\mathcal{G}$ be a Lie superalgebra. The elements $X, Y \in \mathcal{G}$ are called commuting if $[X, Y] = 0$. Obviously every even element of $\mathcal{G}$ commutes with itself (it follows either from (1.1a) or recalling that $\mathcal{G}_{\bar{0}}$ is a Lie algebra). In an abelian Lie superalgebra all odd elements are nilpotent. (Indeed, if $X \in \mathcal{G}_{\bar{1}}$ then $[X, X] = 2X^2$ and if $\mathcal{G}$ is abelian then $X^2 = 0$ ($\mathrm{char}\, F \neq 2$).)

Every associative superalgebra $\mathcal{A}$ is also a Lie superalgebra with respect to the bracket $[X, Y] \equiv X \cdot Y - (-1)^{pp'} Y \cdot X$, where $X \cdot Y$ is the product in $\mathcal{A}$. Thus $[X, Y]$ is the anticommutator if X, Y are both odd and it is the commutator otherwise.

Let $\mathcal{G}$ be a finite-dimensional Lie superalgebra. Then $\mathcal{G}$ contains a unique maximal solvable ideal $\mathcal{R}$ (the solvable radical). The Lie superalgebra $\mathcal{G}/\mathcal{R}$ is semisimple (i. e., it has no solvable ideals). Note that Levi's theorem on $\mathcal{G}$ being a semidirect sum of $\mathcal{R}$ and $\mathcal{G}/\mathcal{R}$ is not true, in general, for Lie superalgebras.

The universal enveloping superalgebra of $\mathcal{G}$ is constructed as follows. Let $T(\mathcal{G})$ be the tensor superalgebra over $\mathcal{G}$ with the induced $\mathbb{Z}_2$ grading, and $\mathcal{R}$ be the ideal of $T(\mathcal{G})$ generated by elements of the form

$$[X, Y] - X \otimes Y + (-1)^{p(X)p(Y)} Y \otimes X.$$

We set $U(\mathcal{G}) = T(\mathcal{G})/\mathcal{R}$. The following holds:

Poincaré–Birkhoff–Witt Theorem. *Let $\mathcal{G} = \mathcal{G}_{\bar{0}} \oplus \mathcal{G}_{\bar{1}}$ be a Lie superalgebra, let $a_1, \dots, a_m$ be a basis of $\mathcal{G}_{\bar{0}}$, and $b_{\bar{1}}, \dots, b_n$ be a basis of $\mathcal{G}_{\bar{1}}$. Then the elements of the form*

$$a_1^{k_1} \dots a_m^{k_m} b_1^{\epsilon_1} \dots b_n^{\epsilon_n}, \quad k_i \geq 0, \quad \epsilon_i = 0, 1,$$

form a basis of $U(\mathcal{G})$. ◇

A *supermatrix* is a matrix where every row (and column) is either even or odd. An even element of a supermatrix is at the intersection of an even (respectively odd) row with an even (respectively odd) column; an odd element of a supermatrix is at the intersection of an even (respectively odd) row with an odd (respectively even) column. A *standard supermatrix*

$$\begin{pmatrix} R & S \\ T & U \end{pmatrix} \tag{1.2}$$

is such that the elements of R, U are even and the elements of S, T are odd.

We shall denote by $\mathrm{Mat}(m/n; \mathcal{A})$ the superspace of supermatrices with m even rows and n odd rows with elements from the superalgebra $\mathcal{A}$. Obviously $\mathrm{Mat}(m/n; \mathcal{A})$ is an associative superalgebra with the ordinary product of matrices.

If $\mathcal{A}$ is an abelian superalgebra then we can introduce in $\mathrm{Mat}(p/q; \mathcal{A})$ the *supertrace* of a standard supermatrix $X = \left(\begin{smallmatrix} R & S \\ T & U \end{smallmatrix}\right)$; it is denoted $\mathrm{str}\, X$ and is defined by $\mathrm{str}\, X \doteq \mathrm{tr}\, R - \mathrm{tr}\, U$; $\mathrm{str}\, X \in \mathcal{A}$.

1.2 Classification of finite-dimensional Lie superalgebras

Examples. The Lie superalgebra

$$\mathrm{gl}(m/n; F) = \mathrm{Mat}(m/n; F) \tag{1.3}$$

is called the *general linear superalgebra*. Note that $\mathrm{gl}(m/n; F)_{\bar{0}} \cong \mathrm{gl}(m, F) \oplus \mathrm{gl}(n, F)$. It is reductive (for $m \neq n$) with one-dimensional center spanned by the unit $(m+n) \times (m+n)$ matrix $I_{m+n} \in \mathrm{gl}(m/n; F)_{\bar{0}}$. Note that $\mathrm{gl}(1/1)$ is four dimensional and solvable.

An important subsuperalgebra of $\mathrm{gl}(m/n; F)$ is

$$\mathrm{sl}(m/n; F) \doteq \{X \in \mathrm{gl}(m/n; F) : \mathrm{str}\, X = 0\}, \tag{1.4}$$

called the *special linear superalgebra*. Note that $\mathrm{sl}(m/n; F)_{\bar{0}} \cong \mathrm{sl}(m, F) \oplus \mathrm{sl}(n, F) \oplus F$ and

$$\begin{aligned} \dim_F \mathrm{sl}(m/n; F)_{\bar{0}} &= m^2 + n^2 - 1 \\ \dim_F \mathrm{sl}(m/n; F)_{\bar{1}} &= 2mn. \end{aligned} \tag{1.5}$$

If $m \neq n$ it is semisimple and is an ideal of $\mathrm{gl}(m/n; F) = FI_{m+n} \oplus \mathrm{sl}(m/n; F)$. If $m = n$ the superalgebra $\mathrm{sl}(n/n; F)$ contains itself the central generator: $I_{2n} \in \mathrm{sl}(n/n; F)$. However, $\dim_F \mathrm{gl}(m/n; F) = \dim_F \mathrm{sl}(m/n; F) + 1$ in all cases. For $m = n$ the extra generator of gl$(n/n;F)$ which does not belong to sl$(n/n;F)$ may be taken to be $\mathcal{K} \equiv \text{diag}(1,\dots,1,-1,\dots,-1)$, with equal number of +1 and −1; thus: str $\mathcal{K} = 2n$.

Note that sl(1/1) is three dimensional and nilpotent. Indeed, take $\mathcal{G} = \text{gl}(1/1)$. Then $\mathcal{G}^{(1)} = [\mathcal{G},\mathcal{G}] = \text{sl}(1/1)$, $\mathcal{G}^{(2)} = [\mathcal{G}^{(1)},\mathcal{G}^{(1)}] = \text{l.s.}\ \{I_2\}$, $\mathcal{G}^{(3)} = [\mathcal{G}^{(2)},\mathcal{G}^{(2)}] = 0$, showing that gl(1/1) is solvable. Take $\mathcal{G} = \text{sl}(1/1) = \mathcal{G}^1$. Then $\mathcal{G}^2 = [\mathcal{G}^1,\mathcal{G}] = \text{l.s.}\ \{I_2\}$, $\mathcal{G}^3 = [\mathcal{G}^2,\mathcal{G}] = 0$, showing that sl(1/1) is nilpotent.

For the rest of the superalgebras sl(m/n) sometimes we shall use the notation of Kac:

$$A(m,n) \equiv \begin{cases} \text{sl}(m+1/n+1;\mathbb{C}), & \text{if } m \neq n,\ m,n \geq 0,\ m+n>0, \\ \text{sl}(n+1/n+1;\mathbb{C})/\mathcal{C}, & \text{if } m=n>0, \end{cases}$$

$$\mathcal{C} \doteq c.l.s.\{I_{2n+2}\}$$

(c.l.s. means complex linear span).

Let $\tilde{b}$ be a nondegenerate bilinear form on the superspace V, dim $V_{\bar{0}} = m$, dim $V_{\bar{1}} = 2n$, such that $V_{\bar{0}}$ and $V_{\bar{1}}$ are orthogonal, $\tilde{b}|_{V_{\bar{0}}}$ is symmetric, and $\tilde{b}|_{V_{\bar{1}}}$ is skew-symmetric. Explicitly $\tilde{b}$ may be given by the matrix B of order $m+2n$:

$$B = \begin{pmatrix} iI_m & 0 & 0 \\ 0 & 0 & I_n \\ 0 & -I_n & 0 \end{pmatrix}.$$

We define in gl$(m/2n;\mathbb{C})$ a subalgebra osp$(m/2n)$ = osp$(m/2n)_{\bar{0}}$+osp$(m/2n)_{\bar{1}}$ by setting

$$\text{osp}(m/2n)_s = \{X \in \text{gl}(m/2n;\mathbb{C})_s : X\,B + i^s\,B\,{}^tX = 0\} \tag{1.6}$$

Explicitly, for $X \in \mathcal{G}_{\bar{0}}$, $s = 0$, we have

$$X = \begin{pmatrix} \alpha & 0 & 0 \\ 0 & \beta & \gamma \\ 0 & \delta & -{}^t\beta \end{pmatrix} \tag{1.7}$$

$$\alpha = -{}^t\alpha, \quad \gamma = {}^t\gamma, \quad \delta = {}^t\delta.$$

Thus osp$(m/2n)_{\bar{0}} \cong \text{so}(m,\mathbb{C}) \oplus \text{sp}(n,\mathbb{C})$ and one has dim osp$(m/2n)_{\bar{0}} = m(m-1)/2 + n(2n+1)$. Because of the above we call osp$(m/2n)$ the *orthosymplectic superalgebra*. For $m = 0$, $n = 0$, respectively, it turns, respectively, into the symplectic, orthogonal, Lie algebra: osp$(0/2n) \cong \text{sp}(n,\mathbb{C})$, osp$(m/0) \cong \text{so}(m,\mathbb{C})$. We note that

$$\text{osp}(2m+1/2n)_{\bar{0}} \cong \text{so}(2m+1,\mathbb{C}) \oplus \text{sp}(n,\mathbb{C}) \cong B_m \oplus C_n, \quad m,n \geq 0,\ m+n>0,$$

$$\text{osp}(2m/2n)_{\bar{0}} \cong \text{so}(2m,\mathbb{C}) \oplus \text{sp}(n,\mathbb{C}) \cong D_m \oplus C_n, \quad m,n \geq 0,\ m+n>0.$$

For $X \in \mathcal{G}_{\bar{1}}$, $s = 1$, we have

$$X = \begin{pmatrix} 0 & \xi & \eta \\ -{}^t\eta & 0 & 0 \\ {}^t\xi & 0 & 0 \end{pmatrix}. \tag{1.8}$$

Thus dim osp$(m/2n)_{\bar{1}} = 2mn$.

Following Kac we introduce the notation

$$B(m,n) \equiv \mathrm{osp}(2m+1/2n), \quad \text{for } m \geq 0,\ n \geq 1,$$
$$D(m,n) \equiv \mathrm{osp}(2m/2n), \quad \text{for } m \geq 2,\ n \geq 1,$$
$$C(n) \equiv \mathrm{osp}(2/2n-2), \quad \text{for } n \geq 2$$

Obviously we have

$$B(m,n)_{\bar{0}} \cong B_m \oplus C_n,$$
$$D(m,n)_{\bar{0}} \cong D_m \oplus C_n,$$
$$C(n)_{\bar{0}} \cong \mathbb{C} \oplus C_{n-1}$$

Note that $C(2) \cong A(1,0) \cong A(0,1)$ but in the considerations below, in particular, for the q-deformations, it makes sense to consider also $C(2)$.

Analogously to the nonsupersymmetric case a bilinear form on a Lie superalgebra $\mathcal{G}$ is called invariant if

$$B(X,Y) = (-1)^{pp'} B(Y,X), \tag{1.9a}$$

(supersymmetry),

$$B(X,Y) = 0, \quad \text{if } X \in \mathcal{G}_{\bar{0}},\ Y \in \mathcal{G}_{\bar{1}}, \tag{1.9b}$$

(consistency),

$$B([X,Y],Z) = B(X,[Y,Z]), \tag{1.9c}$$

(invariance). In the nonsupersymmetric case (1.9a) is called the symmetry property, (1.9b) is trivial (there is no $\mathcal{G}_{\bar{1}}$), and (1.9c) is the same. The Killing form

$$K(X,Y) \equiv \mathrm{str}(ad\,X\ ad\,Y) \tag{1.10}$$

is invariant as before; however, in some cases, e. g., $A(n,n)$, $D(n+1,n)$, it is zero; cf. below.

The list of complex finite-dimensional simple Lie superalgebras consists of two essentially different parts—*classical superalgebras* and *Cartan superalgebras*. For the latter we refer to [241] (cf. also the recent paper [73]).

Classical Lie superalgebras. A Lie superalgebra $\mathcal{G} = \mathcal{G}_{\bar{0}} \oplus \mathcal{G}_{\bar{1}}$ is called a *classical superalgebra* if it is simple and the representation of $\mathcal{G}_{\bar{0}}$ on $\mathcal{G}_{\bar{1}}$ is completely reducible. These algebras are divided into two classes, called 'basic' and 'strange'.

Basic classical Lie superalgebras. A classical superalgebra $\mathcal{G}$ is called a *basic classical superalgebra* if there exists a non-degenerate invariant bilinear $B(\cdot,\cdot)$ form on $\mathcal{G}$.

Kac has proved [241] that the complete list of basic classical (or contragredient) Lie superalgebras is as follows:

1. the simple Lie algebras;

2. $A(m,n)$, $B(m,n)$, $C(n)$, $D(m,n)$, $D(2,1;\bar{\sigma})$, ($\bar{\sigma} = \{\sigma_1, \sigma_2, \sigma_3\}$, $\sum_{i=1}^{3} \sigma_i = 0$), $F(4)$, $G(3)$, where $D(2,1;\bar{\sigma})$, $F(4)$, $G(3)$ are the unique 17-, 40-, 31-dimensional superalgebras, respectively, such that $D(2,1;\bar{\sigma})_{\bar{0}} \cong A_1 \oplus A_1 \oplus A_1$, $F(4)_{\bar{0}} \cong B_3 \oplus A_1$, $G(3)_{\bar{0}} = G_2 \oplus A_1$. The non-degenerate form referred to above is unique (up to a constant factor), and it may be taken to be the Killing form, except in the cases $A(n,n)$, $D(n+1,n)$, $D(2,1;\bar{\sigma})$, for which the Killing form is zero. Actually, in [241] the one-parameter algebra $D(2,1;\lambda)$ ($\lambda \in \mathbb{C}\backslash\{0,-1\}$) is discussed, such that $\sigma_1 = -(1+\lambda)/2$, $\sigma_2 = 1/2$, $\sigma_3 = \lambda/2$.

Strange classical Lie superalgebras. A classical superalgebra is called a *strange classical superalgebra* if there does not exist a non-degenerate invariant bilinear form on $\mathcal{G}$. There are two series of such algebras, $P(n)$ and $Q(n)$ for $n \geq 2$:

$$\begin{aligned}
P(n) &\doteq \left\{X = \begin{pmatrix} \alpha & \beta \\ \gamma & -{}^t\alpha \end{pmatrix} : \operatorname{tr}\alpha = 0,\ \beta = {}^t\beta,\ \gamma = -{}^t\gamma\right\} \subset \mathrm{sl}(n+1, n+1;\mathbb{C}), \\
Q(n) &\doteq \tilde{Q}(n)/\mathcal{C}, \qquad\qquad (1.11) \\
\tilde{Q}(n) &\doteq \left\{X = \begin{pmatrix} \alpha & \beta \\ \beta & \alpha \end{pmatrix} : \operatorname{tr}\beta = 0\right\} \subset \mathrm{sl}(n+1, n+1;\mathbb{C}), \\
&\quad \mathcal{C} = c.l.s.\{I_{2n+2}\}, \\
&\quad \dim P(n) = 2(n+1)^2 - 1, \\
&\quad \dim Q(n) = 2(n+1)^2 - 2.
\end{aligned}$$

Further information as regards this class the reader may find in, e.g., [188, 320, 98, 69, 289] and the references therein.

The classical Lie superalgebras are divided into two types. A classical simple Lie superalgebra $\mathcal{G}$ is said to be of *type I* if the representation of $\mathcal{G}_{\bar{0}}$ on $\mathcal{G}_{\bar{1}}$ is equivalent to the sum of *two* irreducible representations of $\mathcal{G}_{\bar{0}}$, and it is said to be of *type II* if the representation of $\mathcal{G}_{\bar{0}}$ on $\mathcal{G}_{\bar{1}}$ is irreducible.

The superalgebras of type I are $A(m,n)$, $C(n)$, $P(n)$. They admit a $\mathbb{Z}$-grading of the form $\mathcal{G} = \mathcal{G}_{-1} \oplus \mathcal{G}_{\bar{0}} \oplus \mathcal{G}_1$. The $\mathcal{G}_{\bar{0}}$-modules $\mathcal{G}_{\pm}$ are irreducible, and in the cases of $A(m,n)$, $C(n)$ they are contragredient, i.e., conjugate to each other. In more detail: the $\mathcal{G}_{\bar{0}}$-module $\mathcal{G}_{-1}$ is isomorphic to $\mathrm{sl}_{m+1} \otimes \mathrm{sl}_{n+1} \otimes \mathbb{C}$, csp_{2n-2}, $\Lambda^2\, \mathrm{sl}^*_{n+1}$, for $\mathrm{sl}(m+1/n+1)$, $C(n)$, $P(n)$, respectively, while the $\mathcal{G}_{\bar{0}}$-module $\mathcal{G}_1$ is isomorphic to $\mathrm{sl}^*_{m+1} \otimes \mathrm{sl}^*_{n+1} \otimes \mathbb{C}$, csp^*_{2n-2}, $S^2\, \mathrm{sl}^*_{n+1}$, for $\mathrm{sl}(m+1/n+1)$, $C(n)$, $P(n)$, where sl_n and sp_n, respectively, stand for the standard (i.e., matrix) representations of $\mathrm{sl}(n)$ and $\mathrm{sp}(n)$, respectively, csp_n is sp_n plus one-dimensional center, and $*$ denotes the conjugate module.

The superalgebras of type II are $B(m,n)$, $D(m,n)$, $D(2,1;\bar{\sigma})$, $F(4)$, $G(3)$, $Q(n)$. The $\mathcal{G}_{\bar{0}}$-module $\mathcal{G}_{\bar{1}}$ is isomorphic to $\mathrm{so}_{2n+1} \otimes \mathrm{sp}_n$, $\mathrm{so}_{2m} \otimes \mathrm{sp}_n$, $\mathrm{sl}_2 \otimes \mathrm{sl}_2 \otimes \mathrm{sl}_2$, $\mathrm{spin}_7 \otimes \mathrm{sl}_2$, $G_2 \otimes \mathrm{sl}_2$, $\mathrm{ad}\ \mathrm{sl}_{n+1}$, respectively; here spin_7 denotes the spin representation of B_3; so_n and G_2, respectively, stand for the standard representation of $\mathrm{so}(n)$, G_2. Finally, $\mathrm{ad}\ \mathrm{sl}_n$ stands for the adjoint representation of sl_n.

1.3 Root systems

1.3.1 Classical Lie superalgebras

Let $\mathcal{G} = \mathcal{G}_{\bar{0}} \oplus \mathcal{G}_{\bar{1}}$ be a classical Lie superalgebra. We define a *Cartan subalgebra* $\mathcal{H}$ of $\mathcal{G}$ to be a Cartan subalgebra of $\mathcal{G}_{\bar{0}}$. We have the root decomposition as in the even case $\mathcal{G} = \oplus_{\alpha\in\mathcal{H}^*}\mathcal{G}_\alpha$, where $\mathcal{G}_\alpha = \{X \in \mathcal{G} \mid [H,X] = \alpha(H)X, \forall H \in \mathcal{H}\}$, and again the set $\Delta = \{\alpha \in \mathcal{H}^* \mid \alpha \neq 0, \mathcal{G}_\alpha \neq 0\}$ is the root system. Clearly, one has the decomposition $\Delta = \Delta_{\bar{0}} \cup \Delta_{\bar{1}}$, where $\Delta_{\bar{0}}$ is the root system of $\mathcal{G}_{\bar{0}}$, while $\Delta_{\bar{1}}$ is the weight system of the representation of $\mathcal{G}_{\bar{0}}$ in $\mathcal{G}_{\bar{1}}$. The system $\Delta_{\bar{0}}, \Delta_{\bar{1}}$, respectively, is called the *even root system, odd root system*, respectively. A root system $\Pi = \{\alpha_1, \ldots, \alpha_r\}$ is called a *simple root system* if there exist vectors $X_i^+ \in \mathcal{G}_{\alpha_i}, X_i^- \in \mathcal{G}_{-\alpha_i}$, such that $[X_i^+, X_j^-] = \delta_{ij}H_i \in \mathcal{H}$, the vectors X_i^+ and X_i^- generate $\mathcal{G}$, and Π is minimal with these properties. Unlike the even case, isomorphic superalgebras may have different root systems as we shall see below. The simple root system with the minimal number of odd roots is called a *distinguished root system*. As in the even case for each choice of simple roots there exists a Cartan matrix $A = (a_{ij})$, such that

$$[H_i, X_j^\pm] = \pm a_{ij}X_j^\pm. \tag{1.12}$$

For the basic classical Lie superalgebras let $(\cdot,\cdot)$ be the scalar product in $\mathcal{H}^*$ induced from the form $B(\cdot,\cdot)$ restricted to $\mathcal{H}\times\mathcal{H}$ as in the even case.

Considerations similar to the even case lead to the following.

Proposition ([241]). *Let $\mathcal{G}$ be a classical Lie superalgebra with the root decomposition $\mathcal{G} = \oplus_{\alpha\in\mathcal{H}^*}\mathcal{G}_\alpha$ w.r.t. the Cartan subalgebra $\mathcal{H}$. Then:*

a) *$\mathcal{G}_0 = \mathcal{H}$ except for $Q(n)$;*
b) *$\dim \mathcal{G}_\alpha = 1$ when $\alpha \neq 0$ except for $A(1,1)$, $P(2)$, $P(3)$, $Q(n)$;*
c) *if $\mathcal{G}$ is not one of $A(1,1)$, $P(n)$, $Q(n)$, then:*
   1) *$[\mathcal{G}_\alpha, \mathcal{G}_\beta] \neq 0$ iff $\alpha, \beta, \alpha+\beta \in \Delta$;*
   2) *$B(\mathcal{G}_\alpha, \mathcal{G}_\beta) = 0$ for $\alpha \neq -\beta$;*
   3) *$[\mathcal{G}_\alpha, \mathcal{G}_{-\alpha}] = B(\mathcal{G}_\alpha, \mathcal{G}_{-\alpha})H_\alpha$, where H_α is the nonzero vector defined by $B(H_\alpha, H) = \alpha(H), \forall H \in \mathcal{H}$;*
   4) *$B(\cdot,\cdot)$ defines a nondegenerate pairing of $\mathcal{G}_a$ and $\mathcal{G}_{-\alpha}$;*
   5) *$\Delta_{\bar{0}}$ and $\Delta_{\bar{1}}$ are invariant under the action of the Weyl group W of $\mathcal{G}_{\bar{0}}$;*
   6) *if $\alpha \in \Delta$ (respectively $\Delta_{\bar{0}}, \Delta_{\bar{1}}$) then $-\alpha \in \Delta$ (respectively $\Delta_{\bar{0}}, \Delta_{\bar{1}}$);*
   7) *if $\alpha \in \Delta$ then $k\alpha \in \Delta$ iff $k = \pm1$ except for $\alpha \in \Delta_{\bar{1}}$ with $(\alpha,\alpha) \neq 0$ when $k = \pm1, 2$.* ◇

The Weyl group W of the even subalgebra $\mathcal{G}_{\bar{0}}$ may be extended to a larger group by the following *odd reflections* [139, 349]. For $\alpha \in \Delta_{\bar{1}}$ we define

$$s_\alpha\beta = \beta - 2\frac{(\alpha,\beta)}{(\alpha,\alpha)}\alpha, \quad (\alpha,\alpha) \neq 0$$

$$s_\alpha\beta = \beta + \alpha, \quad (\alpha,\alpha) = 0, \quad (\alpha,\beta) \neq 0$$
$$s_\alpha\beta = \beta, \quad (\alpha,\alpha) = 0, \quad (\alpha,\beta) = 0, \quad \alpha \neq \beta$$
$$s_\alpha\alpha = -\alpha \tag{1.13}$$

The difference from the usual reflections is when *isotropic odd roots*, i. e., those with $(\alpha,\alpha) = 0$, are involved.

1.3.2 Basic classical Lie superalgebras

Next we list the root systems of the basic classical Lie superalgebras. For the superalgebras $\mathrm{sl}(m/n)$ and $\mathrm{osp}(m/n)$ the roots will be expressed in terms of the mutually orthogonal linear functionals ϵ_i, δ_j, such that $(\epsilon_i,\epsilon_j) = \delta_{ij}$, $(\delta_i,\delta_j) = -\delta_{ij}$, $(\epsilon_i,\delta_j) = 0$.

$A(m,n)$. The roots are expressed in terms of $\epsilon_1,\dots,\epsilon_{m+1}$, $\delta_1,\dots,\delta_{n+1}$. One has

$$\Delta_{\bar{0}} = \{\epsilon_i - \epsilon_j,\ \delta_i - \delta_j,\ i \neq j\}, \quad \Delta_{\bar{1}} = \{\pm(\epsilon_i - \delta_j)\}. \tag{1.14}$$

Up to W equivalence, all systems of simple roots are determined by two increasing sequences $S = \{s_1 < s_2 < \cdots\}$, $T = \{t_1 < t_2 < \cdots\}$, and a sign:

$$\Pi_{S,T} = \pm\{\epsilon_1 - \epsilon_2, \epsilon_2 - \epsilon_3, \dots, \epsilon_{s_1} - \delta_1, \delta_1 - \delta_2, \dots, \delta_{t_1} - \epsilon_{s_1+1}, \dots\}. \tag{1.15}$$

The distinguished simple root system is obtained for $S = \{m+1\}$, $T = \emptyset$:

$$\Pi = \{\epsilon_1 - \epsilon_2, \epsilon_2 - \epsilon_3, \dots, \epsilon_{m+1} - \delta_1, \delta_1 - \delta_2, \dots, \delta_n - \delta_{n+1}\}, \tag{1.16}$$

the corresponding distinguished positive root system is

$$\Delta_{\bar{0}}^+ = \{\epsilon_i - \epsilon_j, \delta_i - \delta_j, i < j\}, \quad \Delta_{\bar{1}}^+ = \{\epsilon_i - \delta_j\}, \tag{1.17}$$

and the highest distinguished root is the sum of all simple roots:

$$\tilde{\alpha} = \epsilon_1 - \delta_{n+1}.$$

The case $\mathrm{sl}(1,1)$ may be treated similarly (though formally) if we set in the above $m = n = 0$.

$B(m,n)$. The roots are expressed in terms of $\epsilon_1\dots,\epsilon_m$, $\delta_1\dots,\delta_n$. One has

$$\Delta_{\bar{0}} = \{\pm\epsilon_i \pm \epsilon_j, \pm\delta_i \pm \delta_j, i \neq j, \pm\epsilon_i, \pm 2\delta_i\}, \quad \Delta_{\bar{1}} = \{\pm\delta_i, \pm\epsilon_i \pm \delta_j\}. \tag{1.18}$$

Up to W equivalence, all systems of simple roots are determined by the two increasing sequences S and T:

$$\Pi_{S,T} = \{\epsilon_1 - \epsilon_2, \epsilon_2 - \epsilon_3, \dots, \epsilon_{s_1} - \delta_1, \delta_1 - \delta_2, \dots, \delta_{t_1} - \epsilon_{s_1+1}, \dots, \pm\delta_n \text{ (or } \pm\epsilon_m)\}. \tag{1.19}$$

The distinguished simple root system is

$$\Pi = \{\delta_1 - \delta_2, \delta_2 - \delta_3, \ldots, \delta_n - \epsilon_1, \epsilon_1 - \epsilon_2, \ldots, \epsilon_{m-1} - \epsilon_m, \epsilon_m\}, \quad m > 0; \tag{1.20a}$$

$$\Pi = \{\delta_1 - \delta_2, \delta_2 - \delta_3, \ldots, \delta_{n-1} - \delta_n, \delta_n\}, \quad m = 0, \tag{1.20b}$$

the corresponding positive root system is

$$\begin{aligned} &\Delta_{\bar{0}}^+ = \{\epsilon_i \pm \epsilon_j, \delta_i \pm \delta_j, i < j, \epsilon_i, 2\delta_i\}, \quad \Delta_{\bar{1}}^+ = \{\delta_i, \delta_i \pm \epsilon_j\} \quad m > 0; \\ &\Delta_{\bar{0}}^+ = \{\delta_i \pm \delta_j, i < j, 2\delta_i\}, \quad \Delta_{\bar{1}}^+ = \{\delta_i\}, \quad m = 0, \end{aligned} \tag{1.21}$$

the highest distinguished root is twice the sum of all simple roots:

$$\tilde{\alpha} = 2\delta_1.$$

$C(n)$. The roots are expressed in terms of ϵ, $\delta_1 \ldots, \delta_{n-1}$. One has

$$\Delta_{\bar{0}} = \{\pm 2\delta_i, \pm\delta_i \pm \delta_j, i \neq j\}, \quad \Delta_{\bar{1}} = \{\pm\epsilon \pm \delta_j\}. \tag{1.22}$$

Up to W equivalence, we have the following systems of simple roots:

$$\Pi_1^{\pm} = \pm\{\epsilon - \delta_1, \delta_1 - \delta_2, \ldots, \delta_{n-2} - \delta_{n-1}, 2\delta_{n-1}\}, \tag{1.23a}$$

$$\Pi_2^{i\pm} = \pm\{\delta_1 - \delta_2, \delta_2 - \delta_3, \ldots, \delta_i - \epsilon, \epsilon - \delta_{i+1}, \ldots, \delta_{n-2} - \delta_{n-1}, 2\delta_{n-1}\}, \tag{1.23b}$$

$$\Pi_3^{\pm} = \pm\{\delta_1 - \delta_2, \delta_2 - \delta_3, \ldots, \delta_{n-2} - \delta_{n-1}, \delta_{n-1} - \epsilon, \delta_{n-1} + \epsilon\}. \tag{1.23c}$$

The distinguished simple root system is Π_1^+, the corresponding distinguished positive root system is

$$\Delta_{\bar{0}}^+ = \{2\delta_i, \delta_i \pm \delta_j, i < j\}, \quad \Delta_{\bar{1}}^+ = \{\epsilon \pm \delta_j\}, \tag{1.24}$$

and the highest distinguished root is

$$\tilde{\alpha} = \epsilon + \delta_1.$$

$D(m, n)$. The roots are expressed in terms of $\epsilon_1 \ldots, \epsilon_m$, $\delta_1 \ldots, \delta_n$. One has

$$\Delta_{\bar{0}} = \{\pm\epsilon_i \pm \epsilon_j, \pm\delta_i \pm \delta_j, i \neq j, \pm 2\delta_i\}, \quad \Delta_{\bar{1}} = \{\pm\epsilon_i \pm \delta_j\}. \tag{1.25}$$

Up to W equivalence, all systems of simple roots are determined by the two increasing sequences S and$T$, and by a number:

$$\begin{aligned} \Pi_{S,T}^1 &= \{\epsilon_1 - \epsilon_2, \ldots, \epsilon_{s_1} - \delta_1, \delta_1 - \delta_2, \ldots, \delta_{t_1} - \epsilon_{s_1+1}, \ldots, \\ &\qquad \epsilon_{m-1} - \epsilon_m, \epsilon_{m-1} + \epsilon_m, \ (\text{or } \delta_n - \epsilon_m, \delta_n + \epsilon_m)\}, \\ \Pi_{S,T}^2 &= \{\epsilon_1 - \epsilon_2, \ldots, \epsilon_{s_1} - \delta_1, \delta_1 - \delta_2, \ldots, \delta_{t_1} - \epsilon_{s_1+1}, \ldots, \\ &\qquad \delta_{n-1} - \delta_n, 2\delta_n\}. \end{aligned} \tag{1.26}$$

There are two distinguished simple root systems:

$$\Pi^1 = \{\delta_1 - \delta_2, \delta_2 - \delta_3, \dots, \delta_n - \epsilon_1, \epsilon_1 - \epsilon_2, \dots, \epsilon_{m-1} - \epsilon_m, \epsilon_{m-1} + \epsilon_m\}, \tag{1.27a}$$

$$\Pi^2 = \{\epsilon_1 - \epsilon_2, \dots, \epsilon_m - \delta_1, \delta_1 - \delta_2, \dots, \delta_{n-1} - \delta_n, 2\delta_n\}, \tag{1.27b}$$

the distinguished positive root system corresponding to Π^1 is

$$\Delta^+_{\bar{0}} = \{\epsilon_i \pm \epsilon_j, \delta_i \pm \delta_j, i < j, 2\delta_i\}, \quad \Delta^+_{\bar{1}} = \{\delta_i \pm \epsilon_j\}, \tag{1.28}$$

and the highest distinguished root is

$$\tilde{\alpha} = 2\delta_1.$$

$D(2,1;\bar{\sigma})$, $\bar{\sigma} = \{\sigma_1, \sigma_2, \sigma_3\}$, $\sum_{i=1}^{3} \sigma_i = 0$. The roots are expressed in terms of mutually orthogonal functionals ϵ_1, ϵ_2, ϵ_3, such that $(\epsilon_i, \epsilon_j) = \delta_{ij}\sigma_i$. One has

$$\Delta_{\bar{0}} = \{\pm 2\epsilon_i\}, \quad \Delta_{\bar{1}} = \{\pm\epsilon_1 \pm \epsilon_2 \pm \epsilon_3\}. \tag{1.29}$$

Up to W equivalence, there are four systems of simple roots:

$$\Pi_{ij} = \{-2\epsilon_i, \epsilon_1 + \epsilon_2 + \epsilon_3, -2\epsilon_j\}, \quad 1 \le i < j \le 3, \tag{1.30a}$$

$$\Pi_4 = \{\epsilon_1 + \epsilon_2 + \epsilon_3, \epsilon_1 - \epsilon_2 - \epsilon_3, -\epsilon_1 - \epsilon_2 + \epsilon_3\}, \tag{1.30b}$$

of which (1.30a) are distinguished. The positive roots for Π_{ij} are

$$\begin{aligned} \Delta^+_{\bar{0}} &= \{\alpha_1, \alpha_3, \alpha_1 + 2\alpha_2 + \alpha_3\} = \{-2\epsilon_i, -2\epsilon_j, 2\epsilon_k\}, \quad i \neq k \neq j, \\ \Delta^+_{\bar{1}} &= \{\alpha_2, \alpha_1 + \alpha_2, \alpha_2 + \alpha_3, \alpha_1 + \alpha_2 + \alpha_3\}. \end{aligned} \tag{1.31}$$

The highest root for Π_{ij} is even:

$$\tilde{\alpha} = \alpha_1 + 2\alpha_2 + \alpha_3 = 2\epsilon_k.$$

Note that the existence of the highest root requires the non-vanishing of the products:

$$(\alpha_1 + \alpha_2 + \alpha_3, \alpha_2) = 2\sigma_k, \quad (\alpha_1 + \alpha_2, \alpha_2 + \alpha_3) = -2(\sigma_i + \sigma_j) = 2\sigma_k.$$

The positive roots for Π_4 are

$$\begin{aligned} \Delta^+_{\bar{0}} &= \{\alpha_1 + \alpha_2 = 2\epsilon_1, \alpha_2 + \alpha_3 = -2\epsilon_2, \alpha_1 + \alpha_3 = 2\epsilon_3\}, \\ \Delta^+_{\bar{1}} &= \{\alpha_1, \alpha_2, \alpha_3, \alpha_1 + \alpha_2 + \alpha_3\}. \end{aligned} \tag{1.32}$$

The highest root for Π_4 is odd:

$$\tilde{\alpha} = \alpha_1 + \alpha_2 + \alpha_3 = \epsilon_1 - \epsilon_2 + \epsilon_3.$$

$F(4)$. The roots are expressed in terms of mutually orthogonal functionals ϵ_1, ϵ_2, ϵ_3, (corresponding to B_3), and δ (corresponding to A_1), such that $(\epsilon_i, \epsilon_j) = \delta_{ij}$, $(\delta, \delta) = -3$, $(\epsilon_i, \delta) = 0$. One has

$$\Delta_{\bar{0}} = \{\pm\epsilon_i \pm \epsilon_j, i \neq j, \pm\epsilon_i, \pm\delta\}, \quad \Delta_{\bar{1}} = \left\{\frac{1}{2}(\pm\epsilon_1 \pm \epsilon_2 \pm \epsilon_3 \pm \delta)\right\}. \tag{1.33}$$

Up to W equivalence, there are four systems of simple roots:

$$\begin{aligned}
\Pi_1 &= \left\{\frac{1}{2}(\epsilon_1 + \epsilon_2 + \epsilon_3 + \delta), -\epsilon_1, \epsilon_1 - \epsilon_2, \epsilon_2 - \epsilon_3\right\},\\
\Pi_2 &= \left\{\epsilon_1 - \epsilon_2, -\epsilon_1, \frac{1}{2}(\epsilon_1 + \epsilon_2 + \epsilon_3 + \delta), -\delta\right\},\\
\Pi_3 &= \left\{\epsilon_3 - \epsilon_2, \epsilon_2 - \epsilon_1, \frac{1}{2}(\epsilon_1 - \epsilon_2 - \epsilon_3 - \delta), \frac{1}{2}(\epsilon_1 + \epsilon_2 + \epsilon_3 + \delta)\right\},\\
\Pi_4 &= \left\{\epsilon_1 - \epsilon_2, \frac{1}{2}(-\epsilon_1 + \epsilon_2 + \epsilon_3 - \delta), \frac{1}{2}(\epsilon_1 + \epsilon_2 + \epsilon_3 + \delta),\right.\\
&\qquad \left.\frac{1}{2}(-\epsilon_1 - \epsilon_2 - \epsilon_3 + \delta)\right\},
\end{aligned} \tag{1.34}$$

of which Π_1, Π_2 are distinguished. The positive roots for Π_1 are

$$\Delta_{\bar{0}}^+ = \{\pm\epsilon_i - \epsilon_j, i < j, -\epsilon_i, \delta\}, \quad \Delta_{\bar{1}}^+ = \left\{\frac{1}{2}(\delta \pm \epsilon_1 \pm \epsilon_2 \pm \epsilon_3)\right\}; \tag{1.35}$$

the highest root is

$$\tilde{\alpha} = \delta = 2\alpha_1 + 3\alpha_2 + 2\alpha_3 + \alpha_4.$$

The positive roots for Π_2 are

$$\begin{aligned}
\Delta_{\bar{0}}^+ &= \{\epsilon_3 \pm \epsilon_1, \epsilon_3 \pm \epsilon_2, -\epsilon_2 \pm \epsilon_1, -\epsilon_1, -\epsilon_2, \epsilon_3, -\delta\},\\
\Delta_{\bar{1}}^+ &= \left\{\frac{1}{2}(\epsilon_3 \pm \epsilon_1 \pm \epsilon_2 \pm \delta)\right\};
\end{aligned} \tag{1.36}$$

the highest root is

$$\tilde{\alpha} = \epsilon_3 = \alpha_1 + 2\alpha_2 + 2\alpha_3 + \alpha_4.$$

$G(3)$. The roots are expressed in terms of functionals ϵ_1, ϵ_2, ϵ_3, $\epsilon_1 + \epsilon_2 + \epsilon_3 = 0$ (corresponding to G_2), and δ (corresponding to A_1), such that $(\epsilon_i, \epsilon_j) = 3\delta_{ij} - 1$, $(\delta, \delta) = -2$, $(\epsilon_i, \delta) = 0$. One has

$$\Delta_{\bar{0}} = \{\epsilon_i - \epsilon_j, i \neq j, \pm\epsilon_i, \pm 2\delta\}, \quad \Delta_{\bar{1}} = \{\pm\epsilon_i \pm \delta, \pm\delta\}. \tag{1.37}$$

Up to W equivalence, there is a unique system of simple roots:

$$\Pi = \{\delta + \epsilon_1, \epsilon_2, \epsilon_3 - \epsilon_2\}, \tag{1.38}$$

the positive roots are

$$\Delta_{\bar{0}}^{+} = \{\epsilon_i - \epsilon_j, i > j, -\epsilon_1, \epsilon_2, \epsilon_3, 2\delta\}, \quad \Delta_{\bar{1}}^{+} = \{\delta, \delta \pm \epsilon_i\}, \tag{1.39}$$

and the highest distinguished root is

$$\tilde{\alpha} = 2\delta.$$

Now we give the Cartan matrices $A = a_{ij}$ (cf. the definition (1.12)) corresponding to distinguished systems of simple roots. Note the following rule:

$$a_{ij} = \begin{cases} 2(\alpha_i, \alpha_j)/(\alpha_i, \alpha_i) & \text{for } (\alpha_i, \alpha_i) \neq 0 \\ \kappa_j(\alpha_i, \alpha_j) & \text{for } (\alpha_i, \alpha_i) = 0\,, \end{cases} \tag{1.40}$$

where $\kappa_j \in \mathbb{Z}$ is the smallest integer by absolute value so that a_{ij} is integer for all j. Note that these Cartan matrices are symmetrizable, namely, there exists a symmetric Cartan matrix $A^s = (a^s_{ij}) = DA$, where $D = (d_{ij})$ is diagonal: $d_{ij} = \delta_{ij} d_i$, so that

$$a^s_{ik} = \sum_j d_{ij} a_{jk} = d_i a_{ik} = a^s_{ki} = \sum_j d_{kj} a_{ji} = d_k a_{ki} \tag{1.41}$$

The numbers d_i are given below for each Cartan matrix. Let $\mathcal{A}_n$ denote the $n \times n$ Cartan matrix of type A_n.

$A(m, n)$, $m, n \geq 0$, $m + n > 0$. The Cartan matrix corresponding to (1.16) is

$$(a_{ij}) = \begin{pmatrix} \mathcal{A}_m & & & & \\ & & -1 & 0 & \\ & -1 & 0 & 1 & \\ & 0 & -1 & & \\ & & & & \mathcal{A}_n \end{pmatrix}, \quad r = m + n + 1,\ \tau = m + 1 \tag{1.42}$$

$$d_1 = \cdots = d_{m+1} = 1,\ d_{m+2} = \cdots = d_r = -1$$

sl(1, 1). The Cartan matrix corresponding formally to (1.16) (with $m = n = 0$) is

$$(a_{ij}) = (0)\,. \tag{1.43}$$

$B(m, n)$. The Cartan matrix corresponding to (1.20a) with $m, n > 0$ is

$$(a_{ij}) = \begin{pmatrix} \mathcal{A}_{n-1} & & & & & \\ & & -1 & & & \\ & -1 & 0 & 1 & & \\ & & -1 & & & \\ & & & & \mathcal{A}_{m-1} & \\ & & & & & -1 \\ & & & & -2 & 2 \end{pmatrix}, \quad r = m + n,\ \tau = n, \tag{1.44}$$

$$d_1 = \cdots = d_n = 2,\ d_{n+1} = \cdots = d_{r-1} = -2,\ d_r = -1,$$

and the Cartan matrix corresponding (1.20b) with $m = 0$, $n > 0$ is

$$(a_{ij}) = \begin{pmatrix} \mathcal{A}_{n-1} & & \\ & & -1 \\ & -2 & 2 \end{pmatrix}, \quad r = n,\ \tau = n, \tag{1.45}$$

$$d_1 = \cdots = d_{n-1} = 2,\ d_n = 1.$$

$C(n)$. The Cartan matrix corresponding to (1.24) with $n > 2$ is

$$(a_{ij}) = \begin{pmatrix} 0 & 1 & & & \\ -1 & & & & \\ & & \mathcal{A}_{n-2} & & \\ & & & & -2 \\ & & & -1 & 2 \end{pmatrix}, \quad r = n,\ \tau = 1, \tag{1.46}$$

$$d_1 = -1,\ d_2 = \cdots = d_{n-1} = 1,\ d_n = 2,$$

while for $n = 2$ one has

$$(a_{ij}) = \begin{pmatrix} 0 & 2 \\ -1 & 2 \end{pmatrix}, \quad r = 2,\ \tau = 1, \tag{1.47}$$

$$d_1 = -1,\ d_2 = 2.$$

$D(m, n)$, $m \geq 2$, $n \geq 1$. The Cartan matrix corresponding to (1.27a) is

$$(a_{ij}) = \begin{pmatrix} \mathcal{A}_{n-1} & & & & & & \\ & & -1 & & & & \\ & -1 & 0 & 1 & & & \\ & & -1 & & & & \\ & & & & \mathcal{A}_{m-1} & & -1 \\ & & & & & & 0 \\ & & & & -1 & 0 & 2 \end{pmatrix}, \quad r = m + n,\ \tau = n, \tag{1.48}$$

$$d_1 = \cdots = d_n = 1,\ d_{n+1} = \cdots = d_r = -1$$

$D(2, 1; \bar{\sigma})$. The Cartan matrix corresponding to Π_{23} (cf. (1.30a)) is

$$(a_{ij}) = \begin{pmatrix} 2 & -1 & 0 \\ -2\sigma_2 & 0 & -2\sigma_3 \\ 0 & -1 & 2 \end{pmatrix}, \quad r = 3,\ \tau = 2, \tag{1.49}$$

$$d_1 = 2\sigma_2,\ d_2 = 1,\ d_3 = 2\sigma_3.$$

$D(2, 1; \bar{\sigma})'$. The Cartan matrix corresponding to the root system Π_4 (cf. (1.30b)), where all simple roots are odd, is

$$(a_{ij}) = \begin{pmatrix} 0 & 2\sigma_1 & 2\sigma_3 \\ 2\sigma_1 & 0 & 2\sigma_2 \\ 2\sigma_3 & 2\sigma_1 & 0 \end{pmatrix}, \quad r = 3, \tag{1.50}$$

$$d_1 = d_2 = d_3 = 1.$$

$F(4)_1$. The Cartan matrix corresponding to Π_1 (cf. (1.34)) is

$$(a_{ij}) = \begin{pmatrix} 0 & 1 & 0 & 0 \\ -1 & 2 & -2 & 0 \\ 0 & -1 & 2 & -1 \\ 0 & 0 & -1 & 2 \end{pmatrix}, \quad r = 4, \ \tau = 1, \tag{1.51}$$

$$d_1 = -2, \ d_2 = 2, \ d_3 = d_4 = 1.$$

$F(4)_2$. The Cartan matrix corresponding to Π_2 (cf. (1.34)) is

$$(a_{ij}) = \begin{pmatrix} 2 & -1 & 0 & 0 \\ -2 & 2 & -1 & 0 \\ 0 & -1 & 0 & 3 \\ 0 & 0 & -1 & 2 \end{pmatrix}, \quad r = 4, \ \tau = 3, \tag{1.52}$$

$$d_1 = 2, \ d_2 = d_3 = 1, \ d_4 = -3.$$

Note that the (symmetrized) Cartan matrix contains the (symmetrized) Cartan matrix of $D(2,1;\bar{\sigma})$ for the parameters $\sigma_1 = 1$, $\sigma_2 = 1/2$, $\sigma_3 = -3/2$. $F(4)_3$. The Cartan matrix corresponding to Π_3 (cf. (1.34)) is

$$(a_{ij}) = \begin{pmatrix} 2 & -1 & 0 & 0 \\ -1 & 2 & -1 & 0 \\ 0 & -2 & 0 & 1 \\ 0 & 0 & 1 & 0 \end{pmatrix}, \quad r = 4, \tag{1.53}$$

$$d_1 = d_2 = 2, \ d_3 = d_4 = 1.$$

$F(4)_4$. The Cartan matrix corresponding to Π_4 (cf. (1.34)) is

$$(a_{ij}) = \begin{pmatrix} 2 & -1 & 0 & 0 \\ -2 & 0 & 2 & 1 \\ 0 & -2 & 0 & 3 \\ 0 & 1 & -3 & 0 \end{pmatrix}, \quad r = 4, \tag{1.54}$$

$$d_1 = 2, \ d_2 = d_4 = 1, \ d_3 = -1.$$

Note that the symmetrized Cartan matrix contains the Cartan matrix of $D(2,1;\bar{\sigma})'$ for the parameters $\sigma_1 = 1$, $\sigma_2 = -3/2$, $\sigma_3 = 1/2$. $G(3)$. The Cartan matrix is (cf. (1.38))

$$(a_{ij}) = \begin{pmatrix} 0 & 1 & 0 \\ -1 & 2 & -1 \\ 0 & -3 & 2 \end{pmatrix}, \quad r = 3, \ \tau = 1, \tag{1.55}$$

$$d_1 = -3, \ d_2 = 3, \ d_3 = 1.$$

Finally, we give the Dynkin diagrams corresponding to the above Cartan matrices. Nodes ○, ⊗, ●, respectively, are called *white node*, *gray node*, and *black node*, respectively. To each Cartan matrix of rank r there corresponds a Dynkin diagram with r nodes; the ith node is white if α_i is even, and gray, respectively black, if α_i is odd and $a_{ii} = 2$, respectively $a_{ii} = 0$. These three nodes give the three possible Lie (super)algebras of rank 1:

$\mathcal{G}(A,\tau)$	A	τ	diagram	dim.
A_1	(2)	∅	○	3
sl(1, 1)	(0)	{1}	⊗	3
$B(0,1)$	(2)	{1}	●	5

Given two distinct nodes i, j they are not joined if $a_{ij} = a_{ji} = 0$, otherwise they are joined as shown now:

$\mathcal{G}(A,\tau)$	A	τ	diagram	dim.
A_2	$\begin{pmatrix}2 & -1\\ -1 & 2\end{pmatrix}$	∅	○—○	8
B_2	$\begin{pmatrix}2 & -1\\ -2 & 2\end{pmatrix}$	∅	○⟹○	10
G_2	$\begin{pmatrix}2 & -1\\ -3 & 2\end{pmatrix}$	∅	○≡>○	14
$A(1,0)$	$\begin{pmatrix}2 & -1\\ -1 & 0\end{pmatrix}$	{2}	○—⊗	8
$A(0,1)$	$\begin{pmatrix}0 & 1\\ -1 & 2\end{pmatrix}$	{1}	⊗—○	8
$B(1,1)$	$\begin{pmatrix}0 & -1\\ -2 & 2\end{pmatrix}$	{1}	⊗⟹○	12
$B(0,2)$	$\begin{pmatrix}2 & -1\\ -2 & 2\end{pmatrix}$	{2}	○⟹●	14
$C(2)$	$\begin{pmatrix}0 & 2\\ -1 & 2\end{pmatrix}$	{1}	⊗⟸○	8
$A(1,0)$	$\begin{pmatrix}0 & -1\\ -1 & 0\end{pmatrix}$	{1, 2}	⊗—⊗	8
$B(1,1)$	$\begin{pmatrix}0 & -1\\ -2 & 2\end{pmatrix}$	{1, 2}	⊗⟹●	12

For general rank the distinguished Dynkin diagrams are given now (τ is the number of the non-white nodes, r is the rank, giving the number of nodes):

$\mathcal{G}$	**diagram**	τ	r
$A(m,n)$	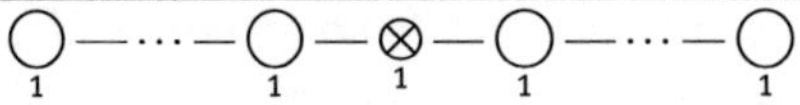	$m+1$	$m+n+1$
$B(m,n)$, $m,n>0$	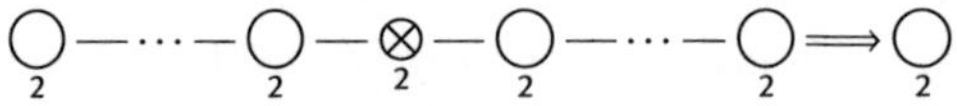	n	$m+n$
$B(0,n)$, $n>0$	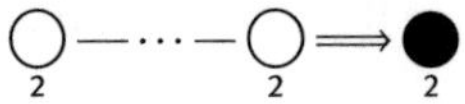	n	n
$C(n)$, $n>2$	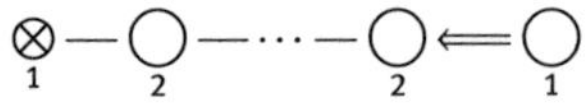	1	n
$D(m,n)$	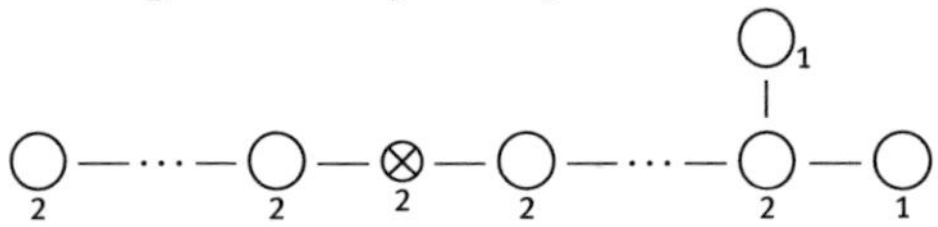	n	$m+n$
$D(2,1;\bar{\alpha})_{23}$	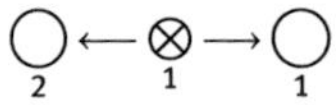	2	3
$F(4)_1$	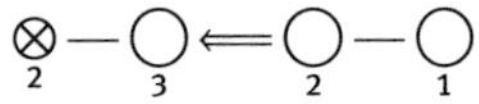	1	4
$F(4)_2$	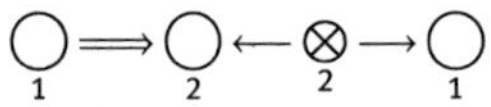	3	4
$G(3)$	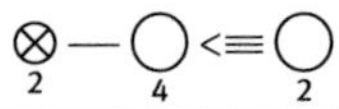	1	3

The number at a node gives the coefficient with which the corresponding simple root enters the decomposition of the highest root.

Some of the different root systems may be written in a unified way [241] as given now via the nondistinguished Dynkin diagrams, where the symbol $\times$ stands for white or gray node (yet in every diagram there is at least one odd node):

$\mathcal{G}$	**diagram**	r
$A(m,n)$		$m+n+1$
$B(m,n)$	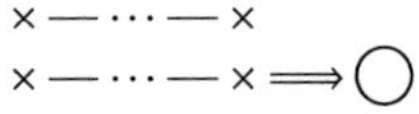	$m+n$, $m,n>0$
$B(0,n)$	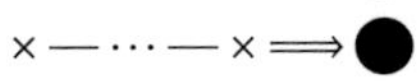	$n>0$
$C(n)$	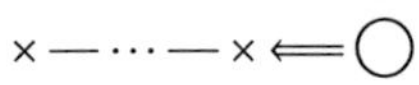	$n>2$
$D(m,n)$	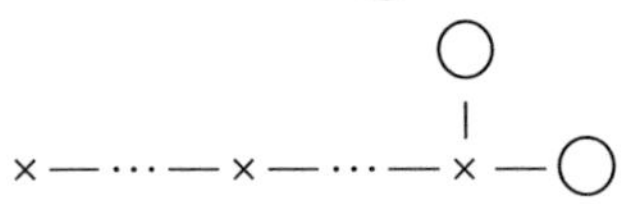	$m+n$
$D(m,n)'$	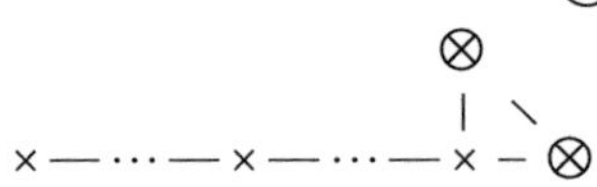	$m+n$
$D(2,1;\bar{\alpha})'$	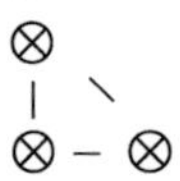	3
$F(4)_3$	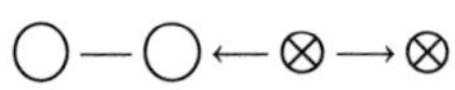	4
$F(4)_4$	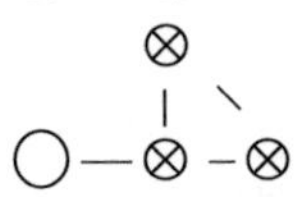	4

According to the results of Yamane [389] the above nondistinguished root systems are equivalent if we extend W with the odd reflections (1.13). This is due to the fact that each such odd reflection transforms the root system Δ into a root system Δ' so that the corresponding Lie superalgebras $\mathcal{G}(\Delta)$ and $\mathcal{G}(\Delta')$ are isomorphic.

1.3.3 Affine basic classical superalgebras

Finally, we give the (extended) distinguished Dynkin diagrams of the affinization $\mathcal{G}^{(1)}$ of the basic classical Lie superalgebras $\mathcal{G}$ (cf. [390]):

$\mathcal{G}^{(1)}$	**diagram**
$A(m,n)^{(1)}$	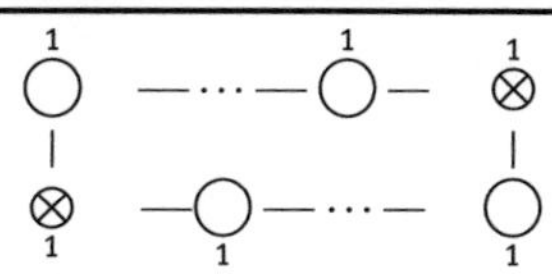
$B(m,n)^{(1)}$	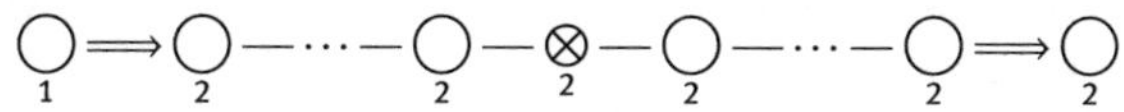
$B(0,n)^{(1)}$	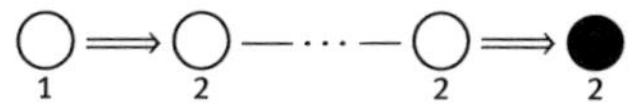
$C(n)^{(1)}$	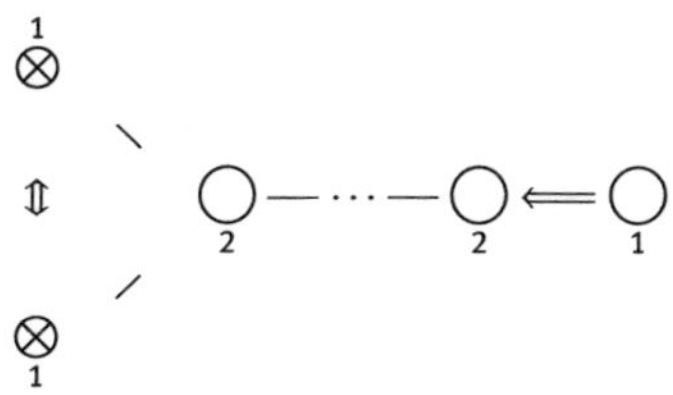
$D(m,n)^{(1)}$	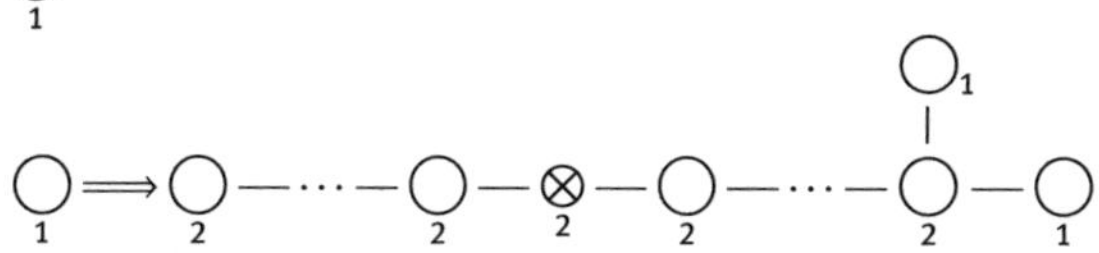
$D(2,1;\bar{\alpha})^{(1)}$	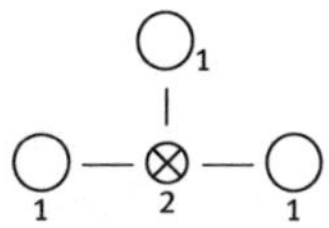
$F(4)^{(1)}$	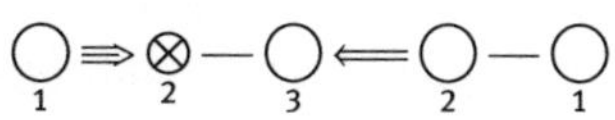
$G(3)^{(1)}$	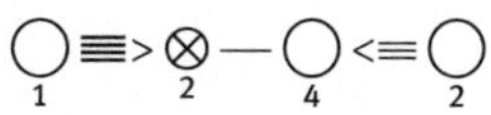

1.4 Representations of simple Lie superalgebras

Highest weight modules, in particular, Verma modules are defined analogously to the even case. Also the following holds.

Ado's theorem. *Every finite-dimensional Lie superalgebra has a finite-dimensional faithful representation.* ◇

Let $\mathcal{G}$ be a basic classical Lie superalgebra. Let

$$\bar{\Delta}_{\bar{0}} \equiv \left\{\alpha \in \Delta_{\bar{0}} \;\middle|\; \frac{1}{2}\alpha \notin \Delta_{\bar{1}}\right\}, \quad \bar{\Delta}_{\bar{1}} \equiv \{\alpha \in \Delta_{\bar{1}} \mid 2\alpha \notin \Delta_{\bar{0}}\}. \tag{1.56}$$

Let $\rho_{\bar{0}} \equiv \frac{1}{2}\sum_{\alpha\in\Delta_{\bar{0}}^+}\alpha$, $\bar{\rho}_{\bar{0}} \equiv \frac{1}{2}\sum_{\alpha\in\bar{\Delta}_{\bar{0}}^+}\alpha$, $\rho_{\bar{1}} = \frac{1}{2}\sum_{\alpha\in\Delta_{\bar{1}}^+}\alpha$, $\rho = \rho_{\bar{0}} - \rho_{\bar{1}}$. Note that $\rho(H_i) = (\alpha_i, \alpha_i)/2$.

Then we have the following.

Proposition ([243]). *The Verma module V^Λ over $\mathcal{G}$ is irreducible iff*

$$2(\Lambda + \rho, \alpha) \neq m(\alpha, \alpha) \tag{1.57}$$

for any $\alpha \in \Delta^+$ and any $m \in \mathbb{N}$. ◇

We shall say that the Verma module V^Λ is reducible w.r.t. the root α with $(\alpha,\alpha) \neq 0$ if there exists $m \in \mathbb{N}$ such that

$$2(\Lambda + \rho, \alpha) = m(\alpha, \alpha), \quad (\alpha,\alpha) \neq 0, \quad m \in \mathbb{N} \tag{1.58}$$

and that V^Λ is reducible w.r.t. the root α with $(\alpha,\alpha) = 0$ if:

$$(\Lambda + \rho, \alpha) = 0, \quad (\alpha, \alpha) = 0. \tag{1.59}$$

If (1.58) holds then there exists a singular vector of weight $\Lambda - m\alpha$ as in the even case, while if (1.59) holds then there exists a singular vector of weight $\Lambda - \alpha$. In particular, if $(\Lambda + \rho, \alpha_i) = 0$ for an odd simple root α_i, then the singular vector is

$$v_s^{\mathrm{odd}} = X_i^- v_0. \tag{1.60}$$

As in the even case there exists a maximal invariant submodule I^Λ, so that the factor module $L_\Lambda = V^\Lambda/I^\Lambda$ is irreducible. The difference with the even case is that it is not true in general, except for $\mathcal{G} = B(0,n)$, that a finite-dimensional representation is completely reducible. Furthermore, $\mathcal{G} \neq B(0,n)$. The irreducible finite-dimensional representations are divided by Kac in two types: *typical representations* and *atypical representations*. The typical finite-dimensional representations are obtained as L_Λ when the Verma module V^Λ is reducible w.r.t. all *even* simple roots, but not w.r.t. any odd root $\alpha \in \bar{\Delta}_{\bar{1}}^+$. (Note here another difference from the even case, where L_Λ is finite-dimensional iff V^Λ is reducible w.r.t. *all* roots.) The atypical finite-dimensional representations are obtained as L_Λ when the Verma module V^Λ is reducible also w.r.t. some odd root $\alpha \in \bar{\Delta}_{\bar{1}}^+$. The interesting feature here is that the factor module $V^\Lambda/\tilde{I}^\Lambda$, where $\tilde{I}^\Lambda$ is the submodule generated by the even singular vectors, is finite dimensional, reducible and indecomposable. It has to be factorized further to obtain L_Λ. Further, we say that an atypical representation is *singly atypical* if $(\Lambda + \rho, \alpha) = 0$ only for one odd root $\alpha \in \bar{\Delta}^+_{\bar{1}}$.

The finite-dimensional representations were classified by Kac.

Theorem ([241, 243]). *Let $\mathcal{G}$ be a basic classical superalgebra, and let V be a finite-dimensional irreducible $\mathcal{G}$-module. Then there exists a vector $v_\Lambda \in V$, $\Lambda \in \mathcal{H}^*$ (unique up to a multiplication by nonzero scalar) such that $X_i^+ v_\Lambda = 0$, $Hv_\Lambda = \Lambda(H)v_\Lambda$. Two $\mathcal{G}$-modules V_1 and V_2 are isomorphic iff $\Lambda_1 = \Lambda_2$. Thus, in our notation $V = L_\Lambda$. The set of numbers $a_i = \Lambda(H_i)$, $i = 1, \ldots, r$, Λ as above, are described by the following conditions (we consider $a_{s,s+1} = 1$ if $a_{ss} = 0$):*

1) *$a_i \in \mathbb{Z}_+$ if $i \neq s$;*
2) *$k \in \mathbb{Z}_+$ where k is given now:*

$\mathcal{G}$	k	b
$B(0,n)$	$\frac{1}{2} a_n$	0
$B(m,n)$, $m > 0$	$a_n - a_{n+1} - \cdots - a_{m+n-1} - \frac{1}{2} a_{m+n}$	m
$D(m,n)$	$a_n - a_{n+1} - \cdots - \frac{1}{2}(a_{m+n-1} + a_{m+n})$	m
$D(2,1;\lambda)$	$(2a_1 - a_2 - \lambda a_3)/(1+\lambda)$	2
$F(4)$	$(2a_1 - 3a_2 - 4a_3 - 2a_4)/3$	4
$G(3)$	$(a_1 - 2a_2 - 3a_3)/2$	3

3) *$k < b$ (from the table) then there are additional conditions:*
$B(m,n)$, $a_{n+k+1} = \cdots = a_{m+n} = 0$;
$D(m,n)$, $a_{n+k+1} = \cdots = a_{m+n} = 0$ if $k \leq m-2$; $a_{m+n-1} = a_{m+n}$ if $k = m-1$;
$D(2,1;\lambda)$, $a_i = 0$ if $k = 0$; $\lambda(a_3 + 1) = \pm(a_2 + 1)$ if $k = 1$;
$F(4)$, $a_i = 0$ if $k = 0$; $k \neq 1$; $a_2 = a_4 = 0$ if $k = 2$; $a_2 = 2a_4 + 1$ if $k = 3$;
$G(3)$, $a_i = 0$ if $k = 0$; $k \neq 1$; $a_2 = 0$ if $k = 2$. ◇

Note that all finite-dimensional irreducible representations (irreps) of $B(0,n)$ are typical. All finite-dimensional irreps of $C(n)$ are either typical or singly atypical.

Let L_Λ be a typical finite-dimensional irrep of $\mathcal{G}$. Then one has

$$\dim L_\Lambda = 2^{\dim \Delta^+_{\bar{1}}} \prod_{\alpha \in \Delta^+_{\bar{0}}} \frac{(\Lambda + \rho, \alpha)}{(\rho_0, \alpha)}, \tag{1.61}$$

$$\begin{aligned} &\dim(L_\Lambda)_{\bar{0}} - \dim(L_\Lambda)_{\bar{1}} = 0, \quad \mathcal{G} \neq B(0,n), \\ &\dim(L_\Lambda)_{\bar{0}} - \dim(L_\Lambda)_{\bar{1}} = \prod_{\alpha \in \bar{\Delta}^+_{\bar{0}}} \frac{(\Lambda + \rho, \alpha)}{(\bar{\rho}_0, \alpha)} = \prod_{\alpha \in \bar{\Delta}^+_{\bar{0}}} \frac{(\Lambda + \rho, \alpha)}{(\rho, \alpha)}, \\ &\qquad \mathcal{G} = B(0,n). \end{aligned} \tag{1.62}$$

The second order Casimir operator is defined analogously to the even case:

$$C_2 = \sum_i (-1)^{p(Y_i)} Y_i Y^i \tag{1.63}$$

where $\{Y_i\}$ and $\{Y^i\}$ are dual homogeneous bases of $\mathcal{G}$.

Let V be a finite-dimensional irreducible $\mathcal{G}$-module with highest weight Λ. Then

$$C_2 u = (\Lambda, \Lambda + 2\rho)u, \quad u \in V. \tag{1.64}$$

1.5 Real forms of the basic classical Lie superalgebras

We go back to the superalgebras $\mathrm{gl}(m,n;F)$, cf. (1.3) with $F = \mathbb{R}, \mathbb{C}, \mathbb{H}$, where $\mathbb{H}$ is the quaternion field, and we fix the standard embeddings: $\mathbb{R} \subset \mathbb{C} \subset \mathbb{H}$. Using this, now we shall consider $\mathrm{gl}(m,n;F)$ as real Lie superalgebras. The *special linear superalgebras* are defined following (1.4), but with a difference for $F = \mathbb{H}$:

$$\begin{aligned} \mathrm{sl}(m/n;F) &\doteq \{X \in \mathrm{gl}(m/n;F) : \mathrm{str}\, X = 0\}, \quad \text{for } F = \mathbb{R}, \mathbb{C} \\ \mathrm{sl}(m/n;\mathbb{H}) &\doteq \{X \in \mathrm{gl}(m/n;\mathbb{H}) : \mathrm{Re}\ \mathrm{str}\, X = 0\}. \end{aligned} \tag{1.65}$$

Furthermore, we introduce the following matrices of order $m + n$:

$$\begin{aligned} S_{p,q} &= \begin{pmatrix} 1_p & 0 & 0 & 0 \\ 0 & -1_{m-p} & 0 & 0 \\ 0 & 0 & 1_q & 0 \\ 0 & 0 & 0 & -1_{n-q} \end{pmatrix}, \\ R_p &= \begin{pmatrix} i1_m & 0 & 0 \\ 0 & 1_p & 0 \\ 0 & 0 & -1_{n-p} \end{pmatrix}, \\ T_p &= \begin{pmatrix} i1_p & 0 & 0 & 0 \\ 0 & i1_{m-p} & 0 & 0 \\ 0 & 0 & 0 & 1_r \\ 0 & 0 & -1_r & 0 \end{pmatrix}, \quad r = \frac{1}{2}n,\ n \in 2\mathbb{N}. \end{aligned} \tag{1.66}$$

Thus, we can define:

- the *special unitary superalgebras* $\mathrm{su}(p, m-p/q, n-q)$ (denoted by Kac $\mathrm{su}(m,n;p,q)$):

$$\mathrm{su}(p, m-p/q, n-q)_s = \{X \in \mathrm{sl}(m/n;\mathbb{C}) \mid S_{p,q}^{-1} X^\dagger S_{p,q} = -i^s X\}; \tag{1.67}$$

- the *real orthosymplectic superalgebras* $\mathrm{osp}(p, m-p/n;\mathbb{R})$ (denoted by Kac $\mathrm{osp}(m,n;p;\mathbb{R})$):

$$\mathrm{osp}(p, m-p/n;\mathbb{R})_s = \{X \in \mathrm{sl}(m/n;\mathbb{R}) \mid T_p^{-1}\, {}^t X T_p = -i^s X\}, \quad n \in 2;\mathbb{N} \tag{1.68}$$

- the *quaternionic orthosymplectic superalgebras* $\mathrm{hosp}(m/p, n-p;\mathbb{H})$ (denoted by Kac $\mathrm{hosp}(m,n;p;\mathbb{H})$):

$$\mathrm{hosp}(m/p, n-p;\mathbb{H})_s = \{X \in \mathrm{sl}(m/n;\mathbb{R}) \mid R_p^{-1} X^\dagger R_p = -i^s X\}. \tag{1.69}$$

- $D(2,1,\alpha;p)$. For every $p = 0,1,2$ there is a representation of $\mathrm{so}(p,4-p) \oplus \mathrm{sl}(2,\mathbb{R})$ in $D(2,1,\alpha)$ which defines a real form of $D(2,1,\alpha)$.
- $F(4;p)$. Each algebra $\mathrm{so}(p,7-p)$, $p = 0,1,2,3$, has a spinor representation $\mathrm{spin}_{p,7-p}$ that is a real form of the B_3-module spin_7. For $p = 1,2,3$ there is a unique real superalgebra $F(4;p)$ with even subalgebra $\mathrm{so}(p,7-p) \oplus \mathrm{so}(3)$, while for $F(4;0)$ the even subalgebra is $\mathrm{so}(7) \oplus \mathrm{so}(2,1)$.
- $G(3;p)$. The standard real forms of G_2 (split and compact), denoted by $G_{2,p}$, $p = 0,1$, respectively, give rise to two real forms of $G(3)$ denoted by $G(3;p)$, $p = 0,1$. The even subalgebra of $G(3;0)$ is $G_{2,0} \oplus \mathrm{su}(2)$, while for $G(3;1)$ the even subalgebra is $G_{2,1} \oplus \mathrm{sl}(2,\mathbb{R})$.

2 Conformal supersymmetry in 4D

Summary

Recently, superconformal field theories in various dimensions have been attracting ever more interest, cf. (for references up to year 2000): [306, 170, 248, 313, 89, 186, 233, 355, 215, 138–140, 49, 193, 141, 142, 347, 72, 160, 168, 258, 273, 309, 9, 58, 102, 189, 171, 177, 311, 317, 4, 76, 101, 103, 164, 167, 172, 174, 173, 175, 105] and the references therein. Particularly important are those for $D \leq 6$, since in these cases the relevant superconformal algebras satisfy Nahm's classification [306] based on the Haag–Lopuszanski–Sohnius theorem [218]. This makes the classification of the UIRs of these superalgebras very important. First such classification was given for the $D = 4$ superconformal algebras $\mathrm{su}(2,2/1)$ [182] and $\mathrm{su}(2,2/N)$ [138–142] (for arbitrary N). Then the classification for $D = 3$ ($\mathrm{osp}(N/4)$ for even N), $D = 5$, and $D = 6$ ($\mathrm{osp}(8^*/2N)$ for $N = 1,2$) was given in [302] (some results being conjectural), and then the $D = 6$ case (for arbitrary N) was finalized in [122] (see Section 3.1).

Once we know the UIRs of a (super-) algebra the next question is to find their characters, since these give the spectrum which is important for the applications. Some results on the spectrum were given in the early papers [233, 355, 215, 140] but it is necessary to have systematic results for which the character formulas are needed. This is the question we address in this chapter for the UIRs of $D = 4$ conformal superalgebras $\mathrm{su}(2,2/N)$. From the mathematical point of view this question is clear only for representations with conformal dimension above the unitarity threshold viewed as irreps of the corresponding complex superalgebra $\mathrm{sl}(4/N)$. But for $\mathrm{su}(2,2/N)$ even the UIRs above the unitarity threshold are truncated for small values of spin and isospin. Moreover, in the applications the most important role is played by the representations with "quantized" conformal dimensions at the unitarity threshold and at discrete points below. In the quantum field or string theory framework some of these correspond to fields with "protected" scaling dimension and therefore imply "non-renormalization theorems" at the quantum level, cf., e. g., [228, 174]. This is intimately related to the super-invariant differential operators and equations satisfied by the superfields at these special representations.

Thus, we need detailed knowledge about the structure of the UIRs from the representation-theoretical point of view. Fortunately, such information is contained in [138–142]. Following these papers we first recall the basic ingredients of the representation theory of the $D = 4$ superconformal algebras. In particular we recall the structure of Verma modules and UIRs. First the general theory for the characters of $\mathrm{su}(2,2/N)$ is developed in great detail. For the general theory we use the (generalized) odd reflections introduced in [139] (see also [349]).[1] We also pin-point the difference between character formulas for $\mathrm{sl}(4,N)$ and $\mathrm{su}(2,2/N)$; for the latter we need to introduce and use the notion of counterterms in the character formulas. The general formulas are valid for arbitrary N and are given for the so-called bare characters (or superfield decompositions). We also summarize our results on the decompositions of long superfields as they descend to the unitarity threshold. To give the character formulas explicitly we need to recall also the character formulas of $\mathrm{su}(2,2)$ and $\mathrm{su}(N)$, for which we give explicitly all formulas that we need. Finally, we give the explicit complete character formulas for $N = 1$ and for a number of important examples for $N = 2,4$.

In this chapter we mostly follow the papers [123, 129, 127, 128] and also using essentially the results of [138–142].

1 For an alternative approach to character formulas, see [149, 57].

https://doi.org/10.1515/9783110527490-002

2.1 Representations of $D = 4$ conformal supersymmetry

2.1.1 The setting

The superconformal algebras in $D = 4$ are $\mathcal{G} = \mathrm{su}(2,2/N)$. The even subalgebra of $\mathcal{G}$ is the algebra $\mathcal{G}_0 = \mathrm{su}(2,2) \oplus u(1) \oplus \mathrm{su}(N)$. We label their physically relevant representations of $\mathcal{G}$ by the signature:

$$\chi = [d;\, j_1, j_2;\, z;\, r_1, \ldots, r_{N-1}] \tag{2.1}$$

where d is the conformal weight, j_1, j_2 are non-negative (half-)integers which are Dynkin labels of the finite-dimensional irreps of the $D = 4$ Lorentz subalgebra so(3, 1) of dimension $(2j_1 + 1)(2j_2 + 1)$, z represents the $u(1)$ subalgebra which is central for $\mathcal{G}_0$ (and for $N = 4$ is central for $\mathcal{G}$ itself), and $r_1, \ldots, r_{N-1}$ are non-negative integers which are Dynkin labels of the finite-dimensional irreps of the internal (or R) symmetry algebra su(N).

We recall that the algebraic approach to $D = 4$ conformal supersymmetry developed in [138–142] involves two related constructions—on function spaces and as Verma modules. The first realization employs the explicit construction of induced representations of $\mathcal{G}$ (and of the corresponding supergroup $G = \mathrm{su}(2,2/N)$) in spaces of functions (superfields) over superspace which are called elementary representations (ER). The UIRs of $\mathcal{G}$ are realized as irreducible components of ERs, and then they coincide with the usually used superfields in indexless notation. The Verma module realization is also very useful as it provides simpler and more intuitive picture for the relation between reducible ERs, for the construction of the irreps, in particular, of the UIRs. For the latter the main tool is an adaptation of the Shapovalov form [353] to the Verma modules [140, 142].

Here we shall use mostly the Verma module construction, however, keeping in mind the construction on function spaces and the corresponding invariant differential operators.

2.1.2 Verma modules

To introduce Verma modules one needs the standard triangular decomposition:

$$\mathcal{G}^{\mathbb{C}} = \mathcal{G}^+ \oplus \mathcal{H} \oplus \mathcal{G}^- \tag{2.2}$$

where $\mathcal{G}^{\mathbb{C}} = \mathrm{sl}(4/N)$ is the complexification of $\mathcal{G}$, $\mathcal{G}^+$, $\mathcal{G}^-$, respectively, are the subalgebras corresponding to the positive, negative, roots of $\mathcal{G}^{\mathbb{C}}$, respectively, and $\mathcal{H}$ denotes the Cartan subalgebra of $\mathcal{G}^{\mathbb{C}}$.

We consider the lowest weight Verma modules, so that $V^{\Lambda} \cong U(\mathcal{G}^+) \otimes v_0$, where $U(\mathcal{G}^+)$ is the universal enveloping algebra of $\mathcal{G}^+$, $\Lambda \in \mathcal{H}^*$ is the lowest weight, and v_0 is the lowest weight vector v_0 such that

$$Xv_0 = 0, \quad X \in \mathcal{G}^-,$$
$$Hv_0 = \Lambda(H)v_0, \quad H \in \mathcal{H}.$$

Furthermore, for simplicity we omit the sign $\otimes$, i. e., we write $Pv_0 \in V^\Lambda$ with $P \in U(\mathcal{G}^+)$.

The lowest weight Λ is characterized by its values on the Cartan subalgebra $\mathcal{H}$, or, equivalently, by its products with the simple roots (given explicitly below). In general, these would be $3 + N$ complex numbers, however, in order to be useful for the representations of the real form $\mathcal{G}$ these values would be restricted to be real and furthermore to correspond to the signatures χ, and we shall write $\Lambda = \Lambda(\chi)$, or $\chi = \chi(\Lambda)$. Note, however, that there are Verma modules to which correspond no ERs, cf. [139] and below.

If a Verma module V^Λ is irreducible then it gives the lowest weight irrep L_Λ with the same weight. If a Verma module V^Λ is reducible then it contains a maximal invariant submodule I^Λ and the lowest weight irrep L_Λ with the same weight is given by factorization: $L_\Lambda = V^\Lambda / I^\Lambda$ [110, 244, 243].

Thus, we need first to know which Verma modules are reducible. The reducibility conditions for highest weight Verma modules over basic classical Lie superalgebra were given by Kac [243]. Translating his conditions to lowest weight Verma modules we have [139]: A lowest weight Verma module V^Λ is reducible only if at least one of the following conditions is true:[2]

$$(\rho - \Lambda, \beta) = m(\beta, \beta)/2, \quad \beta \in \Delta^+, \ (\beta, \beta) \neq 0, \ m \in \mathbb{N} \tag{2.3a}$$
$$(\rho - \Lambda, \beta) = 0, \quad \beta \in \Delta^+, \ (\beta, \beta) = 0, \tag{2.3b}$$

where Δ^+ is the positive root system of $\mathcal{G}^{\mathbb{C}}$, $\rho \in \mathcal{H}^*$ is the very important in representation theory element given by $\rho = \rho_{\bar{0}} - \rho_{\bar{1}}$, where $\rho_{\bar{0}}, \rho_{\bar{1}}$ are the half-sums of the even, odd, respectively, positive roots, $(\cdot, \cdot)$ is the standard bilinear product in $\mathcal{H}^*$.

If a condition from (2.3a) is fulfilled then V^Λ contains a submodule which is a Verma module $V^{\Lambda'}$ with shifted weight given by the pair m, β: $\Lambda' = \Lambda + m\beta$. The embedding of $V^{\Lambda'}$ in V^Λ is provided by mapping the lowest weight vector v_0' of $V^{\Lambda'}$ to the singular vector $v_s^{m,\beta}$ in V^Λ which is completely determined by the conditions

$$Xv_s^{m,\beta} = 0, \quad X \in \mathcal{G}^-,$$
$$Hv_s^{m,\beta} = \Lambda'(H)v_0, \quad H \in \mathcal{H}, \ \Lambda' = \Lambda + m\beta. \tag{2.4}$$

Explicitly, $v_s^{m,\beta}$ is given by an even polynomial in the positive root generators:

$$v_s^{m,\beta} = P^{m,\beta} v_0, \quad P^{m,\beta} \in U(\mathcal{G}^+). \tag{2.5}$$

Thus, the submodule of V^Λ which is isomorphic to $V^{\Lambda'}$ is given by $U(\mathcal{G}^+)P^{m,\beta}v_0$. [More on the even case following the same approach may be found in, e. g., [113, 114].]

2 Many statements below are true for any basic classical Lie superalgebra and would require changes only for the superalgebras osp$(1/2N)$.

If a condition from (2.3b) is fulfilled then V^Λ contains a submodule I^β obtained from the Verma module $V^{\Lambda'}$ with shifted weight $\Lambda' = \Lambda + \beta$ as follows. In this situation V^Λ contains a singular vector

$$\begin{aligned} Xv_s^\beta &= 0, \quad X \in \mathcal{G}^-, \\ Hv_s^\beta &= \Lambda'(H)v_0, \quad H \in \mathcal{H},\ \Lambda' = \Lambda + \beta. \end{aligned} \tag{2.6}$$

Explicitly, v_s^β is given by an odd polynomial in the positive root generators:

$$v_s^\beta = P^\beta v_0, \quad P^\beta \in U(\mathcal{G}^+). \tag{2.7}$$

Then we have

$$I^\beta = U(\mathcal{G}^+)P^\beta v_0, \tag{2.8}$$

which is smaller than $V^{\Lambda'} = U(\mathcal{G}^+)v_0'$, since this polynomial is Grassmannian:

$$(P^\beta)^2 = 0. \tag{2.9}$$

To describe this situation we say that $V^{\Lambda'}$ is *oddly embedded* in V^Λ.

Note, however, that the above formulas describe also more general situations when the difference $\Lambda' - \Lambda = \beta$ is not a root, as used in [139], and below.

The weight shifts $\Lambda' = \Lambda + \beta$, when β is an odd root are called *generalized odd reflections*; see [139]. For future reference we denote them by

$$\hat{\mathcal{S}}_\beta \cdot \Lambda \equiv \Lambda + \beta, \quad (\beta, \beta) = 0,\ (\Lambda - \rho, \beta) = 0. \tag{2.10}$$

Note the difference of this definition with (1.13). Note also that if Λ is as in (2.10) than $\Lambda' = \Lambda + n\beta$ has the same properties. Thus, such generalized odd reflection generates an infinite discrete abelian group:

$$\tilde{W}_\beta \equiv \{(\hat{\mathcal{S}}_\beta)^n \mid n \in \mathbb{Z}\}, \quad \ell((\hat{\mathcal{S}}_\beta)^n) = n, \tag{2.11}$$

where the unit element is obviously obtained for $n = 0$, and $(\hat{\mathcal{S}}_\beta)^{-n}$ is the inverse of $(\hat{\mathcal{S}}_\beta)^n$, and for future use we have also defined the length function $\ell(\cdot)$ on the elements of $\tilde{W}_\beta$. This group acts on the weights Λ extending (2.10):

$$(\hat{\mathcal{S}}_\beta)^n \cdot \Lambda = \Lambda + n\beta, \quad n \in \mathbb{Z},\ (\beta, \beta) = 0,\ (\Lambda - \rho, \beta) = 0. \tag{2.12}$$

This is related to the fact that there is a doubly infinite chain of oddly embedded Verma modules whenever a Verma module is reducible w.r.t. an odd root as in (2.3b). This is explained in detail and used for the classification of the Verma modules in [138], and shall be used below.

Furthermore, to be more explicit we need to recall the root system of $\mathcal{G}^{\mathbb{C}}$—for definiteness—as used in [139]. The positive root system Δ^+ is comprised of α_{ij}, $1 \le i < j \le 4 + N$. The even positive root system $\Delta^+_{\bar{0}}$ is comprised of α_{ij}, with $i, j \le 4$ and α_{ij}, with $i, j \ge 5$, cf. (1.17); the odd positive root system $\Delta^+_{\bar{1}}$ is comprised of α_{ij}, with $i \le 4$, $j \ge 5$, cf. (1.17).

The simple roots are chosen as in (2.4) of [139]:

$$\gamma_1 = \alpha_{12}, \quad \gamma_2 = \alpha_{34}, \quad \gamma_3 = \alpha_{25}, \quad \gamma_4 = \alpha_{4,4+N}, \quad \gamma_k = \alpha_{k,k+1}, \quad 5 \le k \le 3+N. \tag{2.13}$$

Thus, the Dynkin diagram is

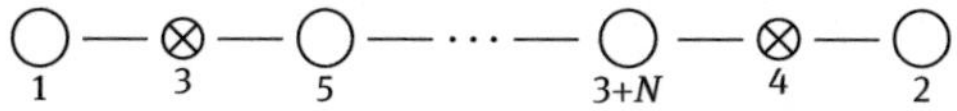

(2.14)

This is a non-distinguished simple root system with two odd simple roots (for the various root systems of the basic classical superalgebras we refer to Kac [241] and to Chapter 1 here).

We choose this diagram since it has a mirror symmetry (conjugation):

$$\gamma_1 \longleftrightarrow \gamma_2, \quad \gamma_3 \longleftrightarrow \gamma_4, \quad \gamma_j \longleftrightarrow \gamma_{N+8-j}, \quad j \ge 5, \tag{2.15}$$

and furthermore it is consistent with the mirror symmetry of the sl(4) and sl(N) root systems by identifying: $\gamma_1 \mapsto \alpha_1$, $\gamma_2 \mapsto \alpha_3$, and $\gamma_j \mapsto \alpha_j$, $j \ge 5$, respectively

Remark. Sometimes we shall use another way of writing the signature related to the above enumeration of simple roots, cf. [139] and (1.16) of [123]:

$$\chi = (2j_1; (\Lambda, \gamma_3); r_1, \dots, r_{N-1}; (\Lambda, \gamma_4); 2j_2), \tag{2.16}$$

(where (Λ, γ_3), (Λ, γ_4) are definite linear combinations of all quantum numbers), or even giving only the Lorentz and su(N) signatures:

$$\chi_N = \{2j_1; r_1, \dots, r_{N-1}; 2j_2\}. \tag{2.17}$$

◇

Let $\Lambda = \Lambda(\chi)$. The products of Λ with the simple roots are [139]:

$$(\Lambda, \gamma_a) = -2j_a, \quad a = 1, 2, \tag{2.18a}$$

$$(\Lambda, \gamma_3) = \frac{1}{2}(d + z') + j_1 - m/N + 1, \tag{2.18b}$$

$$(\Lambda, \gamma_4) = \frac{1}{2}(d - z') + j_2 - m_1 + m/N + 1, \tag{2.18c}$$

$$z' \equiv z(1 - \delta_{N4})$$

$$(\Lambda, \gamma_j) = r_{N+4-j}, \quad 5 \le j \le 3+N. \tag{2.18d}$$

These formulas give the correspondence between signatures χ and lowest weights $\Lambda(\chi)$.

Remark. For $N = 4$ the factor $u(1)$ in $\mathcal{G}_0$ becomes central in $\mathcal{G}$ and $\mathcal{G}^{\mathbb{C}}$. Consequently, the representation parameter z cannot come from the products of Λ with the simple roots, as indicated in (2.18). In that case the lowest weight is actually given by the sum $\Lambda + \tilde{\Lambda}$, where $\tilde{\Lambda}$ carries the representation parameter z. This is explained in detail in [139] and further we shall comment on it no more; the peculiarities for $N = 4$ will be evident in the formulas. ◇

In the case of even roots $\beta \in \Delta_{\bar{0}}^{+}$ there are six roots α_{ij}, $j \leq 4$, coming from the sl(4) factor (which is complexification of su(2, 2)) and $N(N-1)/2$ roots α_{ij}, $5 \leq i$, coming form the sl(N) factor (complexification of su(N)).

The reducibility conditions w.r.t. to the positive roots coming from sl(4)(su(2, 2)) coming from (2.3) (denoting $m \to n_{ij}$ for $\beta \to \alpha_{ij}$) are

$$n_{12} = 1 + 2j_1 \equiv n_1 \tag{2.19a}$$

$$n_{23} = 1 - d - j_1 - j_2 \equiv n_2 \tag{2.19b}$$

$$n_{34} = 1 + 2j_2 \equiv n_3 \tag{2.19c}$$

$$n_{13} = 2 - d + j_1 - j_2 = n_1 + n_2 \tag{2.19d}$$

$$n_{24} = 2 - d - j_1 + j_2 = n_2 + n_3 \tag{2.19e}$$

$$n_{14} = 3 - d + j_1 + j_2 = n_1 + n_2 + n_3. \tag{2.19f}$$

Thus, the reducibility conditions (2.19a,c) are fulfilled automatically for $\Lambda(\chi)$ with χ from (2.1) since we always have $n_1, n_3 \in \mathbb{N}$. There are no such conditions for the ERs since they are induced from the finite-dimensional irreps of the Lorentz subalgebra (parametrized by j_1, j_2). However, to these two conditions correspond differential operators of order $1 + 2j_1$ and $1 + 2j_2$ (as we mentioned above) and these annihilate all functions of the ERs with signature χ.

The reducibility conditions w.r.t. to the positive roots coming from sl(N)(su(N)) are all fulfilled for $\Lambda(\chi)$ with χ from (2.1). In particular, for the simple roots from those condition (2.3) is fulfilled with $\beta \to \gamma_j$, $m = 1 + r_{N+4-j}$, for every $j = 5, 6, \dots, N + 3$. There are no such conditions for the ERs since they are induced from the finite-dimensional UIRs of su(N). However, to these $N-1$ conditions correspond $N-1$ differential operators of orders $1 + r_k$ (as we mentioned) and the functions of our ERs are annihilated by all these operators [139].[3]

For future use we note also the following decompositions:

$$\Lambda = \sum_{j=1}^{N+3} \lambda_j \alpha_{j,j+1} = \Lambda^s + \Lambda^z + \Lambda^u \tag{2.20a}$$

$$\Lambda^s \equiv \sum_{j=1}^{3} \lambda_j \alpha_{j,j+1}, \quad \Lambda^z \equiv \lambda_4 \alpha_{45}, \quad \Lambda^u \equiv \sum_{j=5}^{N+3} \lambda_j \alpha_{j,j+1}, \tag{2.20b}$$

which actually employ the distinguished root system with one odd root α_{45}.

The reducibility conditions for the $4N$ odd positive roots of $\mathcal{G}$ are [138, 139]:

$$d = d^1_{Nk} - z\delta_{N4} \tag{2.21a}$$

3 Note that there are actually as many operators as positive roots of sl(N) but all are expressed in terms of the $N - 1$ above corresponding to the simple roots [139].

$$d^1_{Nk} \equiv 4 - 2k + 2j_2 + z + 2m_k - 2m/N$$

$$d = d^2_{Nk} - z\delta_{N4} \tag{2.21b}$$

$$d^2_{Nk} \equiv 2 - 2k - 2j_2 + z + 2m_k - 2m/N$$

$$d = d^3_{Nk} + z\delta_{N4} \tag{2.21c}$$

$$d^3_{Nk} \equiv 2 + 2k - 2N + 2j_1 - z - 2m_k + 2m/N$$

$$d = d^4_{Nk} + z\delta_{N4} \tag{2.21d}$$

$$d^4_{Nk} \equiv 2k - 2N - 2j_1 - z - 2m_k + 2m/N$$

where in all four cases of (2.21) $k = 1, \dots, N$, $m_N \equiv 0$, and

$$m_k \equiv \sum_{i=k}^{N-1} r_i, \quad m \equiv \sum_{k=1}^{N-1} m_k = \sum_{k=1}^{N-1} k r_k; \tag{2.22}$$

m_k is the number of cells of the kth row of the standard Young tableau, m is the total number of cells. Condition (2.21a.k) corresponds to the root $\alpha_{3,N+5-k}$, (2.21b.k) corresponds to the root $\alpha_{4,N+5-k}$, (2.21c.k) corresponds to the root $\alpha_{1,N+5-k}$, and (2.21d.k) corresponds to the root $\alpha_{2,N+5-k}$.

Note that for a fixed module and fixed $i = 1, 2, 3, 4$, only one of the odd N conditions involving d^i_{Nk} may be satisfied. Thus, no more than four of the conditions (2.21) (two, for $N = 1$) may hold for a given Verma module.

Remark. Note that for $n_2 \in \mathbb{N}$ (cf. (2.19)) the corresponding irreps of su(2,2) are finite dimensional (the necessary and sufficient condition for this is $n_1, n_2, n_3 \in \mathbb{N}$). Then the irreducible LWM L_Λ of su(2,2/N) are also finite dimensional (and non-unitary) independently on whether the corresponding Verma module V^Λ is reducible w.r.t. any odd root. If V^Λ is not reducible w.r.t. any odd root, then these finite-dimensional irreps are called 'typical' [243], otherwise, the irreps are called 'atypical' [243]. In our considerations $n_2 \notin \mathbb{N}$ in all cases, except the trivial 1-dimensional UIR (for which $n_2 = 1$, cf. below). ◇

We shall consider quotients of Verma modules factoring out the *even* submodules for which the reducibility conditions are always fulfilled. Before this we recall the root vectors following [139]. The positive (negative) root vectors corresponding to α_{ij}, $(-\alpha_{ij})$, are denoted by X^+_{ij} (X^-_{ij}). In the $su(2, 2/N)$ matrix notation the convention of [139], (2.7), is

$$X^+_{ij} = \begin{cases} e_{ji} & \text{for } (i,j) = (3,4), (3,j), (4,j),\ 5 \le j \le N+4 \\ e_{ij} & \text{otherwise} \end{cases}$$

$$X^-_{ij} = {}^t(X^+_{ij}) \tag{2.23}$$

where e_{ij} are $(N+4) \times (N+4)$ matrices with all elements zero except the element equal to 1 on the intersection of the ith row and jth column. The simple root vectors X^+_i follow the notation of the simple roots γ_i (2.13):

$$X_1^+ \equiv X_{12}^+, \quad X_2^+ \equiv X_{34}^+, \quad X_3^+ \equiv X_{25}^+, \quad X_4^+ \equiv X_{4,4+N}^+,$$
$$X_k^+ \equiv X_{k,k+1}^+, \quad 5 \leq k \leq 3 + N. \tag{2.24}$$

The mentioned submodules are generated by the singular vectors related to the even simple roots $\gamma_1, \gamma_2, \gamma_5, \ldots, \gamma_{N+3}$ [139]:

$$v_s^1 = (X_1^+)^{1+2j_1} v_0, \tag{2.25a}$$
$$v_s^2 = (X_2^+)^{1+2j_2} v_0, \tag{2.25b}$$
$$v_s^j = (X_j^+)^{1+r_{N+4-j}} v_0, \quad j = 5, \ldots, N+3 \tag{2.25c}$$

(for $N = 1$ (2.25c) being empty). The corresponding submodules are $I_k^\Lambda = U(\mathcal{G}^+) v_s^k$, and the invariant submodule to be factored out is

$$I_c^\Lambda = \bigcup_k I_k^\Lambda. \tag{2.26}$$

Thus, instead of V^Λ we shall consider the factor modules:

$$\widetilde{V}^\Lambda = V^\Lambda / I_c^\Lambda, \tag{2.27}$$

which are closer to the structure of the ERs. In the factorized modules the singular vectors (2.25) become null conditions, i. e., denoting by $\widetilde{|\Lambda\rangle}$ the lowest weight vector of $\widetilde{V}^\Lambda$, we have

$$(X_1^+)^{1+2j_1} \widetilde{|\Lambda\rangle} = 0, \tag{2.28a}$$
$$(X_2^+)^{1+2j_2} \widetilde{|\Lambda\rangle} = 0, \tag{2.28b}$$
$$(X_j^+)^{1+r_{N+4-j}} \widetilde{|\Lambda\rangle} = 0, \quad j = 5, \ldots, N+3. \tag{2.28c}$$

2.1.3 Singular vectors and invariant submodules at the unitary reduction points

We first recall the result of [140] (cf. part (i) of the theorem there) that the following is the complete list of lowest weight (positive energy) UIRs of su(2, 2/N):

$$d \geq d_{\max} = \max(d_{N1}^1, d_{NN}^3), \tag{2.29a}$$
$$d = d_{NN}^4 \geq d_{N1}^1, \quad j_1 = 0, \tag{2.29b}$$
$$d = d_{N1}^2 \geq d_{NN}^3, \quad j_2 = 0, \tag{2.29c}$$
$$d = d_{N1}^2 = d_{NN}^4, \quad j_1 = j_2 = 0, \tag{2.29d}$$

where $d_{\max}$ is the threshold of the continuous unitary spectrum.

Remark. Note that from (2.29a) follows

$$d_{\max} \geq 2 + j_1 + j_2 + m_1,$$

the equality being achieved only when $d^1_{N1} = d^3_{NN}$, while from (2.29b,c) follows

$$d \geq 1 + j_1 + j_2 + m_1, \quad j_1 j_2 = 0,$$

the equality being achieved only when $d^4_{NN} = d^1_{N1}$, or $d^2_{N1} = d^3_{NN}$, for (2.29b) and (2.29c), respectively. We recall the unitarity conditions [291] for the conformal algebra su(2,2):

$$\begin{aligned} d &\geq 2 + j_1 + j_2, \quad j_1 j_2 > 0, \\ d &\geq 1 + j_1 + j_2, \quad j_1 j_2 = 0, \end{aligned} \tag{2.30}$$

i. e., the superconformal unitarity conditions are more stringent that the conformal ones. ◇

Note that in case (d) we have $d = m_1$, $z = 2m/N - m_1$, and that it is trivial for $N = 1$ since then the internal symmetry algebra su(N) is trivial and by definition $m_1 = m = 0$ (the resulting irrep is 1 dimensional with $d = z = j_1 = j_2 = 0$). The UIRs for $N = 1$ (where case (2.29d) is missing) were first given in [182].

Next we note that if $d > d_{\max}$ the factorized Verma modules are irreducible and coincide with the UIRs L_Λ. These UIRs are called **longUIRs** in the modern literature, cf., e. g., [189, 171, 175, 174, 20, 58, 159, 228]. Analogously, we shall use for the cases when $d = d_{\max}$, i. e., (2.29a), the terminology of **semi-short UIRs**, introduced in [189, 174], while the cases (2.29b,c,d) are also called **short UIRs**, cf. [171, 175, 174, 20, 58, 159, 228].

Next we consider in more detail the UIRs the four distinguished reduction points determining the list above:

$$d^1_{N1} = 2 + 2j_2 + z + 2m_1 - 2m/N, \tag{2.31a}$$

$$d^2_{N1} = z + 2m_1 - 2m/N, \quad (j_2 = 0), \tag{2.31b}$$

$$d^3_{NN} = 2 + 2j_1 - z + 2m/N, \tag{2.31c}$$

$$d^4_{NN} = -z + 2m/N, \quad (j_1 = 0). \tag{2.31d}$$

First we recall the singular vectors corresponding to these points. The above reducibilities occur for the following odd roots, respectively:

$$\alpha_{3,4+N} = \gamma_2 + \gamma_4, \quad \alpha_{4,4+N} = \gamma_4, \quad \alpha_{15} = \gamma_1 + \gamma_3, \quad \alpha_{25} = \gamma_3. \tag{2.32}$$

The second and the fourth are the two odd simple roots:

$$\gamma_3 = \alpha_{25}, \quad \gamma_4 = \alpha_{4,4+N} \tag{2.33}$$

and the other two are simply related to these:

$$\alpha_{15} = \alpha_{12} + \alpha_{25} = \gamma_1 + \gamma_3, \quad \alpha_{3,4+N} = \alpha_{34} + \alpha_{4,4+N} = \gamma_2 + \gamma_4. \tag{2.34}$$

Thus, the corresponding singular vectors are

$$v^1_{\text{odd}} = P_{3,4+N}v_0 = (X_4^+X_2^+(h_2-1) - X_2^+X_4^+h_2)v_0 \tag{2.35a}$$
$$= (2j_2X_2^+X_4^+ - (2j_2+1)X_4^+X_2^+)v_0$$
$$= (2j_2X^+_{3,4+N} - X_4^+X_2^+)v_0, \quad d = d^1_{N1} \tag{2.35a$'$}$$
$$v^2_{\text{odd}} = X_4^+v_0, \quad d = d^2_{N1} \tag{2.35b}$$
$$v^3_{\text{odd}} = P_{15}v_0 = (X_3^+X_1^+(h_1-1) - X_1^+X_3^+h_1)v_0 \tag{2.35c}$$
$$= (2j_1X_1^+X_3^+ - (2j_1+1)X_3^+X_1^+)v_0$$
$$= (2j_1X^+_{15} - X_3^+X_1^+)v_0, \quad d = d^3_{NN} \tag{2.35c$'$}$$
$$v^4_{\text{odd}} = X_3^+v_0, \quad d = d^4_{NN}, \tag{2.35d}$$

where $X^+_{3,4+N} = [X_2^+, X_4^+]$ is the odd root vector corresponding to the root $\alpha_{3,4+N}$, $X^+_{15} = [X_1^+, X_3^+]$ is the odd root vector corresponding to the root α_{15}, $h_1, h_2 \in \mathcal{H}$ are Cartan generators corresponding to the roots γ_1, γ_2, (cf. [139]), and passing from the (2.35a) and (2.35c), to the next line we have used the fact that $h_2v_0 = -2j_2v_0$ ($h_1v_0 = -2j_1v_0$), consistently with (2.18b) and (2.18a). These vectors are given in (8.9a),(8.7b),(8.8a),(8.7a), respectively, of [139].

These singular vectors carry over for the factorized Verma modules $\widetilde{V}^\Lambda$:

$$\tilde{\varphi}^1_{\text{odd}} = P_{3,4+N}|\widetilde{\Lambda}\rangle = (X_4^+X_2^+(h_2-1) - X_2^+X_4^+h_2)|\widetilde{\Lambda}\rangle$$
$$= (2j_2X^+_{3,4+N} - X_4^+X_2^+)|\widetilde{\Lambda}\rangle, \quad d = d^1_{N1}, \tag{2.36a}$$
$$\tilde{\varphi}^2_{\text{odd}} = X_4^+|\widetilde{\Lambda}\rangle, \quad d = d^2_{N1}, \tag{2.36b}$$
$$\tilde{\varphi}^3_{\text{odd}} = P_{15}|\widetilde{\Lambda}\rangle = (X_3^+X_1^+(h_1-1) - X_1^+X_3^+h_1)|\widetilde{\Lambda}\rangle$$
$$= (2j_1X^+_{15} - X_3^+X_1^+)|\widetilde{\Lambda}\rangle, \quad d = d^3_{NN}, \tag{2.36c}$$
$$\tilde{\varphi}^4_{\text{odd}} = X_3^+|\widetilde{\Lambda}\rangle, \quad d = d^4_{NN}, \tag{2.36d}$$

where $X^+_{3,4+N} = [X_2^+, X_4^+]$, $X^+_{15} = [X_1^+, X_3^+]$, $h_1, h_2 \in \mathcal{H}$ are Cartan generators corresponding to the roots γ_1, γ_2 (cf. [139]), and passing from the (2.36a) and (2.36c), respectively, to the next line we have used the fact that $h_2v_0 = -2j_2v_0$, $h_1v_0 = -2j_1v_0$, respectively, consistently with (2.18b), (2.18a), respectively

For $j_1 = 0$, $j_2 = 0$, respectively, the vector v^3_{odd}, v^1_{odd}, respectively, is a descendant of the singular vector v^1_s, v^2_s, respectively; cf. (2.25a) and (2.25b), respectively. In the same situations the tilde counterparts $\tilde{\varphi}^1_s$, $\tilde{\varphi}^2_s$ are just zero—cf. (2.28a) and (2.28b), respectively. However, then there is another independent singular vector of $\widetilde{V}^\Lambda$ in both cases. For $j_1 = 0$ it corresponds to the sum of two roots: $\alpha_{15} + \alpha_{25}$ (which sum is not a root!) and is given by equation (D.1) of [139]:

$$\tilde{\varphi}^{34} = X_3^+X_1^+X_3^+|\widetilde{\Lambda}\rangle = X_3^+X^+_{15}|\widetilde{\Lambda}\rangle, \quad d = d^3_{NN}, \ j_1 = 0 \tag{2.37}$$

Checking singularity we see at once that $X_k^-\tilde{\varphi}^{34} = 0$ for $k \neq 3$. It remains to calculate the action of X_3^-:

$$\begin{aligned} X_3^- \tilde{\varphi}^{34} &= h_3 X_1^+ X_3^+ \widetilde{|\Lambda\rangle} - X_3^+ X_1^+ h_3 \widetilde{|\Lambda\rangle} \\ &= X_1^+ X_3^+ (h_3 - 1)\widetilde{|\Lambda\rangle} - X_3^+ X_1^+ h_3 \widetilde{|\Lambda\rangle} = 0, \end{aligned} \tag{2.38}$$

$h_3, h_4 \in \mathcal{H}$ are Cartan generators corresponding to the roots γ_3, γ_4 (cf. [139]), the first term is zero since $\Lambda(h_3) - 1 = \frac{1}{2}(d - d^3_{NN}) = 0$, while the second term is zero due to (2.28a) for $j_1 = 0$.

For $j_2 = 0$ there is a singular vector corresponding to the sum of two roots: $\alpha_{3,4+N} + \alpha_{4,4+N}$ (which sum is not a root) and is given in [139] (cf. the formula before (D.4) there):

$$\tilde{\varphi}^{12} = X_4^+ X_2^+ X_4^+ \widetilde{|\Lambda\rangle} = X_4^+ X_{3,4+N}^+ \widetilde{|\Lambda\rangle}, \quad d = d^1_{N1}, j_2 = 0 \tag{2.39}$$

As above, one checks that $X_k^- v^{12} = 0$ for $k \neq 4$, and then calculates

$$\begin{aligned} X_4^- \tilde{\varphi}^{12} &= h_4 X_2^+ X_4^+ \widetilde{|\Lambda\rangle} - X_4^+ X_2^+ h_4 \widetilde{|\Lambda\rangle} \\ &= X_2^+ X_4^+ (h_4 - 1)\widetilde{|\Lambda\rangle} - X_4^+ X_2^+ h_4 \widetilde{|\Lambda\rangle} = 0 \end{aligned} \tag{2.40}$$

using $\Lambda(h_4) - 1 = \frac{1}{2}(d - d^1_{N1}) = 0$, and (2.28b) for $j_2 = 0$.

To the above two singular vectors in the ER picture correspond second-order super-differential operators given explicitly in formulas (11a,b) of [140], and in formulas (D3) and (D5) of [139].

Remark. Note that w.r.t. V^Λ the analogs of the vectors $\tilde{\varphi}^{34}$ and $\tilde{\varphi}^{12}$ are not singular, but *subsingular vectors* [120, 121]. Indeed, consider the vector in V^Λ given by the same $U(\mathcal{G}^+)$ monomial as $\tilde{\varphi}^{34}$: $v^{34} = X_3^+ X_1^+ X_3^+$. Clearly, $X_k^- v^{34} = 0$ for $k \neq 3$. It remains to calculate the action of X_3^-:

$$\begin{aligned} X_3^- v^{34} &= h_3 X_1^+ X_3^+ v_0 - X_3^+ X_1^+ h_3 v_0 \\ &= X_1^+ X_3^+ (h_3 - 1) v_0 - X_3^+ X_1^+ h_3 v_0 = -X_3^+ X_1^+ v_0 \end{aligned} \tag{2.41}$$

where the first term is zero as above, while the second term is a descendant of the singular vector $v_s^1 = X_1^+ v_0$ (cf. (2.25a) for $j_1 = 0$), which fulfills the definition of subsingular vector. Analogously, for the vector $v^{12} = X_4^+ X_2^+ X_4^+$ we have $X_k^- v^{12} = 0$ for $k \neq 4$, and

$$X_4^- v^{12} = X_4^- X_4^+ X_2^+ X_4^+ = -X_4^+ X_2^+ v_0,$$

(using $\Lambda(h_4) - 1$), which is a descendant of the singular vector $v_s^2 = X_2^+ v_0$, cf. (2.25b) for $j_2 = 0$. ◇

From the expressions of the singular vectors follow, using (2.8), the explicit formulas for the corresponding invariant submodules I^β of the modules $\widetilde{V}^\Lambda$ as follows:

$$\begin{aligned} I^1 &= U(\mathcal{G}^+) P_{3,4+N} \widetilde{|\Lambda\rangle} = U(\mathcal{G}^+)(X_4^+ X_2^+ (h_2 - 1) - X_2^+ X_4^+ h_2)\widetilde{|\Lambda\rangle} \\ &= U(\mathcal{G}^+)(2 j_2 X_{3,4+N}^+ - X_4^+ X_2^+)\widetilde{|\Lambda\rangle}, \quad d = d^1_{N1},\ j_2 > 0, \end{aligned} \tag{2.42a}$$

$$I^2 = U(\mathcal{G}^+) X_4^+ \widetilde{|\Lambda\rangle}, \quad d = d^2_{N1}, \tag{2.42b}$$

$$I^3 = U(\mathcal{G}^+)P_{15}|\widetilde{\Lambda}\rangle = U(\mathcal{G}^+)(X_3^+X_1^+(h_1-1) - X_1^+X_3^+h_1)|\widetilde{\Lambda}\rangle$$
$$= U(\mathcal{G}^+)(2j_1X_{15}^+ - X_3^+X_1^+)|\widetilde{\Lambda}\rangle, \quad d = d^3_{NN}, \, j_1 > 0, \tag{2.42c}$$
$$I^4 = U(\mathcal{G}^+)X_3^+|\widetilde{\Lambda}\rangle, \quad d = d^4_{NN}, \tag{2.42d}$$
$$I^{12} = U(\mathcal{G}^+)\tilde{\varphi}^{12} = X_4^+X_2^+X_4^+|\widetilde{\Lambda}\rangle, \quad d = d^1_{N1}, \, j_2 = 0, \tag{2.42e}$$
$$I^{34} = U(\mathcal{G}^+)\tilde{\varphi}^{34} = X_3^+X_1^+X_3^+|\widetilde{\Lambda}\rangle, \quad d = d^3_{NN}, \, j_1 = 0. \tag{2.42f}$$

Sometimes we shall indicate the signature $\chi(\Lambda)$, writing, e. g., $I^1(\chi)$; sometimes we shall indicate also the resulting signature, writing, e. g., $I^1(\chi,\chi')$ – this is a redundancy since it is determined by what is displayed already: $\chi' = \chi(\Lambda+\beta)$, but will be useful to see immediately in the concrete situations without calculation.

The invariant submodules were used in [140] in the construction of the UIRs, as we shall recall below.

2.1.4 Structure of single-reducibility-condition Verma modules and UIRs

We discuss now the reducibility of Verma modules at the four distinguished points (2.31). We note a partial ordering of these four points:

$$d^1_{N1} > d^2_{N1}, \quad d^3_{NN} > d^4_{NN}, \tag{2.43}$$

or more precisely:

$$d^1_{N1} = d^2_{N1} + 2, (j_2 = 0); \quad d^3_{NN} = d^4_{NN} + 2, (j_1 = 0). \tag{2.44}$$

Due to this ordering at most two of these four points may coincide. Thus, we have two possible situations: of Verma modules (or ERs) reducible at one and at two reduction points from (2.31).

In this section we deal with the situations in which *no two* of the points in (2.31) coincide. According to [140] (Theorem) there are four such situations involving UIRs:

$$d = d_{\max} = d^1_{N1} > d^3_{NN}, \tag{2.45a}$$
$$d = d^2_{N1} > d^3_{NN}, \quad j_2 = 0, \tag{2.45b}$$
$$d = d_{\max} = d^3_{NN} > d^1_{N1}, \tag{2.45c}$$
$$d = d^4_{NN} > d^1_{N1}, \quad j_1 = 0. \tag{2.45d}$$

We shall call these cases *single-reducibility-condition (SRC)* Verma modules or UIRs, depending on the context. In addition, as already stated, we use for the cases when $d = d_{\max}$, i. e., (2.45a,c), the terminology of semi-short UIRs, [189, 174], while the cases (2.45b,d) are also called short UIRs, [171, 175, 174, 20, 58, 159, 228].

As we see the SRC cases have supplementary conditions as specified. And due to the inequalities there are the following additional restrictions:

$$z > j_1 - j_2 - m_1 + 2m/N, \tag{2.46a}$$
$$z > j_1 + 1 - m_1 + 2m/N, \tag{2.46b}$$
$$z < j_1 - j_2 - m_1 + 2m/N, \tag{2.46c}$$
$$z < -1 - j_2 - m_1 + 2m/N. \tag{2.46d}$$

Using these inequalities the unitarity conditions (2.45) may be rewritten more explicitly:

$$d = d_{\max} = d^1_{N1} = d^a \equiv 2 + 2j_2 + z + 2m_1 - 2m/N > d^3_{NN} \tag{2.47a}$$
$$d = d^2_{N1} > d^3_{NN}, \quad j_2 = 0, \tag{2.47b}$$
$$d = d_{\max} = d^3_{NN} = d^c \equiv 2 + 2j_1 - z + 2m/N > d^1_{N1}, \tag{2.47c}$$
$$d = d^4_{NN} > d^1_{N1}, \quad j_1 = 0, \tag{2.47d}$$

where for future use we have introduced notation d^a, d^c.

To finalize the structure we should check the even reducibility conditions (2.19b,d,e,f). It is enough to note that the conditions on d in (2.47a,c):

$$d > 2 + j_1 + j_2 + m_1$$

and in (2.47b,d):

$$d > 1 + j_1 + j_2 + m_1, (j_1 j_2 = 0)$$

are incompatible with (2.19b,d,e,f), except in two cases. The exceptions are in cases (2.47b,d) when $d = 2 + j_1 + j_2 = z$ and $j_1 j_2 = 0$. In these cases we have $n_{14} = 1$ in (2.19f) and there exists a Verma submodule $V^{\Lambda+\alpha_{14}}$. However, the su(2, 2) signature $\chi_0(\Lambda + \alpha_{14})$ is unphysical: $[j_1 - \frac{1}{2}, -\frac{1}{2}; 3 + j_1]$ for $j_2 = 0$, and $[-\frac{1}{2}, j_2 - \frac{1}{2}; 3 + j_1]$ for $j_1 = 0$. Thus, there is no such submodule of $\widetilde{V}^\Lambda$.

Thus, the factorized Verma modules $\widetilde{V}^\Lambda$ with the unitary signatures from (2.45) have only one invariant (odd) submodule which has to be factorized in order to obtain the UIRs. These odd embeddings are given explicitly by

$$\widetilde{V}^\Lambda \to \widetilde{V}^{\Lambda+\beta} \tag{2.48}$$

where we use the convention [138] that arrows point to the oddly embedded module, and we have the following cases for β:

$$\beta = \alpha_{3,4+N}, \quad \text{for (2.45a)}, \ j_2 > 0, \tag{2.49a}$$
$$= \alpha_{4,4+N}, \quad \text{for (2.45b)}, \tag{2.49b}$$
$$= \alpha_{15}, \quad \text{for (2.45c)}, \ j_1 > 0, \tag{2.49c}$$

$$= \alpha_{25}, \quad \text{for (2.45d)}, \tag{2.49d}$$

$$= \alpha_{3,4+N} + \alpha_{4,4+N}, \quad \text{for (2.45a)}, \ j_2 = 0, \tag{2.49e}$$

$$= \alpha_{15} + \alpha_{25}, \quad \text{for (2.45c)}, \ j_1 = 0 \tag{2.49f}$$

This diagram gives the UIR L_Λ contained in $\widetilde{V}^\Lambda$ as follows:

$$L_\Lambda = \widetilde{V}^\Lambda / I^\beta, \tag{2.50}$$

where I^β is given by I^1, I^2, I^3, I^4, I^{12}, I^{34}, respectively, (cf. (2.42)), in the cases (2.49a,b,c,d,e,f), respectively.

It is useful to record the signatures of the shifted lowest weights, i. e., $\chi' = \chi(\Lambda+\beta)$. In fact, for future use we give the signature changes for arbitrary roots. The explicit formulas are [138, 139]:

$$\beta = \alpha_{3,N+5-k}, \quad j_2 > 0, \ r_{k-1} > 0, \tag{2.51a}$$

$$\chi' = \left[d + \frac{1}{2}; \, j_1, j_2 - \frac{1}{2}; \, z + \epsilon_N; \, r_1, \ldots, r_{k-1} - 1, r_k + 1, \ldots, r_{N-1}\right]$$

$$\beta = \alpha_{4,N+5-k}, \quad r_{k-1} > 0, \tag{2.51b}$$

$$\chi' = \left[d + \frac{1}{2}; \, j_1, j_2 + \frac{1}{2}; \, z + \epsilon_N; \, r_1, \ldots, r_{k-1} - 1, r_k + 1, \ldots, r_{N-1}\right]$$

$$\beta = \alpha_{1,N+5-k}, \quad j_1 > 0, \ r_k > 0, \tag{2.51c}$$

$$\chi' = \left[d + \frac{1}{2}; \, j_1 - \frac{1}{2}, j_2; \, z - \epsilon_N; \, r_1, \ldots, r_{k-1} + 1, r_k - 1, \ldots, r_{N-1}\right]$$

$$\beta = \alpha_{2,N+5-k}, \quad r_k > 0, \tag{2.51d}$$

$$\chi' = \left[d + \frac{1}{2}; \, j_1 + \frac{1}{2}, j_2; \, z - \epsilon_N; \, r_1, \ldots, r_{k-1} + 1, r_k - 1, \ldots, r_{N-1}\right]$$

$$\beta_{12} = \alpha_{3,4+N} + \alpha_{4,4+N}, \tag{2.51e}$$

$$\chi'_{12} = [d + 1; \, j_1, 0; \, z + 2\epsilon_N; \, r_1 + 2, r_2, \ldots, r_{N-1}],$$

$$\beta_{34} = \alpha_{15} + \alpha_{25}, \tag{2.51f}$$

$$\chi'_{34} = [d + 1; \, 0, j_2; \, z - 2\epsilon_N; \, r_1, \ldots, r_{N-2}, r_{N-1} + 2],$$

$$\epsilon_N \equiv \frac{2}{N} - \frac{1}{2} \tag{2.52}$$

For each fixed χ the lowest weight $\Lambda(\chi')$ fulfills the same odd reducibility condition as $\Lambda(\chi)$. The lowest weight $\Lambda(\chi'_{12})$ fulfils (2.45b), while the lowest weight $\Lambda(\chi'_{34})$ fulfils (2.45d).

The embedding diagram (2.48) is a piece of a much richer picture [138]. Indeed, notice that if (2.3b) is fulfilled for some odd root β, then it is fulfilled also for an infinite number of Verma modules $V_\ell = V^{\Lambda+\ell\beta}$ for all $\ell \in \mathbb{Z}$. These modules form an infinite chain complex of oddly embedded modules:

$$\cdots \to V_{-1} \to V_0 \to V_1 \to \cdots \tag{2.53}$$

Because of (2.9) this is an exact sequence with one nilpotent operator involved in the whole chain. Of course, once we restrict to the factorized modules $\widetilde{V}^\Lambda$ the diagram will be shortened – this is evident from the signature changes (2.51a,b,c,d). In fact, there are only a finite number of factorized nodules for $N > 1$, while for $N = 1$ the diagram continues to be infinite to the left. Furthermore, when $\beta = \beta_{12}, \beta_{34}$ from the end of the restricted chain one transmutes—via the embeddings (2.42e,f), respectively—to the chain with $\beta = \alpha_{4,N+4}, \alpha_{25}$, respectively. More explicitly, when $\beta = \beta_{12}, \beta_{34}$, then the module V_1 plays the role of V_0 with $\beta = \alpha_{4,N+4}, \alpha_{25}$. All this is explained in detail in [138]. Furthermore, when a factorized Verma module $\widetilde{V}^\Lambda = \widetilde{V}_0^\Lambda$ contains an UIR then not all modules $\widetilde{V}_\ell^\Lambda$ would contain an UIR, [139, 140]. From all this what is important from the view of modern applications can be summarized as follows:

- The semi-short SRC UIRs (cf. (2.45a,c)) are obtained by factorizing a Verma submodule $\widetilde{V}^{\Lambda+\beta}$ containing either another semi-short SRC UIR of the same type (cf. (2.49a,c)) or containing a short SRC UIR of a different type (cf. (2.49e,f)). In contrast, short SRC UIRs (cf. (2.45b,d)) are obtained by factorizing a Verma submodule $\widetilde{V}^{\Lambda+\beta}$ whose irreducible factor module is not unitary (cf. (2.49b,d)).

2.1.5 Structure of double-reducibility-condition Verma modules and UIRs

We consider now the situations in which *two* of the points in (2.31) coincide. According to [140] (Theorem) there are four such situations involving UIRs:

$$d = d_{\max} = d^{ac} \equiv 2 + j_1 + j_2 + m_1 = d^1_{N1} = d^3_{NN}, \tag{2.54a}$$
$$d = d^1_{N1} = d^4_{NN} = 1 + j_2 + m_1, \quad j_1 = 0, \tag{2.54b}$$
$$d = d^2_{N1} = d^3_{NN} = 1 + j_1 + m_1, \quad j_2 = 0, \tag{2.54c}$$
$$d = d^2_{N1} = d^4_{NN} = m_1, \quad j_1 = j_2 = 0. \tag{2.54d}$$

We shall call these *double-reducibility-condition (DRC)* Verma modules or UIRs. As in the previous subsection we shall use for the cases when $d = d_{\max}$, i. e., (2.54a), also the terminology of semi-short UIRs, [189, 174], while the cases (2.54b,c,d) shall also be called short UIRs, [171, 175, 174, 20, 58, 159, 228].

For later use we list more explicitly the values of d and z

$$\begin{aligned} d &= d^{ac} = d^1_{N1} = d^3_{NN} = 2 + j_1 + j_2 + m_1, \\ z &= j_1 - j_2 + 2m/N - m_1; \end{aligned} \tag{2.55a}$$
$$\begin{aligned} d &= d^1_{N1} = d^4_{NN} = 1 + j_2 + m_1, \quad j_1 = 0, \\ z &= -1 - j_2 + 2m/N - m_1; \end{aligned} \tag{2.55b}$$
$$\begin{aligned} d &= d^2_{N1} = d^3_{NN} = 1 + j_1 + m_1, \quad j_2 = 0, \\ z &= 1 + j_1 + 2m/N - m_1; \end{aligned} \tag{2.55c}$$

$$d = d^2_{N1} = d^4_{NN} = m_1, \quad j_1 = j_2 = 0,$$
$$z = 2m/N - m_1. \tag{2.55d}$$

We noted already that for $N = 1$ the last case, **d**, is trivial. Note also that for $N = 2$ we have $2m/N - m_1 = m - m_1 = 0$.

To finalize the structure we should check the even reducibility conditions (2.19b,d,e,f). It is enough to note that the values of d in (2.54), (2.55) are incompatible with (2.19b,d,e,f), except in a few cases. The exceptions are

$$d = d^1_{N1} = d^3_{NN} = 2 + j_1 + j_2, \quad m_1 = 0 \tag{2.56a}$$
$$d = d^1_{N1} = d^4_{NN} = 1 + j_2 + m_1, \quad j_1 = 0,\ m_1 = 0, 1 \tag{2.56b}$$
$$d = d^2_{N1} = d^3_{NN} = 1 + j_1 + m_1, \quad j_2 = 0,\ m_1 = 0, 1 \tag{2.56c}$$
$$d = d^2_{N1} = d^4_{NN} = m_1, \quad j_1 = j_2 = 0,\ m_1 = 0, 1, 2 \tag{2.56d}$$

- In case (2.56a) we have $n_{14} = 1$ in (2.19f) and there exists a Verma submodule $V^{\Lambda+\alpha_{14}}$ with su(2, 2) signature $\chi_0(\Lambda + \alpha_{14}) = [j_1 - \frac{1}{2}, j_2 - \frac{1}{2}; 3 + j_1 + j_2]$. As we can see this signature is unphysical for $j_1 j_2 = 0$. Thus, there is the even submodule $\widetilde{V}^{\Lambda+\alpha_{14}}$ of $\widetilde{V}^\Lambda$ only if $j_1 j_2 \neq 0$.
- In case (2.56b) there are three subcases:

$m_1 = 0, j_2 = \frac{1}{2}$; then $d = \frac{3}{2}$, $n_{24} = 1$, $n_{14} = 2$. The signatures of the embedded submodules of V^Λ are $\chi_0(\Lambda + \alpha_{24}) = [\frac{1}{2}, 0; \frac{5}{2}]$, $\chi_0(\Lambda + 2\alpha_{14}) = [-1, -\frac{1}{2}; \frac{7}{2}]$. Thus, there is only the even submodule $\widetilde{V}^{\Lambda+\alpha_{24}}$ of $\widetilde{V}$.

$m_1 = 0, j_2 = 0$; then $d = 1$, $n_{13} = 1$, $n_{24} = 1$, $n_{14} = 2$. The signatures of the embedded submodules of V^Λ are $\chi_0(\Lambda + \alpha_{13}) = [-\frac{1}{2}, \frac{1}{2}; 2]$, $\chi_0(\Lambda + \alpha_{24}) = [\frac{1}{2}, -\frac{1}{2}; 2]$, $\chi_0(\Lambda + 2\alpha_{14}) = [-1, -1; 3]$, and are all unphysical. However, the Verma module V^Λ has a subsingular vector of weight $\alpha_{23} + \alpha_{14}$, cf. [120], and thus, the factorized Verma module $\widetilde{V}^\Lambda$ has the submodule $\widetilde{V}^{\Lambda+\alpha_{23}+\alpha_{14}}$.

$m_1 = 1$; then $n_{14} = 1$, but as above there is no non-trivial even submodule of $\widetilde{V}^\Lambda$.

- The case (2.56c) is dual to (2.56b) so we list briefly the three subcases:

$m_1 = 0, j_1 = \frac{1}{2}$; then $d = \frac{3}{2}$, $n_{13} = 1$, $n_{14} = 2$. There is only the even submodule $\widetilde{V}^{\Lambda+\alpha_{13}}$ of $\widetilde{V}$.

$m_1 = 0, j_1 = 0$; then $d = 1$, $n_{13} = 1$, $n_{24} = 1$, $n_{14} = 2$. This subcase coincides with the second subcase of (2.56b).

$m_1 = 1$; then $n_{14} = 1$ and as above there is no non-trivial submodule of $\widetilde{V}^\Lambda$.

- In case (2.56d) there are again three subcases:

$m_1 = 0$; then all quantum numbers in the signature are zero and the UIR is the 1-dimensional trivial irrep.

$m_1 = 1$; then $d = 1$, $n_{13} = 1$, $n_{24} = 1$, $n_{14} = 2$. Though this subcase has non-trivial isospin from su(2, 2) point of view it has the same structure as the second subcase of (2.56b) and the factorized Verma module $\widetilde{V}^\Lambda$ has the submodule $\widetilde{V}^{\Lambda+\alpha_{23}+\alpha_{14}}$.

$m_1 = 2$; then $d = 2$ and $n_{14} = 1$; as above there is no non-trivial even submodule of $\widetilde{V}^\Lambda$.

The embedding diagrams for the corresponding modules $\widetilde{V}^\Lambda$ when there are no even embeddings are

$$\begin{array}{ccc} \widetilde{V}^{\Lambda+\beta'} & & \\ \uparrow & & \\ \widetilde{V}^\Lambda & \longrightarrow & \widetilde{V}^{\Lambda+\beta} \end{array} \tag{2.57}$$

$$(\beta,\beta') = (\alpha_{15},\alpha_{3,4+N}), \quad \text{for (2.54a)}, \ j_1 j_2 > 0 \tag{2.58a}$$

$$= (\alpha_{15},\alpha_{3,4+N} + \alpha_{3,4+N}), \quad \text{for (2.54a)}, \ j_1 > 0, \ j_2 = 0 \tag{2.58b}$$

$$= (\alpha_{15} + \alpha_{25},\alpha_{3,4+N}), \quad \text{for (2.54a)}, \ j_1 = 0, \ j_2 > 0 \tag{2.58c}$$

$$= (\alpha_{15} + \alpha_{25},\alpha_{3,4+N} + \alpha_{3,4+N}), \quad \text{for (2.54a)}, \ j_1 = j_2 = 0 \tag{2.58d}$$

$$= (\alpha_{25},\alpha_{3,4+N}), \quad \text{for (2.54b)}, \ j_2 > 0, \tag{2.58e}$$

$$= (\alpha_{25},\alpha_{3,4+N} + \alpha_{4,4+N}), \quad \text{for (2.54b)}, \ j_2 = 0, \tag{2.58f}$$

$$= (\alpha_{15},\alpha_{4,4+N}), \quad \text{for (2.54c)}, \ j_1 > 0, \tag{2.58g}$$

$$= (\alpha_{15} + \alpha_{25},\alpha_{4,4+N}), \quad \text{for (2.54c)}, \ j_1 = 0, \tag{2.58h}$$

$$= (\alpha_{25},\alpha_{4,4+N}), \quad \text{for (2.54d)} \tag{2.58i}$$

This diagram gives the UIR L_Λ contained in $\widetilde{V}^\Lambda$ as follows:

$$L_\Lambda = \widetilde{V}^\Lambda / I^{\beta,\beta'}, \quad I^{\beta,\beta'} = I^\beta \cup I^{\beta'} \tag{2.59}$$

where I^β, $I^{\beta'}$ are given in (2.42), according to the cases in (2.58).

The embedding diagrams for the corresponding modules $\widetilde{V}^\Lambda$ when there are even embeddings are

$$\begin{array}{ccccc} & & \widetilde{V}^{\Lambda+\beta'} & & \\ & & \uparrow & & \\ \widetilde{V}^{\Lambda+\beta_e} & \longleftarrow & \widetilde{V}^\Lambda & \longrightarrow & \widetilde{V}^{\Lambda+\beta} \end{array} \tag{2.60}$$

where

$$(\beta,\beta',\beta_e) = (\alpha_{15},\alpha_{3,4+N},\alpha_{14}), \quad \text{for (2.54a)}, \ j_1 j_2 > 0, \ m_1 = 0 \tag{2.61a}$$

$$= (\alpha_{25},\alpha_{3,4+N},\alpha_{24}), \quad \text{for (2.54b)}, \ j_2 = \frac{1}{2}, \ m_1 = 0 \tag{2.61b}$$

$$= (\alpha_{25},\alpha_{3,4+N} + \alpha_{4,4+N},\alpha_{23} + \alpha_{14}), \quad \text{for (2.54b)}, \ j_2 = m_1 = 0 \tag{2.61c}$$

$$= (\alpha_{15},\alpha_{4,4+N},\alpha_{13}), \quad \text{for (2.54c)}, \ j_1 = \frac{1}{2}, \ m_1 = 0 \tag{2.61d}$$

$$= (\alpha_{15} + \alpha_{25},\alpha_{4,4+N},\alpha_{23} + \alpha_{14}), \quad \text{for (2.54c)}, \ j_1 = m_1 = 0 \tag{2.61e}$$

$$= (\alpha_{25},\alpha_{4,4+N},\alpha_{23} + \alpha_{14}), \quad \text{for (2.54d)}, \ m_1 = 1 \tag{2.61f}$$

This diagram gives the UIR L_Λ contained in $\widetilde{V}^\Lambda$ as follows:

$$L_\Lambda = \widetilde{V}^\Lambda / I^{\beta,\beta',\beta_e}, \quad I^{\beta,\beta'} = I^\beta \cup I^{\beta'} \cup \widetilde{V}^{\Lambda+\beta_e} \tag{2.62}$$

Naturally, the two odd embeddings in (2.57) or (2.60) are the combination of the different cases of (2.48). Similarly, like (2.48) is a piece of the richer picture (2.53), here we have the following analogs of (2.53) [138][4]

$$\begin{array}{ccccccc} & & \vdots & & & & \\ & & \uparrow & & & & \\ & & V_{01} & & & & \\ & & \uparrow & & & & \\ \cdots & \to & V_{00} & \to & V_{10} & \to & \cdots \\ & & \uparrow & & & & \\ & & \vdots & & & & \end{array} \qquad N = 1 \tag{2.63}$$

where $V_{k\ell} \equiv V^{\Lambda+k\beta+\ell\beta'}$, and β, β' are the roots appearing in (2.58a,e,g,i) (or (2.61a,b,d,f))

$$\begin{array}{ccccccc} & & \vdots & & \vdots & & \\ & & \uparrow & & \uparrow & & \\ \cdots & \to & V_{10} & \to & V_{11} & \to & \cdots \\ & & \uparrow & & \uparrow & & \\ \cdots & \to & V_{00} & \to & V_{01} & \to & \cdots \\ & & \uparrow & & \uparrow & & \\ & & \vdots & & \vdots & & \end{array} \qquad N > 1 \tag{2.64}$$

The difference between the cases $N = 1$ and $N > 1$ is due to the fact that if (2.3b) is fulfilled for V_{00} w.r.t. two odd roots β, β' then for $N > 1$ it is fulfilled also for all Verma modules $V_{k\ell}$ again w.r.t. these odd roots β, β', while for $N = 1$ it is fulfilled only for V_{k0} w.r.t. the odd root β and only for $V_{0\ell}$ w.r.t. the odd root β'.

In the cases (2.58b,c,d,f,h) (or (2.61c,e)) we have the same diagrams though their parametrization is more involved [138] (cf. also what we said about transmutation for

4 These diagrams are essential parts of much richer diagrams (which we do not need since we consider only UIRs-related modules) which are explicitly described for any N in [138], and shown there in Figure 1 (for $N = 1$) and Figure 2 (for $N = 2$).

the single chains after (2.53)). However, for the modules with $0 \le k, \ell \le 1$ (which we use) we have simply as before $V_{k\ell} = V^{\Lambda+k\beta+\ell\beta'}$ for the appropriate β, β'.

The richer structure for $N > 1$ has practical consequences for the calculation of the character formulas, as we shall now see.

2.2 Character formulas of positive energy UIRs

2.2.1 Character formulas: generalities

In the beginning of this subsection we follow Dixmier [110]. Let $\hat{\mathcal{G}}$ be a simple Lie algebra of rank ℓ with Cartan subalgebra $\hat{\mathcal{H}}$, root system $\hat{\mathcal{D}}$, simple root system $\hat{\pi}$. Let Γ, (respectively, Γ_+), be the set of all integral (respectively, integral dominant), elements of $\hat{\mathcal{H}}^*$, i. e., $\lambda \in \hat{\mathcal{H}}^*$ such that $(\lambda, \alpha_i^\vee) \in \mathbb{Z}$ (respectively, $\mathbb{Z}_+$), for all simple roots α_i ($\alpha_i^\vee \equiv 2\alpha_i/(\alpha_i, \alpha_i)$). Let V be a lowest weight module with lowest weight Λ and lowest weight vector v_0. It has the following decomposition:

$$V = \bigoplus_{\mu\in\Gamma_+} V_\mu, \quad V_\mu = \{u \in V \mid Hu = (\lambda + \mu)(H)u, \forall\, H \in \mathcal{H}\} \tag{2.65}$$

(Note that $V_0 = \mathbb{C}v_0$.) Let $E(\mathcal{H}^*)$ be the associative abelian algebra consisting of the series $\sum_{\mu\in\mathcal{H}^*} c_\mu e(\mu)$, where $c_\mu \in \mathbb{C}$, $c_\mu = 0$ for μ outside the union of a finite number of sets of the form $D(\lambda) = \{\mu \in \mathcal{H}^* \mid \mu \ge \lambda\}$, using some ordering of $\mathcal{H}^*$, e. g., the lexicographic one; the formal exponents $e(\mu)$ have the properties: $e(0) = 1$, $e(\mu)e(\nu) = e(\mu + \nu)$.

Then the (formal) character of V is defined by

$$\mathrm{ch}_0\, V = \sum_{\mu\in\Gamma_+} (\dim V_\mu) e(\Lambda + \mu) = e(\Lambda) \sum_{\mu\in\Gamma_+} (\dim V_\mu) e(\mu) \tag{2.66}$$

(We shall use subscript '0' for the even case.)

For a Verma module, i. e., $V = V^\Lambda$ one has $\dim V_\mu = P(\mu)$, where $P(\mu)$ is a generalized partition function, $P(\mu) = \#$ of ways μ can be presented as a sum of positive roots β, each root taken with its multiplicity $\dim \mathcal{G}_\beta$ ($= 1$ here), $P(0) \equiv 1$. Thus, the character formula for Verma modules is

$$\mathrm{ch}_0\, V^\Lambda = e(\Lambda) \sum_{\mu\in\Gamma_+} P(\mu) e(\mu) = e(\Lambda) \prod_{\alpha\in\Delta^+} (1 - e(\alpha))^{-1} \tag{2.67}$$

Further we recall the standard reflections in $\hat{\mathcal{H}}^*$:

$$s_\alpha(\lambda) = \lambda - (\lambda, \alpha^\vee)\alpha, \quad \lambda \in \hat{\mathcal{H}}^*,\ \alpha \in \hat{\mathcal{D}} \tag{2.68}$$

The Weyl group W is generated by the simple reflections $s_i \equiv s_{\alpha_i}$, $\alpha_i \in \hat{\pi}$. Thus every element $w \in W$ can be written as the product of simple reflections. It is said that w is written in a reduced form if it is written with the minimal possible number of simple reflections; the number of reflections of a reduced form of w is called the length of w, denoted by $\ell(w)$.

The Weyl character formula for the finite-dimensional irreducible LWM L_Λ over $\hat{\mathcal{G}}$, i. e., when $\Lambda \in -\Gamma_+$, has the form[5]

$$\mathrm{ch}_0\, L_\Lambda = \sum_{w\in W} (-1)^{\ell(w)}\, \mathrm{ch}_0\, V^{w\cdot\Lambda}, \quad \Lambda \in -\Gamma_+ \tag{2.69}$$

where the dot $\cdot$ action is defined by $w \cdot \lambda = w(\lambda - \rho) + \rho$. For future reference we note:

$$s_\alpha \cdot \Lambda = \Lambda + n_\alpha \alpha \tag{2.70}$$

where

$$n_\alpha = n_\alpha(\Lambda) \doteq (\rho - \Lambda, \alpha^\vee) = (\rho - \Lambda)(H_\alpha), \quad \alpha \in \Delta^+ \tag{2.71}$$

In the case of basic classical Lie superalgebras the first character formulas were given by Kac [243, 242].[6] For all such superalgebras (except osp(1/2N)) the character formula for Verma modules is [243, 242]:

$$\mathrm{ch}\, V^\Lambda = e(\Lambda)\Bigg(\prod_{\alpha\in\Delta^+_{\bar{0}}} (1 - e(\alpha))^{-1}\Bigg)\Bigg(\prod_{\alpha\in\Delta^+_{\bar{1}}} (1 + e(\alpha))\Bigg) \tag{2.72}$$

Note that the factor $\prod_{\alpha\in\Delta^+_{\bar{0}}}(1 - e(\alpha))^{-1}$ represents the states of the even sector: $V_0^\Lambda \equiv U((\mathcal{G}_+^{\mathbb{C}})_{(0)})v_0$ (as above in the even case), while $\prod_{\alpha\in\Delta^+_{\bar{1}}}(1 + e(\alpha))$ represents the states of the odd sector: $\hat{V}^\Lambda \equiv (U(\mathcal{G}_+^{\mathbb{C}})/U((\mathcal{G}_+^{\mathbb{C}})_{(0)}))v_0$. Thus, we may introduce a character for $\hat{V}^\Lambda$ as follows:

$$\mathrm{ch}\, \hat{V}^\Lambda \equiv \prod_{\alpha\in\Delta^+_{\bar{1}}} (1 + e(\alpha)). \tag{2.73}$$

In our case, $\hat{V}^\Lambda$ may be viewed as the result of all possible applications of the 4N odd generators $X^+_{a,4+k}$ on v_0, i. e., $\hat{V}^\Lambda$ has 2^{4N} states (including the vacuum). Explicitly, the basis of $\hat{V}^\Lambda$ may be chosen as in [141, 142]:

$$\begin{aligned}\Psi_{\bar{\varepsilon}} &= \Bigg(\prod_{k=N}^{1}(X^+_{1,4+k})^{\varepsilon_{1,4+k}}\Bigg)\Bigg(\prod_{k=N}^{1}(X^+_{2,4+k})^{\varepsilon_{2,4+k}}\Bigg)\\ &\quad\times\Bigg(\prod_{k=1}^{N}(X^+_{3,4+k})^{\varepsilon_{3,4+k}}\Bigg)\Bigg(\prod_{k=1}^{N}(X^+_{4,4+k})^{\varepsilon_{4,4+k}}\Bigg)v_0,\\ \varepsilon_{aj} &= 0, 1\end{aligned} \tag{2.74}$$

5 A more general character formula involves the Kazhdan–Lusztig polynomials $P_{y,w}(u)$, $y, w \in W$ [253].

6 Kac considers highest weight modules but his results are immediately transferable to lowest weight modules.

where $\bar{\varepsilon}$ denotes the set of all ε_{ij}.[7] Thus, the character of $\hat{V}^\Lambda$ may be written as

$$\begin{aligned}
\operatorname{ch} \hat{V}^\Lambda &= \sum_{\bar{\varepsilon}} e(\Psi_{\bar{\varepsilon}}) \\
&= \sum_{\bar{\varepsilon}} \left(\prod_{k=1}^{N} e(\alpha_{1,4+k})^{\varepsilon_{1,4+k}} \right) \left(\prod_{k=1}^{N} e(\alpha_{2,4+k})^{\varepsilon_{2,4+k}} \right) \\
&\quad \times \left(\prod_{k=1}^{N} e(\alpha_{3,4+k})^{\varepsilon_{3,4+k}} \right) \left(\prod_{k=1}^{N} e(\alpha_{4,4+k})^{\varepsilon_{4,4+k}} \right) \\
&= \sum_{\bar{\varepsilon}} e\left(\sum_{k=1}^{N} \sum_{a=1}^{4} \varepsilon_{a,4+k} \alpha_{a,4+k} \right)
\end{aligned} \tag{2.75}$$

(Note that in the above formula there is no actual dependence on Λ.)

We shall use the above to write for the character of V^Λ:

$$\begin{aligned}
\operatorname{ch} V^\Lambda &= \operatorname{ch} \hat{V}^\Lambda \cdot \operatorname{ch}_0 V_0^\Lambda \\
&= \sum_{\bar{\varepsilon}} e\left(\sum_{k=1}^{N} \sum_{a=1}^{4} \varepsilon_{a,4+k} \alpha_{a,4+k} \right) \cdot e(\Lambda) \left(\prod_{\alpha \in \Delta_{\bar{0}}^{+}} (1 - e(\alpha))^{-1} \right) \\
&= \sum_{\bar{\varepsilon}} e\left(\Lambda + \sum_{k=1}^{N} \sum_{a=1}^{4} \varepsilon_{a,4+k} \alpha_{a,4+k} \right) \left(\prod_{\alpha \in \Delta_{\bar{0}}^{+}} (1 - e(\alpha))^{-1} \right) \\
&= \sum_{\bar{\varepsilon}} \operatorname{ch}_0 V_0^{\Lambda + \sum_{k=1}^{N} \sum_{a=1}^{4} \varepsilon_{a,4+k} \alpha_{a,4+k}}
\end{aligned} \tag{2.76}$$

where $\operatorname{ch}_0 V_0^\Lambda$ is the character obtained by restriction of V^Λ to V_0^Λ:

$$\operatorname{ch}_0 V_0^\Lambda = e(\Lambda^z) \cdot \operatorname{ch}_0 V^{\Lambda^s} \cdot \operatorname{ch}_0 V^{\Lambda^u} \tag{2.77}$$

where we use the decomposition $\Lambda = \Lambda^s + \Lambda^z + \Lambda^u$ from (2.20a), and V^{Λ^s}, V^{Λ^u}, respectively, are Verma modules over the complexifications of su(2, 2), su(N), respectively, cf. [130].

Analogously, for the factorized Verma modules $\widetilde{V}^\Lambda$ the character formula is

$$\begin{aligned}
\operatorname{ch} \widetilde{V}^\Lambda &= \operatorname{ch} \hat{V}^\Lambda \cdot \operatorname{ch}_0 \widetilde{V}_0^\Lambda \\
&= \sum_{\bar{\varepsilon}} \operatorname{ch}_0 \widetilde{V}_0^{\Lambda + \sum_{k=1}^{N} \sum_{a=1}^{4} \varepsilon_{a,4+k} \alpha_{a,4+k}}
\end{aligned} \tag{2.78}$$

7 The order chosen in (2.74) was important in the proof of unitarity in [140, 142] and for that purposes one may choose also an order in which the vectors on the first row are exchanged with the vectors on the second row. For our purposes the order is important as far as to avoid impossible states—this is much of the analysis done in the next subsections.

where $\text{ch}_0\,\widetilde{V}_0^\Lambda$ is the character obtained by restriction of $\widetilde{V}^\Lambda$ to $\widetilde{V}_0^\Lambda \equiv U((\mathcal{G}_+^{\mathbb{C}})_{(0)})|\widetilde{\Lambda}\rangle$, or more explicitly:

$$\text{ch}_0\,\widetilde{V}_0^\Lambda = e(\Lambda^z)\cdot \text{ch}_0\,L_{\Lambda^s}\cdot \text{ch}_0\,L_{\Lambda^u} \tag{2.79}$$

where we use the decomposition $\Lambda = \Lambda^s + \Lambda^z + \Lambda^u$ from (2.20a), and character formulas for the irreps of the even subalgebra (cf. [130] and Section 2.3.1 below).

Equation (2.78) represents the expansion of the corresponding superfield in components, and each component has its own even character. We see that this expansion is given exactly by the expansion of the odd character (2.75).

We have already displayed how the UIRs L_Λ are obtained as factor modules of the (even-submodules-factorized) Verma modules $\widetilde{V}^\Lambda$. Of course, this factorization means that the odd singular vectors of $\widetilde{V}^\Lambda$ from (2.36) are becoming null conditions in L_Λ. However, this is not enough to determine the character formulas even when considering our UIRs as irreps of the complexification sl(4/N). The latter is a well known feature even in the bosonic case. Here the situation is much more complicated and much more refined analysis is necessary.

The most important aspect of this analysis is the determination of the superfield content. (This analysis was used in [140, 142] but was not explicated enough.) This is given by the positive norm states $\hat{\Lambda}_\Lambda$ among all states in the odd sector $\hat{V}^\Lambda$. Of course, $\hat{\Lambda}_\Lambda$ may have less than 2^{4N} states.

For future use we introduce notation for the levels of the different chiralities ε_i and the overall level ε

$$\varepsilon_i = \sum_{k=1}^{N} \varepsilon_{i,4+k}, \quad i = 1,2,3,4,\ \varepsilon = \varepsilon_1 + \varepsilon_2 + \varepsilon_3 + \varepsilon_4. \tag{2.80}$$

The odd null conditions entwine with the even null conditions as we shall see. The even null conditions follow from the even singular vectors in (2.25) (alternatively, one may say that they carry over from the even null conditions (2.28) of $\widetilde{V}^\Lambda$). We write down the even null conditions first since they hold for any positive energy UIR:

$$(X_1^+)^{1+2j_1}|\Lambda\rangle = 0, \quad (X_2^+)^{1+2j_2}|\Lambda\rangle = 0, \tag{2.81a}$$

$$(X_j^+)^{1+r_{N+4-j}}|\Lambda\rangle = 0, \quad j = 5,\dots,N+3 \tag{2.81b}$$

((2.81b) being empty for $N = 1$), where by $|\Lambda\rangle$ we shall denote the lowest weight vector of the UIR L_Λ.

2.2.2 Character formulas for the long UIRs

As we mentioned if $d > d_{\max}$ there are no further reducibilities, and the UIRs $L_\Lambda = \widetilde{V}^\Lambda$ are called *long UIRs*, since $\hat{\Lambda}_\Lambda$ may have the maximally possible number of states 2^{4N} (including the vacuum state).

However, the actual number of states may be less than 2^{4N} states due to the fact that – depending on the values of j_a and r_k – not all actions of the odd generators on the vacuum would be allowed. The latter is obvious from equations (2.51). Using the latter we can give the resulting signature of the state $\Psi_{\bar{\varepsilon}}$:

$$\begin{aligned}\chi(\Psi_{\bar{\varepsilon}}) = \Big[& d + \frac{1}{2}\varepsilon;\ j_1 + \frac{1}{2}(\varepsilon_2 - \varepsilon_1), j_2 + \frac{1}{2}(\varepsilon_4 - \varepsilon_3);\\ & z + \epsilon_N(\varepsilon_3 + \varepsilon_4 - \varepsilon_1 - \varepsilon_2);\\ & \dots, r_i + \varepsilon_{1,N+4-i} - \varepsilon_{1,N+5-i} + \varepsilon_{2,N+4-i} - \varepsilon_{2,N+5-i}\\ & - \varepsilon_{3,N+4-i} + \varepsilon_{3,N+5-i} - \varepsilon_{4,N+4-i} + \varepsilon_{4,N+5-i}, \dots \Big]\end{aligned} \tag{2.82}$$

Thus, only if $j_1, j_2 \geq N/2$ and $r_i \geq 4$ (for all i) the number of states is 2^{4N} [140], and the character formula is

$$\operatorname{ch} L_\Lambda = \operatorname{ch} L^0_\Lambda \operatorname{ch} \hat{V}^\Lambda, \tag{2.83a}$$

$$j_1, j_2 \geq N/2,\ r_i \geq 4,\ i = 1, \dots, N-1, \tag{2.83b}$$

where $\operatorname{ch} L^0_\Lambda$ denotes the character of the restriction of L_Λ to the even subalgebra.

The general formula for $\operatorname{ch} L_\Lambda$ shall be written in a similar fashion:

$$\operatorname{ch} L_\Lambda = \operatorname{ch} \hat{\Lambda}_\Lambda \cdot \operatorname{ch}_0 \widetilde{V}^\Lambda_0. \tag{2.84}$$

Moreover, from now on in this section we shall write only the formulas for ch $\hat{\Lambda}_\Lambda$. Thus, equation (2.83) shall be written equivalently as

$$\operatorname{ch} \hat{\Lambda}_\Lambda = \operatorname{ch} \hat{V}^\Lambda, \quad j_1, j_2 \geq N/2,\ r_i \geq 4,\ \forall i. \tag{2.85}$$

As we have noted after (2.78) we do not lose information using this factorized form which has the advantage of brevity. In Section 2.3 the characters written in this factorized form will be called **bare characters.**

If the auxiliary conditions (2.83b) are not fulfilled then a careful analysis is necessary. To simplify the exposition we classify the states by the following quantities:

$$\begin{aligned}&\varepsilon^c_j \equiv \varepsilon_1 - \varepsilon_2, \quad \varepsilon^a_j \equiv \varepsilon_3 - \varepsilon_4,\\ &\varepsilon^i_r \equiv \varepsilon_{1,5+i} + \varepsilon_{2,5+i} + \varepsilon_{3,4+i} + \varepsilon_{4,4+i} - \varepsilon_{1,4+i} - \varepsilon_{2,4+i} - \varepsilon_{3,5+i} - \varepsilon_{4,5+i},\\ &\quad i = 1, \dots, N-1.\end{aligned} \tag{2.86}$$

This gives the following necessary conditions on ε_{ij} for a state to be allowed:

$$\varepsilon^c_j \leq 2j_1, \tag{2.87a}$$

$$\varepsilon^a_j \leq 2j_2, \tag{2.87b}$$

$$\varepsilon^i_r \leq r_{N-i}, \quad i = 1, \dots, N-1. \tag{2.87c}$$

These conditions are also sufficient only for $N = 1$ (when (2.87c) is absent). The exact conditions are

Criterion. The necessary and sufficient conditions for the state $\Psi_{\bar{\varepsilon}}$ of level ε to be allowed are that conditions (2.87) are fulfilled and that the state is a descendant of an allowed state of level $\varepsilon - 1$. ◇

The second part of the criterion will take care first of all of chiral (or anti-chiral) states when some ε_{aj} contribute to opposing sides of the inequalities in (2.87a) and (2.87c) (or (2.87b) and (2.87c)). This happens for $j_1 = r_i = 0$ (or $j_2 = r_i = 0$).

We shall give now the most important such occurrences. Take first the chiral states, i. e., all $\varepsilon_{3,4+k} = \varepsilon_{4,4+k} = 0$. Fix $i = 1, \ldots, N-1$. It is easy to see that the following states are not allowed [123]:

$$\begin{aligned}\psi_{ij} = \phi_{ij}|\Lambda\rangle = X^+_{1,i+4}X^+_{2,i+5}X^+_{a_1,i+6}\cdots X^+_{a_{j-1},i+4+j}|\Lambda\rangle,\quad & a_n = 1,2,\\ j = 1,\ldots,N-i,\quad j_1 = r_{N-i} = \cdots = r_{N-i-j+1} = 0,&\\ \text{in addition, for } N > 2,\ i > 1 \text{ holds } r_{N-i+1} \neq 0.&\end{aligned} \tag{2.88}$$

Consider now anti-chiral states, i. e., such that $\varepsilon_{1,4+k} = \varepsilon_{2,4+k} = 0$, for all $k = 1, \ldots, N$. Fix $i = 1, \ldots, N-1$. Then the following anti-chiral states are not allowed:

$$\begin{aligned}\psi'_{ij} = \phi'_{ij}|\Lambda\rangle = X^+_{3,i+5}X^+_{4,i+4}X^+_{b_1,i+3}\cdots X^+_{b_{j-1},i+5-j}|\Lambda\rangle,\quad & b_n = 3,4,\\ j = 1,\ldots,i,\quad j_2 = r_{N-i} = \cdots = r_{N-i+j-1} = 0,&\\ \text{in addition, for } N > 2,\ i > 1 \text{ holds } r_{N-i-1} \neq 0.&\end{aligned} \tag{2.89}$$

Furthermore, any combinations of ϕ_{ij} and $\phi'_{i'j'}$ are not allowed.

Note that for $N \geq 4$ the states in (2.88) and (2.89) do not exhaust the states forbidden by our criterion. For example, for $N = 4$ there are the following forbidden states:

$$\begin{aligned}\psi_4 = \phi_4|\Lambda\rangle = X^+_{28}X^+_{17}X^+_{16}X^+_{25}|\Lambda\rangle,\quad & j_1 = r_1 = r_2 = r_3 = 0,\\ \psi'_4 = \phi'_4|\Lambda\rangle = X^+_{45}X^+_{36}X^+_{37}X^+_{48}|\Lambda\rangle,\quad & j_2 = r_1 = r_2 = r_3 = 0\end{aligned} \tag{2.90}$$

Summarizing the discussion so far, the general character formula may be written as follows:

$$\operatorname{ch}\hat{\Lambda}_\Lambda = \operatorname{ch}\hat{V}^\Lambda - \mathcal{R},\quad d > d_{\max}, \tag{2.91}$$

$$\mathcal{R} = e(\hat{V}^\Lambda_{\text{excl}}) = \sum_{\substack{\text{excluded}\\ \text{states}}} e(\Psi_{\bar{\varepsilon}}),$$

$$\begin{aligned}e(\Psi_{\bar{\varepsilon}}) = & \left(\prod_{k=N}^{1} e(\alpha_{1,4+k})^{\varepsilon_{1,4+k}}\right)\left(\prod_{k=N}^{1} e(\alpha_{2,4+k})^{\varepsilon_{2,4+k}}\right)\\ & \times \left(\prod_{k=1}^{N} e(\alpha_{3,4+k})^{\varepsilon_{3,4+k}}\right)\left(\prod_{k=1}^{N} e(\alpha_{4,4+k})^{\varepsilon_{4,4+k}}\right),\end{aligned}$$

where the counterterms denoted by $\mathcal{R}$ are determined by $\hat{V}^\Lambda_{\text{excl}}$ which is the collection of all states (i. e., collection of ε_{jk}) which violate the conditions (2.87), or are impossible in the sense of (2.88) and/or (2.89). Of course, each excluded state is accounted for only once even if it is not allowed for several reasons.

Finally, we consider two important conjugate special cases.

First, the chiral sector of R-symmetry scalars with $j_1 = 0$. Taking into account (2.87a,c) ((2.87b) is trivially satisfied for chiral states) and our criterion it is easy to see that the appearance of the generators $X^+_{1,4+k}$ is restricted as follows. The generator X^+_{15} may appear only in the state

$$X^+_{15}X^+_{25}|\Lambda\rangle \tag{2.92}$$

and its descendants. The generator X^+_{16} may only appear either in states descendant to the state (2.92) or in the state

$$X^+_{16}X^+_{25}|\Lambda\rangle \tag{2.93}$$

and its descendants including only generators $X^+_{a,5+\ell}$, $a = 1, 2$, $\ell > 1$. Furthermore, the restrictions are described recursively, namely, fix ℓ such that $1 < \ell \leq N - 1$. The generator $X^+_{1,5+\ell}$ may only appear either in states containing generators $X^+_{1,5+j}$, where $0 \leq j < \ell$, or in the state

$$X^+_{1,5+\ell}X^+_{2,4+\ell}X^+_{2,3+\ell}\cdots X^+_{2,5}|\Lambda\rangle \tag{2.94}$$

and its descendants including only generators $X^+_{a,5+\ell'}$, $a = 1, 2$, $\ell' > \ell$.

The chiral part of the basis is further restricted. Namely, there are only N chiral states that can be built from the generators $X^+_{2,4+k}$ alone, given as follows:

$$X^+_{2,4+k}\cdots X^+_{25}|\Lambda\rangle, \quad k = 1, \dots, N, \ j_1 = r_i = 0, \ \forall i. \tag{2.95}$$

This follows from (2.87c) which in this case is reduced to $\varepsilon_{1i} \leq \varepsilon_{1,i+1}$ for $i = 1, \dots, N - 1$.

Secondly, the anti-chiral sector of R-symmetry scalars with $j_2 = 0$. Taking into account (2.87b,c) and our criterion it is easy to see that the appearance of the generators $X^+_{3,4+k}$ is restricted as follows. The generator $X^+_{3,4+N}$ may appear only in the state

$$X^+_{3,4+N}X^+_{4,4+N}|\Lambda\rangle \tag{2.96}$$

and its descendants. The generator $X^+_{3,3+N}$ may only appear either in states descendant to the state (2.96) or in the state

$$X^+_{3,3+N}X^+_{4,4+N}|\Lambda\rangle \tag{2.97}$$

and its descendants including only generators $X^+_{a,4+N-\ell}$, $a = 3, 4$, $\ell > 1$. Furthermore, fix ℓ such that $1 < \ell \leq N - 1$. The generator $X^+_{3,4+N-\ell}$ may only appear either in states containing generators $X^+_{3,4+N-j}$, where $0 \leq j < \ell$, or in the state

$$X^+_{3,4+N-\ell}X^+_{4,5+N-\ell}X^+_{4,6+N-\ell}\cdots X^+_{4,4+N}|\Lambda\rangle \tag{2.98}$$

and its descendants including only generators $X^+_{a,4+N-\ell'}$, $a = 3, 4$, $\ell' > \ell$.

The anti-chiral part of the basis is further restricted. Namely, there are only N anti-chiral states that can be built from the generators $X^+_{4,4+k}$ alone:

$$X^+_{4,5+N-k}X^+_{4,6+N-k}\cdots X^+_{4,4+N}|\Lambda\rangle, \quad k = 1,\dots,N,\ j_2 = r_i = 0,\ \forall i. \tag{2.99}$$

This follows from (2.87c), which for such states becomes $\varepsilon_{4,4+N-i} \le \varepsilon_{4,5+N-i}$ for $i = 1,\dots,N-1$.

2.2.3 Character formulas of SRC UIRs

Here we consider the four SRC cases.

• **a** $d = d^1_{N1} = d^a \equiv 2 + 2j_2 + z + 2m_1 - 2m/N > d^3_{NN}$.

• Let first $j_2 > 0$.

In these semi-short SRC cases we have the odd null condition (following from the singular vector (8.9a) of [139]; cf. also (2.35a'), (2.36a), and (2.42a)):

$$\begin{aligned} P_{3,4+N}|\Lambda\rangle &= (X^+_4X^+_2(h_2-1) - X^+_2X^+_4h_2)|\Lambda\rangle \\ &= (2j_2X^+_{3,4+N} - X^+_4X^+_2)|\Lambda\rangle = 0 \end{aligned} \tag{2.100}$$

where $X^+_{3,4+N} = [X^+_2, X^+_4]$. Clearly, condition (2.100) means that the generator $X^+_{3,4+N}$ is eliminated from the basis that is built on the lowest weight vector $|\Lambda\rangle$. Thus, for $N = 1$ and if $r_1 > 0$ for $N > 1$ the character formula is

$$\operatorname{ch}\hat{\Lambda}_\Lambda = \prod_{\substack{\alpha\in\Delta^+_{\bar{1}} \\ \alpha\neq\alpha_{3,4+N}}} (1 + e(\alpha)) - \mathcal{R}, \quad j_2r_1 > 0. \tag{2.101}$$

There are no counterterms when $j_1 \ge N/2$, $j_2 \ge (N-1)/2$, and $r_i \ge 4$ (for all i), and then the number of states is 2^{4N-1}.

Remark. For the finite-dimensional irreps of sl(4/N) (in fact, of all basic classical Lie superalgebras) such situations are called 'singly atypical' and the character formulas look exactly as (2.101) with $\mathcal{R} = 0$, cf. [47, 375].

For character formulas of finite-dimensional irreps beyond the singly atypical case cf. [350, 375, 68, 364], and the references therein. ◇

When there are no counterterms (also for the complex sl$(4/N)$ case) this formula follows easily from (2.50). Indeed, in the case at hand $I^\beta = I^1$ (cf. (2.42a)); then from $L_\Lambda = \widetilde{V}^\Lambda/I^1$ follow $\hat{\Lambda}_\Lambda = \hat{V}^\Lambda/\hat{I}^1$ and

$$\operatorname{ch}\hat{\Lambda}_\Lambda = \operatorname{ch}\hat{V}^\Lambda - \operatorname{ch}\hat{I}^1, \tag{2.102}$$

where $\hat{I}^1$ is the projection of I^1 to the odd sector. Naively, the character of $\hat{I}^1$ should be given by the character of $\hat{V}^{\Lambda+\alpha_{3,4+N}}$, however, as discussed in general—cf. (2.8), I^1 is smaller than $\hat{V}^{\Lambda+\alpha_{3,4+N}}$ and its character is given with a prefactor:

$$\mathrm{ch}\,\hat{I}^1 = \frac{1}{1+e(\alpha_{3,4+N})}\,\mathrm{ch}\,\hat{V}^{\Lambda+\alpha_{3,4+N}} = \frac{e(\alpha_{3,4+N})}{1+e(\alpha_{3,4+N})}\,\mathrm{ch}\,\hat{V}^{\Lambda}. \tag{2.103}$$

Now (2.101) (with $\mathcal{R} = 0$) follows from the combination of (2.102) and (2.103).

Equation (2.101) may also be described by using the generalized odd reflection (2.10) with $\beta = \alpha_{3,4+N}$:

$$\mathrm{ch}\,\hat{\Lambda}_\Lambda = \mathrm{ch}\,\hat{V}^\Lambda - \frac{1}{1+e(\alpha_{3,4+N})}\,\mathrm{ch}\,\hat{V}^{\hat{S}_{\alpha_{3,4+N}}\cdot\Lambda} - \mathcal{R} \tag{2.104a}$$

$$= \sum_{\hat{S}\in\widehat{W}_{\alpha_{3,4+N}}} (-1)^{\ell(\hat{S})}\hat{S}\cdot\mathrm{ch}\,\hat{V}^\Lambda - \mathcal{R}, \tag{2.104b}$$

where $\widehat{W}_\beta \equiv \{1, \hat{S}_\beta\}$ is a two-element semi-group restriction of $\tilde{W}_\beta$, and we have made a further formalization by introducing notation for the action of a generalized odd reflection on the characters:

$$\hat{S}_\beta\cdot\mathrm{ch}\,V^\Lambda = \frac{1}{1+e(\beta)}\,\mathrm{ch}\,V^{\hat{S}_\beta\cdot\Lambda} = \frac{1}{1+e(\beta)}\,\mathrm{ch}\,V^{\Lambda+\beta} = \frac{e(\beta)}{1+e(\beta)}\,\mathrm{ch}\,V^\Lambda. \tag{2.105}$$

In particular, we shall show that in many cases character formulas of equations (2.101) and (2.104) may be written as follows:

$$\mathrm{ch}\,\hat{\Lambda}_\Lambda = \sum_{\hat{S}\in\widehat{W}_\beta} (-1)^{\ell(\hat{S})}\hat{S}\cdot(\mathrm{ch}\,\hat{V}^\Lambda - \mathcal{R}_{\mathrm{long}}), \tag{2.106}$$

where $\mathcal{R}_{\mathrm{long}}$ represents the counterterms for the long superfields for the same values of j_1 and r_i as Λ, while the value of j_2 is zero when j_2 from Λ is zero, otherwise it has to be the generic value $j_2 \geq N/2$.

Writing (2.101) as (2.104) (or (2.106)) may look as a complicated way to describe the cancellation of a factor from the character formula for $\hat{V}^\Lambda$, however, first of all it is related to the structure of $\widetilde{V}^\Lambda$ given by (2.50), and furthermore may be interpreted—when there are no counterterms—as the following decomposition:

$$\hat{V}^\Lambda = \hat{\Lambda}_\Lambda \oplus \hat{\Lambda}_{\Lambda+\beta}, \tag{2.107}$$

for $\beta = \alpha_{3,4+N}$. Indeed, for generic signatures $\hat{\Lambda}_{\Lambda+\beta}$ is isomorphic to $\hat{\Lambda}_\Lambda$ as a vector space (this is due to the fact that $V^{\Lambda+\beta}$ has the same reducibilities as V^Λ); they differ only by the vacuum state. Thus, when there are no counterterms, $\hat{\Lambda}_\Lambda$ and $\hat{\Lambda}_{\Lambda+\beta}$ have the same 2^{4N-1} states.

It is more important that there is a similar decomposition valid for many cases beyond the generic, i. e., we have

$$(\hat{\Lambda}_{\mathrm{long}})_{|_{d=d^a}} = \hat{\Lambda}_\Lambda \oplus \hat{\Lambda}_{\Lambda+\alpha_{3,4+N}}, \quad N = 1 \text{ or } r_1 > 0 \text{ for } N > 1, \tag{2.108}$$

where $\hat{\Lambda}_{\text{long}}$ is a long superfield with the same values of j_1 and r_i as Λ, while the value of j_2 has to be specified, and equality is as vector spaces.

For $N > 1$ there are possible additional truncations of the basis. To make the exposition easier we need additional notation. Let i_0 be an integer such that $0 \le i_0 \le N-1$, and $r_i = 0$ for $i \le i_0$, and if $i_0 < N-1$ then $r_{i_0+1} > 0$.[8]

Let now $N > 1$ and $i_0 > 0$; then the generators $X^+_{3,4+N-i}$, $i = 1, \dots, i_0$, are eliminated from the basis. This follows from the following recursive null conditions:

$$P_{3,4+N-i}|\Lambda\rangle = (2j_2 X^+_{3,4+N-i} - X^+_{4,4+N-i} X^+_2)|\Lambda\rangle = 0, \quad i \le i_0. \tag{2.109}$$

From the above it follows that for $i_0 > 0$ the decomposition (2.108) cannot hold. Indeed, the generators $X^+_{3,4+N-i}$, $i = 1, \dots, i_0$, are eliminated from the irrep $\hat{\Lambda}_\Lambda$ due to the fact that we are at a reducibility point, but there is no reason for them to be eliminated from the long superfield. Certainly, some of these generators are present in the second term $\hat{\Lambda}_{\Lambda+\alpha_{3,4+N}}$ in (2.108), but that would be only those which in the long superfield were in states of the kind: $\Phi X^+_{3,4+N}|\Lambda\rangle$, and, certainly, such states do not exhaust the occurrence of the discussed generators in the long superfield. Symbolically, instead of the decomposition (2.108) we shall write

$$(\hat{\Lambda}_{\text{long}})|_{d=d^a} = \hat{\Lambda}_\Lambda \oplus \hat{\Lambda}_{\Lambda+\alpha_{3,4+N}} \oplus \hat{\Lambda}'_\Lambda, \quad N > 1,\ i_0 > 0, \tag{2.110}$$

where we have represented the excess states by the last term with prime. With the prime we stress that this is not a genuine irrep, but just a book-keeping device. Formulas as (2.110) in which not all terms are genuine irreps shall be called *quasi-decompositions.*

The corresponding character formula is

$$\operatorname{ch} \hat{\Lambda}_\Lambda = \prod_{\substack{\alpha \in \Delta^+_{\bar 1} \\ \alpha \ne \alpha_{3,5+N-k} \\ k=1,\dots,1+i_0}} (1 + e(\alpha)) - \mathcal{R} \tag{2.111a}$$

$$= \sum_{\hat{S} \in \widehat{W}^a_{i_0}} (-1)^{\ell(\hat{S})} \hat{S} \cdot \operatorname{ch} \hat{V}^\Lambda - \mathcal{R} \tag{2.111b}$$

$$= \sum_{\hat{S} \in \widehat{W}^a_{i_0}} (-1)^{\ell(\hat{S})} \hat{S} \cdot (\operatorname{ch} \hat{V}^\Lambda - \mathcal{R}_{\text{long}}), \tag{2.111c}$$

$$\widehat{W}^a_{i_0} \equiv \widehat{W}_{\alpha_{3,N+4}} \times \widehat{W}_{\alpha_{3,N+3}} \times \cdots \times \widehat{W}_{\alpha_{3,N+4-i_0}}. \tag{2.111d}$$

The restrictions (2.87) used to determine the counterterms are, of course, with $\varepsilon_{3,5+N-k} = 0$, $k = 1, \dots, 1 + i_0$. Equations (2.101), (2.104), and (2.106) are special cases

8 This is formally valid for $N = 1$ with $i_0 = 0$ since $r_0 \equiv 0$ by convention. This shall be used to make certain statements valid for general N.

of (2.111a,b,c), respectively, for $i_0 = 0$. The maximal number of states in $\hat{\Lambda}_\Lambda$ is 2^{4N-1-i_0}. This is the number of states that is obtained from the action of the Weyl group $\widehat{W}^a_{i_0}$ on $\operatorname{ch}\hat{V}^\Lambda$, while the actual counterterm is obtained from the action of the Weyl group on $\mathcal{R}_{\text{long}}$.

In the extreme case of R-symmetry scalars, $i_0 = N-1$, i. e., $r_i = 0$, $i = 1, \dots, N-1$, or, equivalently, $m_1 = 0 = m$, all the N generators $X^+_{3,4+k}$ are eliminated. The character formula is again (2.111) taken with $i_0 = N-1$.

- Let now $j_2 = 0$. Then all null conditions above follow from (2.28b), so these conditions do not mean elimination of the mentioned vectors. As we know in this situation we have the singular vector (2.39) which leads to the following null condition:

$$X^+_{3,4+N}X^+_{4,4+N}|\Lambda\rangle = X^+_4 X^+_2 X^+_4|\Lambda\rangle = 0. \tag{2.112}$$

The state in (2.112) and all of its 2^{4N-2} descendants are zero for any N. Thus, the character formula is similar to (2.104), but with $\alpha_{3,4+N}$ replaced by $\beta_{12} = \alpha_{3,4+N} + \alpha_{4,4+N}$:

$$\operatorname{ch}\hat{\Lambda}_\Lambda = \sum_{\hat{\mathcal{S}}\in\widehat{W}_{\beta_{12}}} (-1)^{\ell(\hat{\mathcal{S}})}\hat{\mathcal{S}}\cdot(\operatorname{ch}\hat{V}^\Lambda - \mathcal{R}_{\text{long}}), \quad N = 1 \text{ or } r_1 > 0, \tag{2.113}$$

where $\widehat{W}_{\beta_{12}} \equiv \{1, \beta_{12}\}$. Note that for $N = 1$ equation (2.113) is equivalent to (2.101).

Here we have a decomposition similar to (2.108):

$$(\hat{\Lambda}_{\text{long}})_{|_{d=d^a}} = \hat{\Lambda}_\Lambda \oplus \hat{\Lambda}_{\Lambda+\beta_{12}}, \quad N = 1 \text{ or } r_1 > 0 \text{ for } N > 1, \tag{2.114}$$

where $\hat{\Lambda}_{\text{long}}$ has the same values of $j_1, j_2 (= 0), r_i$ as Λ. Note, however, that the UIR $\hat{\Lambda}_{\Lambda+\beta_{12}}$ belongs to type **b** below.

There are more eliminations for $N > 1$ when $i_0 > 0$. For instance we can show that all states obtained as in (2.98) considered for $\ell = 1, \dots, i_0$ are not allowed [123]. From this it follows that if $i_0 > 0$ the decomposition (2.114) does not hold. Instead, there is a quasi-decomposition similar to (2.110).

We can be more explicit in the case when all $r_i = 0$. In that case all the vectors $X^+_{3,5+N-k}$ are eliminated from all anti-chiral states [123].

The anti-chiral part of the basis is further restricted. As we know, when $j_2 = r_i = 0$, $\forall i$, there are only N anti-chiral states that can be built from the generators $X^+_{4,4+k}$ alone, given in (2.99). Thus the corresponding character formula is

$$\operatorname{ch}\hat{\Lambda}_\Lambda = \sum_{k=1}^{N}\prod_{i=1}^{k} e(\alpha_{4,5+N-i}) + \prod_{\substack{\alpha\in\Delta^+_{\bar{1}}\\ \varepsilon_1+\varepsilon_2>0}} (1 + e(\alpha)) - \mathcal{R}, \quad j_2 = r_i = 0,\ \forall i. \tag{2.115}$$

- **b** $d = d^2_{N1} = z + 2m_1 - 2m/N > d^3_{NN}$, $j_2 = 0$.

In these short single-reducibility-condition cases we have the odd null condition (following from the singular vector (2.36b))

$$X^+_4|\Lambda\rangle = X^+_{4,4+N}|\Lambda\rangle = 0. \tag{2.116}$$

Since $j_2 = 0$ from (2.28b) and (2.116) follows the additional null condition,

$$X^+_{3,4+N}|\Lambda\rangle = [X^+_2, X^+_4]|\Lambda\rangle = 0 \tag{2.117}$$

For $N > 1$ and $r_1 > 2$ each of these UIRs enters as the second term in decomposition (2.114), when the first term is an UIR of type **a** with $j_2 = 0$, as explained above.

Furthermore, for $N > 1$ there are additional recursive null conditions if $r_i = 0$, $i \le i_0$, which follow from (2.28c) and (2.117):

$$\begin{aligned} X^+_{3,4+N-i}|\Lambda\rangle &= [X^+_{3,5+N-i}, X^+_{4+N-i}]|\Lambda\rangle = 0, \quad r_j = 0,\ 1 \le j \le i \le i_0 \\ X^+_{4,4+N-i}|\Lambda\rangle &= [X^+_{4,5+N-i}, X^+_{4+N-i}]|\Lambda\rangle = 0, \quad r_j = 0,\ 1 \le j \le i \le i_0 \end{aligned} \tag{2.118}$$

Thus, $2(1+i_0)$ generators $X^+_{3,5+N-k}, X^+_{4,5+N-k}$, $k = 1, \dots, 1+i_0$, are eliminated. The maximal number of states in $\hat{\Lambda}_\Lambda$ is $2^{4N-2-2i_0}$.

The corresponding character formula is

$$\operatorname{ch} \hat{\Lambda}_\Lambda = \sum_{\hat{\mathcal{S}} \in \widehat{W}^b_{i_0}} (-1)^{\ell(\hat{\mathcal{S}})} \hat{\mathcal{S}} \cdot \operatorname{ch} \hat{V}^\Lambda - \mathcal{R}, \tag{2.119a}$$

$$\begin{aligned} \widehat{W}^b_{i_0} \equiv \widehat{W}_{\alpha_{3,N+4}} &\times \widehat{W}_{\alpha_{3,N+3}} \times \cdots \times \widehat{W}_{\alpha_{3,N+4-i_0}} \times \widehat{W}_{\alpha_{4,N+4}} \\ &\times \widehat{W}_{\alpha_{4,N+3}} \times \cdots \times \widehat{W}_{\alpha_{4,N+4-i_0}}, \quad j_2 = r_i = 0,\ i \le i_0 \end{aligned} \tag{2.119b}$$

where determining the counterterms we use $\varepsilon_{a,4+k} = 0$, $a = 3, 4$, $k = 1, \dots, 1 + i_0$.

In the case of R-symmetry scalars ($i_0 = N - 1$) we have

$$X^+_{3,4+k}|\Lambda\rangle = 0, \quad X^+_{4,4+k}|\Lambda\rangle = 0, \quad k = 1, \dots, N,\ r_i = 0,\ \forall i. \tag{2.120}$$

The character formula is (2.119) taken with $1 + i_0 = N$. These UIRs should be called chiral since all anti-chiral generators are eliminated.

The next two cases are conjugates of the first two and the exposition will be compact.

- **c** $d = d^3_{NN} = d^c \equiv 2 + 2j_1 - z + 2m/N > d^1_{N1}$.
- **d** $d = d^4_{NN} = -z + 2m/N > d^1_{N1}$, $j_1 = 0$.

These cases are conjugate to the cases **a**, **b**, respectively. All results may be obtained by the substitutions: $j_1 \to j_2$, $r_i \to r_{N-i}$, $z \to -z$, $\alpha_{a,4+k} \to \alpha_{4-a,N+5-k}$, $a = 1, 2$, $k = 1, \dots, N$, and so we shall omit them here, cf. [123].

2.2.4 Character formulas of DRC UIRs

Let first $N > 1$ and $r_1 r_{N-1} > 0$ (i. e., $i_0 = i'_0 = 0$). Then we have the following character formula:

$$\operatorname{ch} \hat{\Lambda}_\Lambda = \sum_{\hat{\mathcal{S}} \in \widehat{W}_{\beta,\beta'}} (-1)^{\ell(\hat{\mathcal{S}})} \hat{\mathcal{S}} \cdot \operatorname{ch} \hat{V}^\Lambda - \mathcal{R} \tag{2.121a}$$

$$= \mathrm{ch}\,\hat{V}^{\Lambda} - \frac{1}{1+e(\beta)}\,\mathrm{ch}\,\hat{V}^{\Lambda+\beta} - \frac{1}{1+e(\beta')}\,\mathrm{ch}\,\hat{V}^{\Lambda+\beta'}$$

$$+ \frac{1}{(1+e(\beta))(1+e(\beta'))}\,\mathrm{ch}\,\hat{V}^{\Lambda+\beta+\beta'} - \mathcal{R}, \tag{2.121b}$$

$$\widehat{W}_{\beta,\beta'} \equiv \widehat{W}_{\beta} \times \widehat{W}_{\beta'} \tag{2.121c}$$

The above formula is proved similarly to what we had in the SRC cases, however, it takes into account the richer structure given explicitly already in the paper [138]. The proof is as follows [123]: The two terms with minus sign on the first line of (2.121b) take into account the factorization of the oddly embedded submodules $I^{\beta}, I^{\beta'}$, cf. (2.59). The non-trivial moment is the contribution of the module $\hat{V}^{\Lambda+\beta+\beta'}$. It is oddly embedded in $\hat{V}^{\Lambda}$ via both submodules $\hat{V}^{\Lambda+\beta}$ and $\hat{V}^{\Lambda+\beta'}$, [138], and its contribution is taken out *two times*—once with I^{β}, and a second time with $I^{\beta'}$. Thus, we need the term with plus sign on the second line of (2.121b) to restore its contribution once. We can not apply the same kind of arguments for $N = 1$, nevertheless, equation (2.121) holds also then for the case (2.58a), cf. Appendix A.1. of [123] and Section 2.3.2.

• **ac** $d = d_{\max} = d^1_{N1} = d^3_{NN} = d^{ac} \equiv 2 + j_1 + j_2 + m_1$.

In these semi-short DRC cases hold the null condition (2.100) and its conjugate. In addition, for $N > 1$ if $r_i = 0$, $i = 1, \ldots, i_0$, we have (2.109) and if $r_{N-i} = 0$, $i = 1, \ldots, i'_0$, we have the conjugate to (2.109).

There are two basic situations. The first is when $i_0 + i'_0 \leq N - 2$. This means that not all r_i are zero and all eliminations are as described separately for cases • **a** and • **c**. These semi-short UIRs may be called Grassmann analytic following [174], since odd generators from different chiralities are eliminated. The maximal number of states in $\hat{\Lambda}_{\Lambda}$ is $2^{4N-2-i_0-i'_0}$.

The second is when $i_0 + i'_0 \leq N - 2$ does not hold which means that all r_i are zero, and in fact we have $i_0 = i'_0 = N - 1$ and all generators $X^+_{1,4+k}$ and $X^+_{3,4+k}$ are eliminated. The maximal number of states in $\hat{\Lambda}_{\Lambda}$ is 2^{2N}.

• For $j_1 j_2 > 0$ the corresponding character formulas are combinations of (2.111) and its conjugate [123]:

$$\mathrm{ch}\,\hat{\Lambda}_{\Lambda} = \sum_{\hat{\mathcal{S}} \in \widehat{W}^{ac}_{i_0,i'_0}} (-1)^{\ell(\hat{\mathcal{S}})} \hat{\mathcal{S}} \cdot (\mathrm{ch}\,\hat{V}^{\Lambda} - \mathcal{R}_{\mathrm{long}}), \tag{2.122}$$

$$\begin{aligned} \widehat{W}^{ac}_{i_0,i'_0} &\equiv \widehat{W}^{a}_{i_0} \times \widehat{W}^{c}_{i'_0}, \quad j_1 j_2 > 0, \\ &\text{either } i_0 + i'_0 \leq N - 2, \\ &r_i = 0,\ i = 1, 2, \ldots, i_0, N - i'_0, N - i'_0 + 1, \ldots, N - 1, \\ &r_i > 0,\ i = i_0 + 1, N - i'_0 - 1, \\ &\text{or } i_0 = i'_0 = N - 1,\ r_i = 0,\ \forall i. \end{aligned}$$

The last subcase is of R-symmetry scalars. It is also the only formula in the case under consideration – **ac** – valid for $N = 1$ (where there are no counterterms).

For $N > 1$ and $i_0 = i'_0 = 0$ equation (2.122) is equivalent to (2.121) with $\beta = \alpha_{15}$, $\beta' = \alpha_{3,4+N}$. Also we have the following decomposition:

$$(\hat{\Lambda}_{\text{long}})_{|_{d=d^{ac}}} = \hat{\Lambda}_\Lambda \oplus \hat{\Lambda}_{\Lambda+\alpha_{15}} \oplus \hat{\Lambda}_{\Lambda+\alpha_{3,4+N}} \oplus \hat{\Lambda}_{\Lambda+\alpha_{15}+\alpha_{3,4+N}}, \quad r_1 r_{N-1} > 0, \tag{2.123}$$

$\hat{\Lambda}_{\text{long}}$ being a long superfield with the same values of r_i as Λ and with $j_1, j_2 \geq N/2$.

- For $j_1 > 0$, $j_2 = 0$ the corresponding character formulas are combinations of (2.113) and the conjugate to (2.111) [123]:

$$\begin{aligned} \operatorname{ch} \hat{\Lambda}_\Lambda &= \sum_{\hat{\mathcal{S}} \in \widehat{W}^{a'c}_{i'_0}} (-1)^{\ell(\hat{\mathcal{S}})} \hat{\mathcal{S}} \cdot \operatorname{ch} \hat{V}^\Lambda - \mathcal{R} \\ &= \sum_{\hat{\mathcal{S}} \in \widehat{W}^{a'c}_{i'_0}} (-1)^{\ell(\hat{\mathcal{S}})} \hat{\mathcal{S}} \cdot (\operatorname{ch} \hat{V}^\Lambda - \mathcal{R}_{\text{long}}), \quad r_1 > 0, \\ \widehat{W}^{a'c}_{i'_0} &\equiv \widehat{W}_{\beta_{12}} \times \widehat{W}^c_{i'_0}, \quad \beta_{12} = \alpha_{3,4+N} + \alpha_{4,4+N}. \end{aligned} \tag{2.124}$$

For $i_0 = i'_0 = 0$ we have the decomposition

$$(\hat{\Lambda}_{\text{long}})_{|_{d=d^{ac}}} = \hat{\Lambda}_\Lambda \oplus \hat{\Lambda}_{\Lambda+\alpha_{15}} \oplus \hat{\Lambda}_{\Lambda+\beta_{12}} \oplus \hat{\Lambda}_{\Lambda+\alpha_{15}+\beta_{12}}, \quad r_1 r_{N-1} > 0, \tag{2.125}$$

where $\hat{\Lambda}_{\text{long}}$ is a long superfield with the same values of $j_2 (= 0)$, r_i as Λ and with $j_1 \geq N/2$. Note that the UIR $\hat{\Lambda}_{\Lambda+\alpha_{15}}$ is also of the type **ac** under consideration, while the last two UIRs are short from type **bc** considered below.

For R-symmetry scalars we combine (2.115) and the conjugate to (2.111a):

$$\operatorname{ch} \hat{\Lambda}_\Lambda = \sum_{k=1}^{N} \prod_{i=1}^{k} e(\alpha_{4,5+N-i}) + \prod_{\substack{\alpha \in \Delta_{\bar{1}}^+ \\ \alpha \neq \alpha_{1,4+k}, \\ k=1,\dots,N \\ \varepsilon_2 > 0}} (1 + e(\alpha)) - \mathcal{R}, \quad r_i = 0, \ \forall i. \tag{2.126}$$

- The case $j_1 = 0$, $j_2 > 0$ is obtained from the previous one by conjugation. Here for $i_0 = i'_0 = 0$ we have the decomposition

$$(\hat{\Lambda}_{\text{long}})_{|_{d=d^{ac}}} = \hat{\Lambda}_\Lambda \oplus \hat{\Lambda}_{\Lambda+\alpha_{3,4+N}} \oplus \hat{\Lambda}_{\Lambda+\beta_{34}} \oplus \hat{\Lambda}_{\Lambda+\alpha_{3,4+N}+\beta_{34}}, \quad r_1 r_{N-1} > 0 \tag{2.127}$$

where $\hat{\Lambda}_{\text{long}}$ is a long superfield with the same values of $j_1 (= 0)$, r_i as Λ and with $j_2 \geq N/2$. Note that the UIR $\hat{\Lambda}_{\Lambda+\alpha_{3,4+N}}$ is again of the type **ac** under consideration, while the last two UIRs are actually from type **ad** considered below.

- For $j_1 = j_2 = 0$ the corresponding character formulas are combinations of (2.113) and its conjugate:

$$\operatorname{ch} \hat{\Lambda}_\Lambda = \sum_{\hat{\mathcal{S}} \in \widehat{W}^{a'c'}_{i'_0}} (-1)^{\ell(\hat{\mathcal{S}})} \hat{\mathcal{S}} \cdot (\operatorname{ch} \hat{V}^\Lambda - \mathcal{R}_{\text{long}}), \tag{2.128}$$

$$\widehat{W}^{a'c'}_{i'_0} \equiv \widehat{W}_{\beta_{12}} \times \widehat{W}_{\beta_{34}}, \quad r_1 r_{N-1} > 0.$$

For $i_0 = i'_0 = 0$ we have the decomposition

$$(\hat{\Lambda}_{\text{long}})_{|_{d=d^{ac}}} = \hat{\Lambda}_\Lambda \oplus \hat{\Lambda}_{\Lambda+\beta_{12}} \oplus \hat{\Lambda}_{\Lambda+\beta_{34}} \oplus \hat{\Lambda}_{\Lambda+\beta_{12}+\beta_{34}}, \quad r_1 r_{N-1} > 0, \tag{2.129}$$

where $\hat{\Lambda}_{\text{long}}$ is a long superfield with the same values of $j_1(=0)$, $j_2(=0)$, r_i as Λ. Note that the UIR $\hat{\Lambda}_{\Lambda+\beta_{12}}$ is of the type **bc**, $\hat{\Lambda}_{\Lambda+\beta_{34}}$ is of the type **ad**, $\hat{\Lambda}_{\Lambda+\beta_{12}+\beta_{34}}$ is of the type **bd**, these three being considered below.

For R-symmetry scalars we combine (2.115) and its conjugate:

$$\begin{aligned} \operatorname{ch} \hat{\Lambda}_\Lambda = & \sum_{k=1}^{N} \prod_{i=1}^{k} e(\alpha_{2,4+i}) + \sum_{k=1}^{N} \prod_{i=1}^{k} e(\alpha_{4,5+N-i}) \\ & + \prod_{\substack{\alpha\in\Delta_{\bar{1}}^+ \\ \varepsilon_1+\varepsilon_2>0 \\ \varepsilon_3+\varepsilon_4>0}} (1 + e(\alpha)) - \mathcal{R}, \quad r_i = 0, \ \forall i. \end{aligned} \tag{2.130}$$

• **ad** $\quad d = d^1_{N1} = d^4_{NN} = 1 + j_2 + m_1, j_1 = 0.$

In these short DRC cases hold the three null conditions (2.100), and the conjugates to (2.116) and (2.117). In addition, for $N > 1$ if $r_i = 0$, $i = 1, \dots, i_0$, hold (2.109) and if $r_{N-i} = 0$, $i = 1, \dots, i'_0$, hold the conjugate of (2.118).

If $i_0 + i'_0 \le N-2$ all eliminations are as described separately for cases • **a** and • **d**. All these are Grassmann-analytic UIRs. The maximal number of states in $\hat{\Lambda}_\Lambda$ is $2^{4N-3-i_0-2i'_0}$. Interesting subcases are the so-called BPS states, cf., [9, 172, 173, 176, 174, 159, 338, 106, 22, 104]. They are characterized by the number κ of odd generators which annihilate them—then the corresponding case is called $\frac{\kappa}{4N}$-BPS state. For example consider $N = 4$ and $\frac{1}{4}$-BPS cases with $z = 0 \Rightarrow d = 2m/N$. One such case is obtained for $i_0 = 1$, $i'_0 = 0$, $j_2 > 0$, then $d = \frac{1}{2}(2r_2 + 3r_3)$, $r_1 = 0$, $r_2 > 0$, $r_3 = 2(1 + j_2)$.

For $j_2 m_1 > 0$ the corresponding character formula is a combination of (2.111) and the conjugate of (2.119):

$$\operatorname{ch} \hat{\Lambda}_\Lambda = \sum_{\hat{S}\in\widehat{W}^{ad}_{i_0,i'_0}} (-1)^{\ell(\hat{S})} \hat{S} \cdot \operatorname{ch} \hat{V}^\Lambda - \mathcal{R}, \tag{2.131}$$

$$\begin{aligned} \widehat{W}^{ad}_{i_0,i'_0} \equiv \widehat{W}^{a}_{i_0} \times \widehat{W}^{d}_{i'_0}, \quad & j_2 m_1 > 0, \\ & r_i = 0, \ i = 1, 2, \dots, i_0, N - i'_0, N - i'_0 + 1, \dots, N-1, \\ & r_i > 0, \ i = i_0 + 1, N - i'_0 - 1. \end{aligned}$$

For $i_0 = i'_0 = 0$ some of these UIRs appear (up to two times) in the decomposition (2.127) [123].

For $j_2 = 0$, $m_1 > 0$ the corresponding character formula is a combination of (2.113) and the conjugate of (2.119a):

$$\operatorname{ch} \hat{\Lambda}_\Lambda = \sum_{\hat{S}\in\widehat{W}^{a'd}_{i'_0}} (-1)^{\ell(\hat{S})} \hat{S} \cdot \operatorname{ch} \hat{V}^\Lambda - \mathcal{R}, \tag{2.132}$$

$$\widehat{W}^{a'd}_{i'_0} \equiv \widehat{W}_{\beta_{12}} \times \widehat{W}^{d}_{i'_0},$$

where $\beta_{12} = \alpha_{3,4+N} + \alpha_{4,4+N}$. For $i_0 = i'_0 = 0$ some of these UIRs appear in the decomposition (2.129) or (2.127) [123].

In the case of R-symmetry scalars we have $i_0 = i'_0 = N - 1$, $\kappa = 3N$, and all generators $X^+_{1,4+k}, X^+_{2,4+k}, X^+_{3,4+k}$ are eliminated. Here we have $d = -z = 1 + j_2$. These anti-chiral irreps form one of the three series of *massless UIRs*; they are denoted χ^+_s, $s = j_2 = 0, \frac{1}{2}, 1, \ldots$, in Section 3 of [140]. Besides the vacuum they contain only N states in $\hat{\Lambda}_\Lambda$ given by (2.99) for $k = 1, \ldots, N$. These should be called ultrashort UIRs. The character formula can be written most explicitly:

$$\mathrm{ch}\,\hat{\Lambda}_\Lambda = 1 + \sum_{k=1}^{N} \prod_{i=1}^{k} e(\alpha_{4,5+N-i}), \quad j_1 = r_i = 0,\ \forall i, \tag{2.133}$$

and it is valid for any j_2. In the case under consideration – **ad** – only the last character formula is valid for $N = 1$.

• **bc** $d = d^2_{N1} = d^3_{NN} = 1 + j_1 + m_1, j_2 = 0, z = 2m/N - m_1 + 1 + j_1$.

This case is conjugate to the previous one and all results may be obtained by the substitutions given for the SRC conjugate cases.

• **bd** $d = d^2_{N1} = d^4_{NN} = m_1, j_1 = j_2 = 0, z = 2m/N - m_1$.

In these short DRC cases hold the four null conditions (2.116) and (2.117), and their conjugates.

For $N = 1$ this is the trivial irrep with $d = z = 0$. This follows from the fact that, since $d = j_1 = j_2 = 0$, we also have the even reducibility condition (2.19b) (and consequently (2.19d,e,f)). Thus, we have the null conditions: $X^+_k|\Lambda\rangle = 0$ for all simple root generators (and consequently for all generators) and the irrep consists only of the vacuum $|\Lambda\rangle$.

For $N > 1$ the situation is non-trivial. In addition to the mentioned conditions, and if $r_i = 0, i = 1, \ldots, i_0$, hold (2.118) and if $r_{N-i} = 0, i = 1, \ldots, i'_0$, hold the conjugates of (2.118).

If $i_0 + i'_0 \leq N - 2$ all eliminations are as described separately for cases • **b** and • **d.** These are also Grassmann-analytic UIRs. The maximal number of states in $\hat{\Lambda}_\Lambda$ is $2^{4N-4-2i_0-2i'_0}$. For $N = 4$ for the BPS cases we take $z = \frac{1}{2}(r_3 - r_1) = 0 \Rightarrow d = 2r_1 + r_2$. In the $\frac{1}{4}$-BPS case we have $i_0 = i'_0 = 0, r_1 = r_3 > 0$.

For $i_0 = i'_0 = 0$ some of these UIRs appear in the decomposition (2.129) [123].

Most interesting is the case $i_0 + i'_0 = N - 2$, then there is only one non-zero r_i, namely, $r_{1+i_0} = r_{N-1-i'_0} > 0$, while the rest r_i are zero. Thus, the Young tableau parameters are $m_1 = r_{1+i_0}, m = (1 + i_0)r_{1+i_0}$.

An important subcase is when $d = m_1 = 1$, then $m = i_0 + 1 = N - 1 - i'_0, r_i = \delta_{mi}$, and these irreps form the third series of *massless* UIRs. In Section 3 of [140] they are denoted $\chi'_n, n = m \geq \frac{1}{2}N\ (z = 2n/N - 1), \chi'^+_n, n = N - m \geq \frac{1}{2}N\ (z = 1 - 2n/N)$. Note that for even N we have the coincidence $\chi'_n = \chi'^+_n$, where $n = m = N - m = N/2$. Here we shall parametrize these UIRs by the parameter $i_0 = 0, 1, \ldots, N - 1$.

Another subcase here are $\frac{1}{2}$-BPS states for even N with $z = 0 \Rightarrow d = m_1 = 2m/N \Rightarrow i_0 = i'_0 = N/2 - 1 \Rightarrow m_1 = r_{N/2}$, $m = \frac{N}{2} r_{N/2}$. These are also massless only if $r_{N/2} = 1$, which is the self-conjugate case: χ'_n, $n = N/2$. For $N = 4$ we have $i_0 = i'_0 = 1$, $r_1 = r_3 = 0$, $r_2 > 0$, which is also massless if $r_2 = 1$.

Finally, in the case of R-symmetry scalars we have $i_0 = i'_0 = N - 1$ and all $4N$ odd generators $X^+_{1,4+k}$, $X^+_{2,4+k}$, $X^+_{3,4+k}$, $X^+_{4,4+k}$, are eliminated. More than this, all quantum numbers are zero (cf. (2.54d)), and this is the trivial irrep. The latter follows exactly as explained above for the case $N = 1$.

For $m_1 > 0$ the corresponding character formula is a combination of (2.119) and its conjugate:

$$\begin{aligned}
\operatorname{ch} \hat{\Lambda}_\Lambda &= \sum_{\hat{S} \in \widehat{W}^{bd}_{i_0,i'_0}} (-1)^{\ell(\hat{S})} \hat{S} \cdot \operatorname{ch} \hat{V}^\Lambda - \mathcal{R}, \\
\widehat{W}^{bd}_{i_0,i'_0} &\equiv \widehat{W}^b_{i_0} \times \widehat{W}^d_{i'_0}, \\
& r_i = 0,\ i = 1, 2, \ldots, i_0, N - i'_0, N - i'_0 + 1, \ldots, N - 1, \\
& r_i > 0,\ i = i_0 + 1, N - i'_0 - 1,
\end{aligned} \tag{2.134}$$

where $\mathcal{R}$ designates the counterterms due to our criterion, in particular, due to (2.87) taken with $\varepsilon_{a,N+1-k} = 0$, $a = 1, 2$, $k = 1, \ldots, 1 + i'_0$, $\varepsilon_{bj} = 0$, $b = 3, 4$, $k = j, \ldots, 1 + i_0$.

Also for the third series of massless UIRs we can give a much more explicit character formula without counterterms. Fix the parameter $i_0 = 0, 1, \ldots, \ldots, N-2$. Then we have only the following states in $\hat{\Lambda}_\Lambda$:

$$\begin{aligned}
& X^+_{2,N+4-j} \cdots X^+_{2,N+4-i_0} |\Lambda\rangle, \quad j = 0, 1, \ldots, i_0, \\
& X^+_{4,4+k} \cdots X^+_{4,N+3-i_0} |\Lambda\rangle, \quad k = 1, \ldots, N - 1 - i_0,
\end{aligned} \tag{2.135}$$

altogether N states besides the vacuum [123]. The corresponding character formula for the massless UIRs of this series is therefore

$$\begin{aligned}
\operatorname{ch} \hat{\Lambda}_\Lambda &= 1 + \sum_{j=0}^{i_0} \prod_{i=j}^{i_0} e(\alpha_{2,N+4-i}) + \sum_{k=1}^{N-1-i_0} \prod_{i=k}^{N-1-i_0} e(\alpha_{4,4+i}), \\
& i_0 = 0, 1, \ldots, N - 2,\ r_i = \delta_{i,i_0+1}.
\end{aligned} \tag{2.136}$$

Remark. In this text we use mostly the Verma (factor-)module realization of the UIRs. We give here a short remark on what happens with the elementary representations (ERs) realization of the UIRs. We first recall that in the even case the ERs are the most fundamental objects in the representation theory of semisimple groups. Namely, as shown first by Langlands [270] and Knapp-Zuckerman [259], every irreducible admissible representation of a real connected semisimple Lie group G with finite centre is equivalent to a sub- representation of an elementary representation of G. In the even case the ERs may be defined as C^∞ functions on the local coordinates of G/P, where

P are suitable parabolic subgroups of G. In the superconformal context, cf. [139], the ERs are superfields depending on Minkowski space-time and on $4N$ Grassmann coordinates θ^i_a, $\bar{\theta}^k_b$, $a, b = 1, 2$, $i, k = 1, \ldots, N$.[9] There is 1-to-1 correspondence in these dependencies and the odd null conditions. Namely, if the condition $X^+_{a,4+k}|\Lambda\rangle = 0$, $a = 1, 2$, holds, then the superfields of the corresponding ER do not depend on the variable θ^k_a, while if the condition $X^+_{a,4+k}|\Lambda\rangle = 0$, $a = 3, 4$, holds, then the superfields of the corresponding ER do not depend on the variable $\bar{\theta}^k_{a-2}$. These statements were used in the proof of unitarity for the ERs picture, cf. [142], but were not explicated. They were analyzed in detail in the papers [9, 171, 175, 172, 173, 176, 174], using the notions of 'harmonic superspace analyticity' and Grassmann analyticity [197]. ◇

2.2.5 Summary and discussion

First we recall the results on decompositions of long irreps as they descend to the unitarity threshold. In the SRC cases in subsection 2.2.3 we have established that for $d = d_{\max}$ we have the two-term decompositions given in (2.108), (2.114), and their conjugates. In the DRC cases in subsection 2.2.4 we have established that for $N > 1$ and $d = d_{\max} = d^{ac}$ we have the four-term decompositions given in (2.123), (2.125), (2.127), (2.129).

Next we note that for $N = 1$ all SRC cases enter some decomposition, while no DRC cases enter any decomposition. For $N > 1$ the situation is more diverse and so we give the list of UIRs that do **not** enter decompositions:

- **SRC cases:**
- **a** $d = d_{\max} = d^a = d^1_{N1} = 2 + 2j_2 + z + 2m_1 - 2m/N > d^3_{NN}$, j_1, j_2 arbitrary, $r_1 = 0$.
- **b** $d = d^2_{N1} = z + 2m_1 - 2m/N > d^3_{NN}$, $j_2 = 0$, j_1 arbitrary, $r_1 \leq 2$.
- **c** $d = d_{\max} = d^c = d^3_{NN} = 2 + 2j_1 - z + 2m/N > d^1_{N1}$, j_1, j_2 arbitrary, $r_{N-1} = 0$.
- **d** $d = d^4_{NN} = -z + 2m/N > d^1_{N1}$, $j_1 = 0$, j_1 arbitrary, $r_{N-1} \leq 2$.

- **DRC cases:** all non-trivial cases for $N = 1$, while for $N > 1$ the list is
- **ac** $d = d_{\max} = d^{ac} = d^1_{N1} = d^3_{NN} = 2 + j_1 + j_2 + m_1$, j_1, j_2 arbitrary, $r_1 r_{N-1} = 0$.
- **ad** $d = d^1_{N1} = d^4_{NN} = 1 + j_2 + m_1$, $j_1 = 0$, j_2 arbitrary, $r_{N-1} \leq 2$, $r_1 = 0$ for $N > 2$.
- **bc** $d = d^2_{N1} = d^3_{NN} = 1 + j_1 + m_1$, $j_2 = 0$, j_1 arbitrary, $r_1 \leq 2$, $r_{N-1} = 0$ for $N > 2$.
- **bd** $d = d^2_{N1} = d^4_{NN} = m_1$, $j_1 = j_2 = 0$, $r_1, r_{N-1} \leq 2$ for $N > 2$, $r_1 \leq 4$ for $N = 2$.

We would like to point out possible application of our results to important developments in conformal field theory. Recently, there has been growing interest in superfields with conformal dimensions which are protected from renormalization in the sense that they cannot develop anomalous dimensions [174, 19, 58, 20, 228, 150]. Initially, the idea was that this happens because the representations under which they

9 A mathematically precise formulation is given in [139], while for the even case we refer to [113, 114].

transform determine these dimensions uniquely. Later, it was argued that one can tell which operators will be protected in the quantum theory simply by looking at the representations they transform under and whether they can be written in terms of single trace 1/2 BPS operators (chiral primaries or CPOs) on analytic superspace [228]. In [150] it was shown how, at the unitarity threshold, a long multiplet can be decomposed into four semi-short multiplets, and decompositions similar to ours, i. e., involving the modules given in (2.121) and [123] (which follow from the odd embeddings given in [138]), were considered for $N = 2, 4$. However, the decompositions of [150] are justified on the dimensions of the finite-dimensional irreps of the Lorentz and su(N) subalgebras inducing the superfields involved in the decompositions, and in particular, the latter hold also when $r_1 r_{N-1} = 0$.

Independently of the above, we would like to make a *mathematical* remark. As a by-product of our analysis we have obtained character formulas for the complex Lie superalgebras sl($4/N$). The point is that our character formulas have as starting point character formulas of Verma modules and factor modules over sl($4/N$). Thus, almost all character formulas in Section 3, more precisely, formulas (2.91), (2.101), (2.104), (2.106), (2.111), (2.113), (2.119), (2.121), (2.122), (2.124), (2.128), (2.131), (2.132), (2.133), and (2.134) and their conjugates become character formulas for sl($4/N$) for the same values of the representation parameters by just discarding the counterterms $\mathcal{R}$, $\mathcal{R}_{\text{long}}$, respectively

2.3 Explicit character formulas for $N = 1, 2, 4$

2.3.1 Characters of the even subalgebra

For the characters of the even subalgebra: $\mathcal{G}_0^{\mathbb{C}} = \text{sl}(4) \oplus \text{gl}(1) \oplus \text{sl}(N)$ of $\mathcal{G}^{\mathbb{C}}$, we use formulas (2.77), (2.79). In fact, since the subalgebra $\mathcal{G}_0^{\mathbb{C}}$ is reductive the corresponding character formulas will be given by the products of the character formulas of the two simple factors sl(4) and sl(N).

We start with the sl(4) case. We denoted the six positive roots of sl(4) by α_{ij}, $1 \leq i < j \leq 4$. (Of course, they are part of the root system of sl($4/N$).) For the simplification of the character formulas we use notation for the formal exponents corresponding to the sl(4) simple roots: $t_j \equiv e(\alpha_j)$, $j = 1, 2, 3$; then for the three non-simple roots we have $e(\alpha_{13}) = t_1 t_2$, $e(\alpha_{24}) = t_2 t_3$, $e(\alpha_{14}) = t_1 t_2 t_3$. In terms of these the character formula for a Verma module over sl(4) is

$$\text{ch}_0\, V^{\Lambda^s} = \frac{e(\Lambda^s)}{(1-t_1)(1-t_2)(1-t_3)(1-t_1t_2)(1-t_2t_3)(1-t_1t_2t_3)} \tag{2.137}$$

where by Λ^s we denote the sl(4) lowest weight.

The representations of sl(4) which we consider are infinite dimensional. When $d > d_{\max}$ all the numbers: n_2, n_{12}, n_{23}, n_{13} from (2.19) cannot be positive integers. Then the only reducibilities of the sl(4) Verma module are related to the complexification of the Lorentz subalgebra of su(2, 2), i. e., with sl(2) ⊕ sl(2), and the character formula is given by the product of the two character formulas for finite-dimensional sl(2) irreps. In short, the sl(4) character formula is

$$\begin{aligned}\operatorname{ch} L^2_{d;j_1,j_2} &= \operatorname{ch}_0 V^{\Lambda^s} - \operatorname{ch}_0 V^{\Lambda^s+n_1\alpha_{12}} \\ &\quad - \operatorname{ch}_0 V^{\Lambda^s+n_3\alpha_{34}} + \operatorname{ch}_0 V^{\Lambda^s+n_1\alpha_{12}+n_3\alpha_{34}} \\ &= \frac{e(\Lambda^s)(1-t_1^{n_1})(1-t_3^{n_3})}{(1-t_1)(1-t_2)(1-t_3)(1-t_1t_2)(1-t_2t_3)(1-t_1t_2t_3)} \\ &= \frac{e(\Lambda^s)}{(1-t_2)(1-t_1t_2)(1-t_2t_3)(1-t_1t_2t_3)} \\ &\quad \times \left(\sum_{j=0}^{n_1-1} t_1^j\right)\left(\sum_{k=0}^{n_3-1} t_3^k\right) \\ &= \frac{e(\Lambda^s)}{(1-t_2)(1-t_1t_2)(1-t_2t_3)(1-t_1t_2t_3)} \mathcal{Q}_{n_1,n_3}(t_1,t_3), \\ &\quad n_1 = 2j_1+1,\ n_3 = 2j_2+1,\ d > d_{\max},\end{aligned} \tag{2.138}$$

and we have introduced for later use notation $\mathcal{Q}_{n_1,n_2}$ for the character factorized by $e(\Lambda^s)/((1-t_2)(1-t_1t_2)(1-t_2t_3)(1-t_1t_2t_3))$. The above formula obviously has the form (2.69) replacing $W \mapsto W_2 \times W_2$, where W_2 is the two-element Weyl group of sl(2).

When $d \le d_{\max}$ there are additional even reducibilities, cf. (2.60) and (2.61).

First we consider the case when $d = 2 + j_1 + j_2$, i. e., the unitarity threshold when $j_1j_2 \neq 0$. Using the definitions in (2.19) we have

$$\begin{aligned} &n_1 = 1+2j_1, \quad n_2 = -1-2j_1-2j_2, \quad n_3 = 1+2j_2, \\ &n_{12} = -2j_2, \quad n_{23} = -2j_1, \quad n_{13} = 1. \end{aligned} \tag{2.139}$$

The corresponding character formula is given in [121] (equation (4.32c)) (one has to make the changes $m_{23} \mapsto n_1$, $m_{12} \mapsto n_3$, since that formula is parametrized w.r.t. some referent dominant weight, and set $m_2 = 1$):

$$\begin{aligned}\operatorname{ch} L_{d^{ac};j_1,j_2} &= \operatorname{ch} V^{\Lambda_{ac}}(1 - t_1^{n_1} - t_3^{n_3} - t_{13} + t_1^{n_1}t_3^{n_3} \\ &\quad + t_1^{n_1}t_{23} + t_{12}t_3^{n_3} - t_1^{n_1}t_2t_3^{n_3}) \\ &= \operatorname{ch} V^{\Lambda_{ac}}((1-t_1^{n_1})(1-t_3^{n_3}) - t_{13}(1-t_1^{n_1-1})(1-t_3^{n_3-1})) \\ &= \frac{e(\Lambda_{ac})}{(1-t_2)(1-t_1t_2)(1-t_2t_3)(1-t_1t_2t_3)} \mathcal{P}_{n_1,n_3}(t_1,t_2,t_3), \\ &\quad d^{ac} = 2+j_1+j_2,\ n_1 = 1+2j_1,\ n_3 = 1+2j_2, \\ \mathcal{P}_{n_1,n_3} &= \mathcal{Q}_{n_1,n_3} - t_{13}\mathcal{Q}_{n_1-1,n_3-1}.\end{aligned} \tag{2.140}$$

Note that this formula is valid also for $j_1 = 0$ ($n_1 = 1$) and/or $j_2 = 0$ ($n_3 = 1$) when the second term disappears ($\mathcal{Q}_{0,n} = \mathcal{Q}_{n,0} = 0$), and then the formula coincides with (2.138).

This may be explained with the fact that when $j_1 j_2 = 0$, then the value $d = 2 + j_1 + j_2$ is not a threshold, instead $d = 1 + j_1 + j_2$ is the threshold.

Next we consider the case when $d = 1+j_1+j_2$, i. e., the massless unitarity threshold when $j_1 j_2 = 0$. First we take $j_1 = 0$. Using the definitions in (2.19) we have

$$\begin{aligned} &n_1 = 1, \quad n_2 = -2j_2, \quad n_3 = 1 + 2j_2, \\ &n_{12} = 1 - 2j_2, \quad n_{23} = 1, \quad n_{13} = 2. \end{aligned} \tag{2.141}$$

The corresponding character formula is given in [121] (equation (4.32b)) (one has to make the changes $m_{13} \mapsto n_3$, $m_2 \mapsto n_1 = 1$, $m_3 \mapsto n_{23} = 1$, $m_{23} \mapsto n_{13} = 2$, since that formula is parametrized w.r.t. some more general referent dominant weight):

$$\begin{aligned} \operatorname{ch} L_{d^{ad};0,j_2} &= \operatorname{ch} V^{\Lambda_{ad}}(1 - t_1 - t_3^{n_3} + t_1 t_3^{n_3} \\ &\quad - t_{23} + t_2 t_3^{n_3} + t_1^2 t_{23} - t_1^2 t_2 t_3^{n_3} \\ &\quad + t_1 t_{23}^2 - t_1 t_2^2 t_3^{n_3} - t_{13}^2 + t_{12}^2 t_3^{n_3}) \\ &= \operatorname{ch} V^{\Lambda_{ad}}(1 - t_1)(1 - t_3^{n_3} - t_{23}(1 + t_1)(1 - t_3^{n_3-1}) \\ &\quad + t_2 t_{13}(t_3 - t_3^{n_3-1})) \\ &= \frac{e(\Lambda_{ad})}{(1 - t_2)(1 - t_1 t_2)(1 - t_2 t_3)(1 - t_1 t_2 t_3)} \mathcal{P}_{n_3}(t_1, t_2, t_3), \end{aligned} \tag{2.142}$$

$$\mathcal{P}_{n_3} = \begin{cases} \mathcal{Q}_{n_3} - t_{23}(1 + t_1)\mathcal{Q}_{n_3-1} + t_{23} t_{13} \mathcal{Q}_{n_3-2}, & n_3 \geq 2 \\ 1 - t_2 t_{13}, & n_3 = 1 \end{cases}$$

This formula simplifies considerably for $j_2 = 0$ ($n_3 = 1$) and $j_2 = \frac{p+q-2}{2}$ ($n_3 = 2$):

$$\operatorname{ch} L_{d^{ad}=1;0,0} = \frac{e(\Lambda_{ad})(1 - t_2 t_{13})}{(1 - t_2)(1 - t_{12})(1 - t_{23})(1 - t_{13})}. \tag{2.143}$$

$$\operatorname{ch} L_{d^{ad}=\frac{3}{2};0,\frac{p+q-2}{2}} = \frac{e(\Lambda_{ad})(1 + t_3 - t_{23} - t_{13})}{(1 - t_2)(1 - t_{12})(1 - t_{23})(1 - t_{13})}. \tag{2.144}$$

The case $j_2 = 0$ is obtained from the above by the changes $n_3 \mapsto n_1$ and $t_1 \longleftrightarrow t_3$, and the character formula is

$$\begin{aligned} \operatorname{ch} L_{d^{bc};j_1,0} &= \operatorname{ch} V^{\Lambda_{ad}}(1 - t_3 - t_1^{n_1} + t_1^{n_1} t_3 \\ &\quad - t_{12} + t_1^{n_1} t_2 + t_{12} t_3^2 - t_1^{n_1} t_2 t_3^2 \\ &\quad + t_{12}^2 t_3 - t_1^{n_1} t_2^2 t_3 - t_{13}^2 + t_1^{n_1} t_{23}^2) \\ &= \operatorname{ch} V^{\Lambda_{bc}}(1 - t_3)(1 - t_1^{n_1} - t_{12}(1 + t_3)(1 - t_1^{n_1-1}) \\ &\quad + t_2 t_{13}(t_1 - t_1^{n_1-1})) \\ &= \frac{e(\Lambda_{bc})}{(1 - t_2)(1 - t_1 t_2)(1 - t_2 t_3)(1 - t_1 t_2 t_3)} \mathcal{P}'_{n_1}(t_1, t_2, t_3), \end{aligned} \tag{2.145}$$

$$\mathcal{P}'_{n_1} = \begin{cases} \mathcal{Q}'_{n_1} - t_{12}(1 + t_3)\mathcal{Q}'_{n_1-1} + t_{12} t_{13} \mathcal{Q}'_{n_1-2}, & n_1 \geq 2 \\ 1 - t_2 t_{13}, & n_1 = 1 \end{cases} \tag{2.146}$$

This formula simplifies considerably for $j_1 = 0$ ($n_1 = 1$) when it coincides with (2.143) since $\mathcal{P}_1' = \mathcal{P}_1$, while in the case $j_1 = \frac{p+q-2}{2}$ ($n_1 = 2$) we have

$$\operatorname{ch} L_{d^{bc}=\frac{3}{2};\frac{p+q-2}{2},0} = \frac{e(\Lambda_{ad})(1 + t_1 - t_{12} - t_{13})}{(1-t_2)(1-t_{12})(1-t_{23})(1-t_{13})}. \tag{2.147}$$

Remark. It is not surprising that the three cases (2.143), (2.144) and (2.147) (when $j_1 + j_2 \leq \frac{p+q-2}{2}$) are special since (unlike the other massless irreps) they are not related to finite-dimensional irreps. The easiest way to see this is by the value of the Casimir, which in terms of the parameters n_{ik} is given as follows:

$$C_2 = \frac{1}{2}\left(n_{13}^2 + n_2^2 + \frac{1}{2}(n_1 - n_3)^2\right) - 5 \tag{2.148}$$

and is normalized so that for each finite-dimensional irrep (when $n_k \in \mathbb{N}$) it is non-negative and zero only for the 1-dimensional irrep ($n_k = 1$). It is easy to see that for the massless cases given in (2.141) one has

$$C_2 = 3(j_2^2 - 1), \quad j_2 \in \frac{p+q-2}{2}\mathbb{Z}_+, \tag{2.149}$$

which is indeed negative for $j_2 = 0, \frac{p+q-2}{2}$ and non-negative for $j_2 \geq 1$. In fact, the finite-dimensional irrep, related to a massless case with $j_2 \geq 1$, has dimension [111]:

$$\dim \Lambda_{\mathrm{fdir},j_2} = \frac{1}{3}j_2(4j_2^2 - 1), \quad j_2 \in 1 + \frac{p+q-2}{2}\mathbb{Z}_+. \tag{2.150}$$

For the conjugate massless case we have similarly

$$C_2 = 3(j_1^2 - 1), \quad j_1 \in \frac{p+q-2}{2}\mathbb{Z}_+, \tag{2.151}$$

$$\dim \Lambda_{\mathrm{fdir}',j_1} = \frac{1}{3}j_1(4j_1^2 - 1), \quad j_1 \in 1 + \frac{p+q-2}{2}\mathbb{Z}_+, \tag{2.152}$$

with the same conclusions.

These three representations are the so-called minimal UIRs of su(2, 2), cf., e. g., [261], and for other of their properties we refer to [120, 125]. ◇

In the case of sl(N) the representations are finite dimensional, since we induce from UIRs of su(N). The character formula is (2.69), which we repeat in order to introduce the corresponding notation:

$$\mathcal{S}_{r_1,\ldots,r_{N-1}} = \sum_{w \in W_u} (-1)^{\ell(w)} \operatorname{ch}_0 V^{w\cdot\Lambda^u}, \quad \Lambda^u \in -\Gamma_+^u \tag{2.153}$$

The index u is to distinguish the quantities pertinent to the case.

We shall write down explicitly the cases that we shall need, namely, sl(2) and sl(4). In the sl(2) case the Weyl group has only two elements and the character formula is very simple:

$$\begin{aligned} \mathcal{S}_r &= \operatorname{ch} V^{\Lambda}(1 - t^{r+1}) = e(\Lambda)\frac{1 - t^{r+1}}{1 - t} \\ &= e(\Lambda)(1 + t + t^2 + \cdots + t^r), \\ e(\Lambda) &= t^{-r/2}, \quad r \in \mathbb{Z}_+. \end{aligned} \tag{2.154}$$

In the sl(4) case the Weyl group has 24 elements and the character formula is:

$$\begin{aligned} \mathcal{S}_{r_1,r_2,r_3} = {} & \operatorname{ch} V^{\Lambda^u}(1 - t_1^{n_1} - t_2^{n_2} - t_3^{n_3} + t_1^{n_1}t_3^{n_3} + t_1^{n_1}t_2^{n_{12}} + t_2^{n_{23}}t_3^{n_3} \\ & + t_1^{n_{12}}t_2^{n_2} + t_2^{n_2}t_3^{n_{23}} - t_1^{n_1}t_2^{n_{13}}t_3^{n_3} - t_1^{n_{12}}t_2^{n_2}t_3^{n_{23}} - t_1^{n_{13}}t_2^{n_{23}}t_3^{n_3} \\ & - t_1^{n_1}t_2^{n_{12}}t_3^{n_{13}} - (t_1t_2)^{n_{12}} - (t_2t_3)^{n_{23}} + t_1^{n_{12}}t_2^{n_2+n_{13}}t_3^{n_{23}} \\ & + t_1^{n_1}(t_2t_3)^{n_{13}} + (t_1t_2)^{n_{12}}t_3^{n_{13}} + t_1^{n_{13}}(t_2t_3)^{n_{23}} + (t_1t_2)^{n_{13}}t_3^{n_3} \\ & - t_1^{n_{12}}t_2^{n_2+n_{13}}t_3^{n_{13}} - t_1^{n_{13}}t_2^{n_2+n_{13}}t_3^{n_{23}} - (t_1t_2t_3)^{n_{13}} + (t_1t_2t_3)^{n_{13}}t_2^{n_2}), \\ & n_k = r_k + 1,\ n_{12} = r_1 + r_2 + 2, n_{23} = r_2 + r_3 + 2, \\ & n_{13} = r_1 + r_2 + r_3 + 3 \end{aligned} \tag{2.155}$$

The expression for Λ^u in terms of the sl(4) simple roots is

$$\begin{aligned} \Lambda^u &= -\frac{p+q-2}{2}\left(r_2 + \frac{p+q-2}{2}(3r_1 + r_3)\right)\beta_1 - \left(r_2 + \frac{p+q-2}{2}(r_1 + r_3)\right)\beta_2 \\ &\quad - \frac{p+q-2}{2}\left(r_2 + \frac{p+q-2}{2}(r_1 + 3r_3)\right)\beta_3 \\ &= \sum_{k=1}^{3} \lambda_k \beta_k, \end{aligned} \tag{2.156}$$

where the minus signs are due to the fact that Λ^u is assumed to be lowest weight, for highest weight the minuses become pluses. Furthermore, the notation for the simple roots depends on the application, e. g., when applied to the compact part of the even subalgebra for $N = 4$: $\mathcal{G}_0 = \mathrm{su}(2,2) \oplus u(1) \oplus \mathrm{su}(4)$ then the roots are mapped: $\beta_1 \mapsto \alpha_7$, $\beta_2 \mapsto \alpha_6$, $\beta_3 \mapsto \alpha_5$.

We give some explicit examples that would be actually used:

$$\begin{aligned} \mathcal{S}_{0,0,0} &= 1, \\ \mathcal{S}_{1,0,0} &= e(\Lambda^u)(1 + t_1 + t_{12} + t_{13}), \\ \mathcal{S}_{0,0,1} &= e(\Lambda^u)(1 + t_3 + t_{23} + t_{13}), \\ \mathcal{S}_{0,1,0} &= e(\Lambda^u)(1 + t_2 + t_{12} + t_{23} + t_{13} + t_{13}t_2), \\ \mathcal{S}_{2,0,0} &= e(\Lambda^u)(1 + t_1 + t_1^2 + t_{12} + t_{12}^2 + t_{13} + t_{13}^2 + t_{12}t_1 + t_{13}t_1 + t_{13}t_{12}), \end{aligned} \tag{2.157}$$

$$\mathcal{S}_{0,0,2} = e(\Lambda^u)(1 + t_3 + t_3^2 + t_{23} + t_{23}^2 + t_{13} + t_{13}^2 + t_{23}t_3 + t_{13}t_3 + t_{13}t_{23}),$$
$$e(\Lambda^u) = \sum_{k=1}^{3} t_k^{\lambda_k}.$$

Naturally, the number of terms in the character formula is equal to the dimension of the corresponding irrep:

$$\dim_{r_1,r_2,r_3} = \frac{1}{12}(r_1+1)(r_2+1)(r_3+1)(r_1+r_2+2)(r_2+r_3+2)(r_1+r_2+r_3+3). \quad (2.158)$$

2.3.2 $N = 1$

• Long superfields

If $d > d_{\max}$, $j_1 j_2 > 0$ then $\hat{\Lambda}_\Lambda$ has the maximum possible number of states, 16.

The bare character formula following from (2.84) is (see also [123]):

$$\begin{aligned} \operatorname{ch}\hat{\Lambda}_\Lambda &= \prod_{\alpha\in\Delta_{\bar{1}}^+} (1 + e(\alpha)) \\ &= (1 + e(\alpha_{15}))(1 + e(\alpha_{25}))(1 + e(\alpha_{35}))(1 + e(\alpha_{45})), \end{aligned} \quad (2.159)$$

where we use (2.13):

$$\begin{aligned} \alpha_{15} &= \gamma_1 + \gamma_3 = \alpha_1 + \gamma_3, \quad \alpha_{25} = \gamma_3, \\ \alpha_{35} &= \gamma_2 + \gamma_4 = \alpha_3 + \gamma_4, \quad \alpha_{45} = \gamma_4 \end{aligned} \quad (2.160)$$

The mirror symmetry of (2.159) follows from (2.15):

$$\gamma_1 \longleftrightarrow \gamma_2, \quad \gamma_3 \longleftrightarrow \gamma_4$$

For the characters we shall need also the following notation related to the odd roots:

$$\vartheta \equiv e(\alpha_{15}), \quad \theta \equiv e(\alpha_{25}), \quad \bar{\vartheta} \equiv e(\alpha_{35}), \quad \bar{\theta} \equiv e(\alpha_{45}), \quad (2.161)$$

(the bar respecting the mirror symmetry), and for the sl(4)-related variables: $t_k \equiv e(\alpha_k)$, $k = 1, 2, 3$; cf. Section 2.2. Note the relation

$$\vartheta = t_1\theta, \quad \bar{\vartheta} = t_3\bar{\theta}, \quad (2.162)$$

which is also mirror symmetric.

The full character formula takes into account the characters of the conformal algebra entries, i. e., we have

$$\operatorname{ch} L_\Lambda = e(\Lambda)\{\operatorname{ch}' L^2_{d;j_1,j_2} + e(\alpha_{15})\operatorname{ch}' L^2_{d+\frac{p+q-2}{2};j_1-\frac{p+q-2}{2},j_2}$$

$$
\begin{aligned}
&+ e(\alpha_{25})\,\mathrm{ch}'\,L^2_{d+\frac{p+q-2}{2};j_1+\frac{p+q-2}{2},j_2} \\
&+ e(\alpha_{35})\,\mathrm{ch}'\,L^2_{d+\frac{p+q-2}{2};j_1,j_2-\frac{p+q-2}{2}} + e(\alpha_{45})\,\mathrm{ch}'\,L^2_{d+\frac{p+q-2}{2};j_1,j_2+\frac{p+q-2}{2}} \\
&+ e(\alpha_{15})e(\alpha_{25})\,\mathrm{ch}'\,L^2_{d+1;j_1,j_2} + e(\alpha_{15})e(\alpha_{35})\,\mathrm{ch}'\,L^2_{d+1;j_1-\frac{p+q-2}{2},j_2-\frac{p+q-2}{2}} \\
&+ e(\alpha_{15})e(\alpha_{45})\,\mathrm{ch}'\,L^2_{d+1;j_1-\frac{p+q-2}{2},j_2+\frac{p+q-2}{2}} \\
&+ e(\alpha_{25})e(\alpha_{35})\,\mathrm{ch}'\,L^2_{d+1;j_1+\frac{p+q-2}{2},j_2-\frac{p+q-2}{2}} \\
&+ e(\alpha_{25})e(\alpha_{45})\,\mathrm{ch}'\,L^2_{d+1;j_1+\frac{p+q-2}{2},j_2+\frac{p+q-2}{2}} + e(\alpha_{35})e(\alpha_{45})\,\mathrm{ch}'\,L^2_{d+1;j_1,j_2} \\
&+ e(\alpha_{15})e(\alpha_{25})e(\alpha_{35})\,\mathrm{ch}'\,L^2_{d+\frac{3}{2};j_1,j_2-\frac{p+q-2}{2}} \\
&+ e(\alpha_{15})e(\alpha_{25})e(\alpha_{45})\,\mathrm{ch}'\,L^2_{d+\frac{3}{2};j_1,j_2+\frac{p+q-2}{2}} \\
&+ e(\alpha_{15})e(\alpha_{35})e(\alpha_{45})\,\mathrm{ch}'\,L^2_{d+\frac{3}{2};j_1-\frac{p+q-2}{2},j_2} \\
&+ e(\alpha_{25})e(\alpha_{35})e(\alpha_{45})\,\mathrm{ch}'\,L^2_{d+\frac{3}{2};j_1+\frac{p+q-2}{2},j_2} \\
&+ e(\alpha_{15})e(\alpha_{25})e(\alpha_{35})e(\alpha_{45})\,\mathrm{ch}'\,L^2_{d+2;j_1,j_2}\Big\}
\end{aligned}
\tag{2.163}
$$

where $\mathrm{ch}'\,L^2_{d;j_1,j_2}$ is the conformal (actually sl(4)) character formula of equation (2.138), the prime denoting that we have omitted the prefactor $e(\Lambda^s)$ since above it is distributed among the other prefactors. Substituting the explicit expressions for $\mathrm{ch}'\,L^2_{d;j_1,j_2}$ we obtain

$$
\begin{aligned}
\mathrm{ch}\,L_\Lambda = {}& \frac{e(\Lambda)}{(1-t_2)(1-t_1t_2)(1-t_2t_3)(1-t_1t_2t_3)} \\
&\times \big\{(1+\vartheta\theta)(1+\bar\vartheta\bar\theta)\mathcal{Q}_{n_1,n_3} + \vartheta\mathcal{Q}_{n_1-1,n_3} + \theta\mathcal{Q}_{n_1+1,n_3} \\
&+ \bar\vartheta\mathcal{Q}_{n_1,n_3-1} + \bar\theta\mathcal{Q}_{n_1,n_3+1} + \vartheta\bar\vartheta\mathcal{Q}_{n_1-1,n_3-1} \\
&+ \vartheta\bar\theta\mathcal{Q}_{n_1-1,n_3+1} + \theta\bar\vartheta\mathcal{Q}_{n_1+1,n_3-1} + \theta\bar\theta\mathcal{Q}_{n_1+1,n_3+1} \\
&+ \vartheta\theta\bar\vartheta\mathcal{Q}_{n_1,n_3-1} + \vartheta\theta\bar\theta\mathcal{Q}_{n_1,n_3+1} \\
&+ \vartheta\bar\vartheta\bar\theta\mathcal{Q}_{n_1-1,n_3} + \theta\bar\vartheta\bar\theta\mathcal{Q}_{n_1+1,n_3}\big\}
\end{aligned}
\tag{2.164}
$$

where the polynomials $\mathcal{Q}$ are in terms of the sl(4)-related variables t_1, t_2, t_3.

When $j_1j_2 = 0$ there are less terms in the character formula, since $\mathcal{Q}_{0,\cdot} = 0 = \mathcal{Q}_{\cdot,0}$, and further the entries simplify.

If $j_1 = 0, j_2 > 0$ then the generator X^+_{15} can appear only together with the generator X^+_{25}, and $\hat\Lambda_\Lambda$ has 12 states = 3(chiral)×4(anti-chiral) states.[10] The bare character formula is (2.91), where the counterterm $\mathcal{R}$ in our case is

$$
\mathcal{R} = e(\alpha_{15})(1+e(\alpha_{35}))(1+e(\alpha_{45}))
\tag{2.165}
$$

10 In statements like this each sector includes the vacuum.

and the full character formula is

$$\operatorname{ch} L_\Lambda = \frac{e(\Lambda)}{(1-t_2)(1-t_1t_2)(1-t_2t_3)(1-t_1t_2t_3)} \times \{(1+\vartheta\theta)(1+\bar\vartheta\bar\theta)\mathcal{Q}_{n_3} + \theta(1+\bar\vartheta\bar\theta)(1+t_1)\mathcal{Q}_{n_3} + \bar\vartheta(1+\vartheta\theta)\mathcal{Q}_{n_3-1} + \bar\theta(1+\vartheta\theta)\mathcal{Q}_{n_3+1} + \theta\bar\vartheta(1+t_1)\mathcal{Q}_{n_3-1} + \theta\bar\theta(1+t_1)\mathcal{Q}_{n_3+1}\} \tag{2.166}$$

where we use:

$$\mathcal{Q}_{1,n_3} = \sum_{k=0}^{n_3-1} t_3^k \equiv \mathcal{Q}_{n_3}, \quad \mathcal{Q}_{2,n_3}(t_1,t_3) = (1+t_1)\mathcal{Q}_{n_3}(t_3).$$

The next case is conjugate. If $j_1 > 0, j_2 = 0$ then the generator X_{35}^+ can appear only together with the generator X_{45}^+, and $\hat\Lambda_\Lambda$ has 12 states. The bare character formula is again (2.91) with counterterm:

$$\mathcal{R} = e(\alpha_{35})(1+e(\alpha_{15}))(1+e(\alpha_{25})) \tag{2.167}$$

and the full character formula is

$$\operatorname{ch} L_\Lambda = \frac{e(\Lambda)}{(1-t_2)(1-t_1t_2)(1-t_2t_3)(1-t_1t_2t_3)} \times \{(1+\vartheta\theta)(1+\bar\vartheta\bar\theta)\mathcal{Q}'_{n_1} + \bar\theta(1+\vartheta\theta)(1+t_3)\mathcal{Q}'_{n_1} + \vartheta(1+\bar\vartheta\bar\theta)\mathcal{Q}'_{n_1-1} + \theta(1+\bar\vartheta\bar\theta)\mathcal{Q}'_{n_1+1} + \vartheta\bar\theta(1+t_3)\mathcal{Q}'_{n_1-1} + \theta\bar\theta(1+t_3)\mathcal{Q}'_{n_1+1}\} \tag{2.168}$$

where we use:

$$\mathcal{Q}_{n_1,1} = \sum_{j=0}^{n_1-1} t_1^j \equiv \mathcal{Q}'_{n_1}, \quad \mathcal{Q}_{n_1,2}(t_1,t_3) = (1+t_3)\mathcal{Q}'_{n_1}(t_1).$$

The next case combines the previous two. If $j_1 = j_2 = 0$ then the generator X_{15}^+ can appear only together with the generator X_{25}^+, the generator X_{35}^+ can appear only together with the generator X_{45}^+, and $\hat\Lambda_\Lambda$ has 9 states = 3(chiral)×3(anti-chiral) states. The character formula is (2.91) with:

$$\mathcal{R} = \vartheta(1+e(\alpha_{35}))(1+e(\alpha_{45})) + e(\alpha_{35})(1+e(\alpha_{15}))(1+e(\alpha_{25})) - e(\alpha_{15})e(\alpha_{35}), \tag{2.169}$$

i. e., we combine the counterterms of the previous two cases, but need to subtract a counterterm that is counted twice. The bare character formula is

$$\operatorname{ch} \hat\Lambda_\Lambda = (1+e(\alpha_{25})+\vartheta e(\alpha_{25}))(1+e(\alpha_{45})+e(\alpha_{35})e(\alpha_{45})) \tag{2.170}$$

and the full character formula may be obtained from (2.166) setting $n_3 = 1$ (or from (2.168) setting $n_1 = 1$):

$$\begin{aligned}\mathrm{ch}\, L_\Lambda = {} & \frac{e(\Lambda)}{(1-t_2)(1-t_1t_2)(1-t_2t_3)(1-t_1t_2t_3)} \\ & \times \{(1+\vartheta\theta)(1+\bar{\vartheta}\bar{\theta}) + \theta(1+\bar{\vartheta}\bar{\theta})(1+t_1) \\ & + \bar{\theta}(1+\vartheta\theta)(1+t_3) + \theta\bar{\theta}(1+t_1)(1+t_3)\}\end{aligned} \tag{2.171}$$

using $\mathcal{Q}_1 = 1$, $\mathcal{Q}_2 = 1 + t_3$ (or $\mathcal{Q}'_1 = 1$, $\mathcal{Q}'_2 = 1 + t_1$).

• **SRC cases**

• **a** $d = d_{\max} = d^1_{11} = 2 + 2j_2 + z > d^3_{11}$.

The generator X^+_{35} is eliminated [123] and for $z = 0$ (then $j_2 > j_1$) these are $\frac{1}{4}$-BPS cases [127].

These are called semi-conserved superfields in [76]. For $j_2 > 0$ they obey the first-order super-differential operator given explicitly in equations (7a) of [140]. When $j_2 = 0$ that first-order super-differential operator has trivial kernel and is replaced by second-order super-differential operator given in (11b) of [140].

• $j_1 > 0$. Here there are only 8 states.[11] The bare character formula is (2.101) (or equivalently (2.104)) without counterterms (see also [123]):

$$\mathrm{ch}\, \hat{\Lambda}_\Lambda = \prod_{\substack{\alpha\in\Delta_{\bar{1}}^+ \\ \alpha\neq\alpha_{35}}} (1 + e(\alpha)). \tag{2.172}$$

The full character formula follows from (2.164):

$$\begin{aligned}\mathrm{ch}\, L_\Lambda = {} & \frac{e(\Lambda)}{(1-t_2)(1-t_1t_2)(1-t_2t_3)(1-t_1t_2t_3)} \\ & \times \{(1+\vartheta\theta)\mathcal{Q}_{n_1,n_3} + \vartheta\mathcal{Q}_{n_1-1,n_3} \\ & + \theta\mathcal{Q}_{n_1+1,n_3} + \bar{\theta}\mathcal{Q}_{n_1,n_3+1} + \vartheta\bar{\theta}\mathcal{Q}_{n_1-1,n_3+1} \\ & + \theta\bar{\theta}\mathcal{Q}_{n_1+1,n_3+1} + \vartheta\theta\bar{\theta}\mathcal{Q}_{n_1,n_3+1}\}\end{aligned} \tag{2.173}$$

• $j_1 = 0$. The generator X^+_{15} can appear only together with the generator X^+_{25} and there are only 6 states. Then the bare character formula is (2.101) (or equivalently (2.104)) with counterterm:

$$\begin{aligned}\mathrm{ch}\, \hat{\Lambda}_\Lambda &= \prod_{\substack{\alpha\in\Delta_{\bar{1}}^+ \\ \alpha\neq\alpha_{35}}} (1 + e(\alpha)) - \mathcal{R} \\ &= (1 + e(\alpha_{25}) + \vartheta e(\alpha_{25}))(1 + e(\alpha_{45})), \\ \mathcal{R} &= \vartheta(1 + e(\alpha_{45})).\end{aligned} \tag{2.174}$$

11 For brevity, here and often below we shall say "there are M states" meaning "there are M states in $\hat{\Lambda}_\Lambda$".

The full character formula follows from (2.173):

$$\operatorname{ch} L_\Lambda = \frac{e(\Lambda)}{(1-t_2)(1-t_1t_2)(1-t_2t_3)(1-t_1t_2t_3)} \times \{(1+\vartheta\theta)\mathcal{Q}_{n_3} + \theta(1+t_1)(\mathcal{Q}_{n_3} + \bar\theta\mathcal{Q}_{n_3+1}) + (1+\vartheta\theta)\bar\theta\mathcal{Q}_{n_3+1}\} \tag{2.175}$$

• **b** $d = d^2_{11} = z > d^3_{11}, j_2 = 0$.

These UIRs are called chiral since all anti-chiral generators are eliminated. They obey the first-order super-differential operator given explicitly in equations (7b) of [140].

• $j_1 > 0$. The generators X^+_{35} and X^+_{45} are eliminated and there are only 4 states. The bare character formula is (2.119) (for $i_0 = 0$) without counterterms:

$$\operatorname{ch} \hat\Lambda_\Lambda = (1+e(\alpha_{15}))(1+e(\alpha_{25})) \tag{2.176}$$

The full character formula follows from (2.173):

$$\operatorname{ch} L_\Lambda = \frac{e(\Lambda)}{(1-t_2)(1-t_1t_2)(1-t_2t_3)(1-t_1t_2t_3)} \times \{(1+\vartheta\theta)\mathcal{Q}'_{n_1} + \vartheta\mathcal{Q}'_{n_1-1} + \theta\mathcal{Q}'_{n_1+1}\} \tag{2.177}$$

• $j_1 = 0$. The generators X^+_{35} and X^+_{45} are eliminated, the generator X^+_{15} can appear only together with the generator X^+_{25}, and there are only 3 states. The bare character formula is (2.119) (for $i_0 = 0$) with counterterm $\mathcal{R} = e(\alpha_{15})$:

$$\operatorname{ch} \hat\Lambda_\Lambda = 1 + e(\alpha_{25}) + e(\alpha_{15})e(\alpha_{25}) \tag{2.178}$$

The full character formula follows from (2.177):

$$\operatorname{ch} L_\Lambda = \frac{e(\Lambda)}{(1-t_2)(1-t_1t_2)(1-t_2t_3)(1-t_1t_2t_3)} \times \{1 + \vartheta\theta + \theta(1+t_1)\} \tag{2.179}$$

The next two cases **c, d** are conjugate to the above **a, b**. The character formulas are obtainable by the changes $j_1 \longleftrightarrow j_2$, $n_1 \longleftrightarrow n_3$, $t_1 \longleftrightarrow t_3$, $\vartheta \longleftrightarrow \bar\vartheta$, $\theta \longleftrightarrow \bar\theta$, ($\alpha_{15} \longleftrightarrow \alpha_{35}$, $\alpha_{25} \longleftrightarrow \alpha_{45}$). Thus, we list the character formulas without explanations.

• **c** $d = d_{max} = d^3_{11} = 2 + 2j_1 - z > d^1_{11}$.

The generator X^+_{15} is eliminated [123] and for $z = 0$ (then $j_1 > j_2$) these are $\frac{1}{4}$-BPS cases [127].

These are called semi-conserved superfields in [76]. For $j_1 > 0$ they obey the first-order super-differential operator given explicitly in equations (7c) of [140]. When $j_1 = 0$ that first-order super-differential operator has trivial kernel and is replaced by second-order super-differential operator given in (11a) of [140].

- $j_2 > 0$. The character formulas are

$$\mathrm{ch}\,\hat{\Lambda}_\Lambda = \prod_{\substack{\alpha\in\Delta_{\bar{1}}^+ \\ \alpha\neq\alpha_{15}}} (1 + e(\alpha)). \tag{2.180}$$

$$\mathrm{ch}\,L_\Lambda = \frac{e(\Lambda)}{(1-t_2)(1-t_1t_2)(1-t_2t_3)(1-t_1t_2t_3)} \tag{2.181}$$

$$\begin{aligned} &\times \{(1+\bar{\vartheta}\bar{\theta})\mathcal{Q}_{n_1,n_3} + \bar{\vartheta}\mathcal{Q}_{n_1,n_3-1} \\ &+ \bar{\theta}\mathcal{Q}_{n_1,n_3+1} + \theta\mathcal{Q}_{n_1+1,n_3} + \theta\bar{\vartheta}\mathcal{Q}_{n_1+1,n_3-1} \\ &+ \theta\bar{\theta}\mathcal{Q}_{n_1+1,n_3+1} + \theta\bar{\vartheta}\bar{\theta}\mathcal{Q}_{n_1+1,n_3}\} \end{aligned} \tag{2.182}$$

- $j_2 = 0$. The character formulas are

$$\mathrm{ch}\,\hat{\Lambda}_\Lambda = (1 + e(\alpha_{25}))(1 + e(\alpha_{45}) + e(\alpha_{35})e(\alpha_{45})). \tag{2.183}$$

$$\begin{aligned} \mathrm{ch}\,L_\Lambda = &\frac{e(\Lambda)}{(1-t_2)(1-t_1t_2)(1-t_2t_3)(1-t_1t_2t_3)} \\ &\times \{(1+\bar{\vartheta}\bar{\theta})\mathcal{Q}'_{n_1} + \bar{\theta}(1+t_3)(\mathcal{Q}'_{n_1} + \theta\mathcal{Q}'_{n_1+1}) \\ &+ (1+\theta\bar{\vartheta})\bar{\theta}\mathcal{Q}'_{n_1+1}\} \end{aligned} \tag{2.184}$$

• **d** $d = d^4_{11} = -z > d^1_{11}, j_1 = 0$.

These UIRs are called anti-chiral since all chiral generators are eliminated. They obey the first-order super-differential operator given explicitly in equations (7d) of [140].

- $j_2 > 0$. The character formulas are

$$\mathrm{ch}\,\hat{\Lambda}_\Lambda = (1 + e(\alpha_{35}))(1 + e(\alpha_{45})) \tag{2.185}$$

$$\begin{aligned} \mathrm{ch}\,L_\Lambda = &\frac{e(\Lambda)}{(1-t_2)(1-t_1t_2)(1-t_2t_3)(1-t_1t_2t_3)} \\ &\times \{(1+\bar{\vartheta}\bar{\theta})\mathcal{Q}_{n_3} + \bar{\vartheta}\mathcal{Q}_{n_3-1} + \bar{\theta}\mathcal{Q}_{n_3+1}\} \end{aligned} \tag{2.186}$$

- $j_2 = 0$. The character formulas are

$$\mathrm{ch}\,\hat{\Lambda}_\Lambda = 1 + e(\alpha_{45}) + e(\alpha_{35})e(\alpha_{45}) \tag{2.187}$$

$$\begin{aligned} \mathrm{ch}\,L_\Lambda = &\frac{e(\Lambda)}{(1-t_2)(1-t_1t_2)(1-t_2t_3)(1-t_1t_2t_3)} \\ &\times \{1 + \bar{\vartheta}\bar{\theta} + \bar{\theta}(1+t_3)\} \end{aligned} \tag{2.188}$$

• **DRC cases**

• **ac** $d = d^{ac} = d_{\max} = d^1_{11} = d^3_{11} = d^{ac} = 2 + j_1 + j_2, z = z_{ac} = j_1 - j_2$.

These are the conserved superfields. For $j_1j_2 \neq 0$ they obey the two first-order super-differential operators given explicitly in equations (7a,c) of [140]. (As we noted above these semi-short UIRs may be called Grassmann analytic following [174].)

When $j_1 = 0$ ($j_2 = 0$) the first-order super-differential operator from (7c) of [140] ((7a) of [140], respectively) has trivial kernel and is replaced by second -order super-differential operator given in (11a) of [140] ((11b) of [140], respectively).

The generators X_{15}^+ and X_{35}^+ are eliminated (though for different reasons for $j_1 > 0$ and $j_1 = 0$, respectively, for $j_2 > 0$ and $j_2 = 0$). For $j_2 = j_2$ (then $z = 0$) these are $\frac{p+q-2}{2}$-BPS cases [127].[12] There are only 4 states and the bare character formula is (2.122) (see also (3.84) from [123]) for $i_0 = i_0' = 0$ without counterterms:

$$\begin{aligned}\operatorname{ch} \hat{\Lambda}_\Lambda &= (1 + e(\alpha_{25}))(1 + e(\alpha_{45}))\\ &= \operatorname{ch} \hat{V}^\Lambda - \frac{1}{1 + e(\alpha_{15})} \operatorname{ch} \hat{V}^{\Lambda+\alpha_{15}} - \frac{1}{1 + e(\alpha_{35})} \operatorname{ch} \hat{V}^{\Lambda+\alpha_{35}}\\ &\quad + \frac{1}{(1 + e(\alpha_{15}))(1 + e(\alpha_{35}))} \operatorname{ch} \hat{V}^{\Lambda+\alpha_{15}+\alpha_{35}},\end{aligned} \tag{2.189}$$

where the terms with minus may be interpreted as taking out states, while the last term indicates adding back what was taken two times. The corresponding decomposition of L_Λ is given by

$$\begin{aligned}L_\Lambda = L_{d^{ac};j_1,j_2} \otimes L^z_{z_{ac}} + L_{d^{ac}+\frac{p+q-2}{2};j_1+\frac{p+q-2}{2},j_2} \otimes L^z_{z_{ac}-\frac{3}{2}}\\ + L_{d^{ac}+\frac{p+q-2}{2};j_1,j_2+\frac{p+q-2}{2}} \otimes L^z_{z_{ac}+\frac{3}{2}} + L_{d^{ac}+1;j_1+\frac{p+q-2}{2},j_2+\frac{p+q-2}{2}} \otimes L^z_{z_{ac}}\end{aligned} \tag{2.190}$$

Note that for all four conformal entries is fulfilled the relation $d = 2 + j_1 + j_2$, which for $j_1 j_2 \neq 0$ is the conformal unitarity threshold. Thus, for the conformal characters we have to use equation (2.140), and then the full character formula is

$$\begin{aligned}\operatorname{ch} L_\Lambda &= e(\Lambda)\{\operatorname{ch}' L_{d^{ac};j_1,j_2} + e(\alpha_{25})\operatorname{ch}' L_{d^{ac}+\frac{p+q-2}{2};j_1+\frac{p+q-2}{2},j_2}\\ &\quad + e(\alpha_{45})\operatorname{ch}' L_{d^{ac}+\frac{p+q-2}{2};j_1,j_2+\frac{p+q-2}{2}}\\ &\quad + e(\alpha_{25})e(\alpha_{45})\operatorname{ch}' L_{d^{ac}+1;j_1+\frac{p+q-2}{2},j_2+\frac{p+q-2}{2}}\}\\ &= \frac{e(\Lambda)}{(1-t_1)(1-t_2)(1-t_3)(1-t_{12})(1-t_{23})(1-t_{13})}\\ &\quad \times \{\mathcal{P}_{n_1,n_3} + \theta\mathcal{P}_{n_1+1,n_3} + \bar{\theta}\mathcal{P}_{n_1,n_3+1} + \theta\bar{\theta}\mathcal{P}_{n_1+1,n_3+1}\}\end{aligned} \tag{2.191}$$

where ch′ is the character equation (2.140) without the prefactor $e(\Lambda_{ac})$—this prefactor is subsumed in the overall prefactor $e(\Lambda)$ since the relative difference weights between the four terms of (2.191) are taken into account by the prefactors $e(\alpha_{k5})$.

• **ad** $d = d^{ad} = d^1_{11} = d^4_{11} = 1 + j_2 = -z,\ j_1 = 0.$

The generators X_{15}^+, X_{25}^+ and X_{35}^+ are eliminated (for the latter for different reasons for $j_2 > 0$ and $j_2 = 0$). These are the first series of massless UIRs, and everything is

12 There are only three BPS cases for $N = 1$, the other two were mentioned above in cases **a, c**.

already explicit in the general formulas. There are only 2 states and the bare character formula is (2.133) for $N = 1$:

$$\mathrm{ch}\,\hat{\Lambda}_\Lambda = 1 + e(\alpha_{45}). \tag{2.192}$$

The corresponding decomposition of L_Λ is given by

$$L_\Lambda = L_{d^{ad};0,j_2} \otimes L^z_{z_{ad}} + L_{d^{ad}+\frac{p+q-2}{2};0,j_2+\frac{p+q-2}{2}} \otimes L^z_{z_{ad}+\frac{3}{2}} \tag{2.193}$$

Note that for both conformal entries the relation $d = 1 + j_1 + j_2$ is fulfilled, which for $j_1 j_2 = 0$ is the conformal unitarity threshold. For the characters we have to use equation (2.142), (or its reductions for $j_2 = 0, \frac{p+q-2}{2}$), and then the full character formula is

$$\begin{aligned} \mathrm{ch}L_\Lambda &= e(\Lambda)\{\mathrm{ch}' L_{d^{ad};0,j_2} + e(\alpha_{45})\mathrm{ch}' L_{d^{ad}+\frac{p+q-2}{2};0,j_2+\frac{p+q-2}{2}}\} \\ &= e(\Lambda)\{\mathcal{P}_{n_3} + \bar{\theta}\mathcal{P}_{n_3+1}\}. \end{aligned} \tag{2.194}$$

where ch' is the character equation (2.142) without the prefactor $e(\Lambda_{ad})$—cf. the explanation above. This formula simplifies for $j_2 = 0$ $(n_3 = 1)$ (using both (2.143) and (2.144)):

$$\begin{aligned} \mathrm{ch}L_\Lambda &= \frac{e(\Lambda)}{(1-t_2)(1-t_{12})(1-t_{23})(1-t_{13})} \\ &\quad \times (1 - t_1 t_2^2 t_3 + \bar{\theta}(1 + t_3 - t_{23} - t_{13})) \end{aligned} \tag{2.195}$$

The next case is conjugate.

- **bc** $d = d^{bc} = d^2_{11} = d^3_{11} = 1 + j_1 = z,\ j_2 = 0.$

The generators X^+_{15}, X^+_{35}, and X^+_{45} are eliminated (for the first for different reasons for $j_1 > 0$ and $j_1 = 0$). These are the second series of massless UIRs. There are only 2 states and the bare character formula is (3.100) from [123] for $N = 1$:

$$\mathrm{ch}\,\hat{\Lambda}_\Lambda = 1 + e(\alpha_{25}). \tag{2.196}$$

For the full characters we have to use equation (2.145) (or its reductions for $j_1 = 0, \frac{p+q-2}{2}$), and then we have

$$\begin{aligned} \mathrm{ch}L_\Lambda &= e(\Lambda)\{\mathrm{ch}' L_{d^{bc};j_1,0} + e(\alpha_{25})\mathrm{ch}' L_{d^{bc}+\frac{p+q-2}{2};j_1+\frac{p+q-2}{2},0}\} \\ &= e(\Lambda)\{\mathcal{P}'_{n_1} + \theta\mathcal{P}'_{n_1+1}\}. \end{aligned} \tag{2.197}$$

This formula simplifies for $j_1 = 0$ $(n_1 = 1)$ (using (2.143) and (2.147)):

$$\begin{aligned} \mathrm{ch}L_\Lambda &= \frac{e(\Lambda)}{(1-t_2)(1-t_{12})(1-t_{23})(1-t_{13})} \\ &\quad \times (1 - t_1 t_2^2 t_3 + \theta(1 + t_1 - t_{12} - t_{13})) \end{aligned} \tag{2.198}$$

- **bd** $d = d^2_{11} = d^4_{11} = j_1 = j_2 = z = 0$

As we explained this is the trivial 1-dimensional irrep consisting of the vacuum.

2.3.3 $N = 2$

For $N = 2$ we consider only the **DRC cases.**

First we introduce notation for the odd roots when $N = 2$ using (2.13):

$$\begin{aligned} &\alpha_{15} = \alpha_1 + \gamma_3, \quad \alpha_{16} = \alpha_1 + \gamma_3 + \alpha_5, \\ &\alpha_{25} = \gamma_3, \quad \alpha_{26} = \gamma_3 + \alpha_5, \\ &\alpha_{35} = \alpha_2 + \gamma_4 + \alpha_5, \quad \alpha_{36} = \alpha_2 + \gamma_4, \\ &\alpha_{45} = \gamma_4 + \alpha_5, \quad \alpha_{46} = \gamma_4. \end{aligned} \tag{2.199}$$

Note that as a consequence of our mirror symmetry (2.15) we have

$$\begin{aligned} &\alpha_{15} \longleftrightarrow \alpha_{36}, \quad \alpha_{16} \longleftrightarrow \alpha_{35} \\ &\alpha_{25} \longleftrightarrow \alpha_{46}, \quad \alpha_{26} \longleftrightarrow \alpha_{45} \end{aligned} \tag{2.200}$$

For the characters we shall need also the following notation:

$$\begin{aligned} &\vartheta_k \equiv e(\alpha_{1,7-k}), \quad \theta_k \equiv e(\alpha_{2,7-k}), \quad k = 1,2 \\ &\bar\vartheta_k \equiv e(\alpha_{3,4+k}), \quad \bar\theta_k \equiv e(\alpha_{4,4+k}), \quad k = 1,2, \end{aligned} \tag{2.201}$$

the bar respecting the mirror symmetry of (2.200).

• **ac** $d = d^{ac} = d_{\max} = d^1_{21} = d^3_{22} = 2 + j_1 + j_2 + r, z = j_1 - j_2$.

The maximal number of states is 64 = 8(chiral)×8(anti-chiral), achieved for $r \geq 4$. The 8 anti-chiral, chiral, states are as described in • **a**, • **c**, respectively, (differing for $j_2 > 0$ and $j_2 = 0$, and $j_1 > 0$ and $j_1 = 0$, respectively).

• $j_1 j_2 > 0$.

Here we have the bare character formulas of equations (2.122) (without counterterms for $r \geq 4$). The states $X^+_{15}|\Lambda\rangle$, $X^+_{36}|\Lambda\rangle$, and their descendants are eliminated. (As we noted these semi-short UIRs may be called Grassmann analytic following [174].) For $r = 0$ also the generators X^+_{35} and X^+_{16} are eliminated. Thus, when $j_1 = j_2$ (then $z = 0$) for $r > 0$ we have $\frac{1}{4}$-BPS cases, and for $r = 0$ we have $\frac{p+q-2}{2}$-BPS cases. We recall from [123] that corresponding to the values of r: $r \geq 4$, $r = 3$, $r = 2$, $r = 1$, $r = 0$, there are, respectively, 64, 63, 57, 42, 11 terms in the superfield.

We shall present the character only for the last case which is the shortest semi-short $N = 2$ superfield. The 11 corresponding states are

$$\begin{aligned} &|\Lambda\rangle, \quad X^+_{25}|\Lambda\rangle, \quad X^+_{46}|\Lambda\rangle, \quad X^+_{26}X^+_{25}|\Lambda\rangle, \quad X^+_{45}X^+_{46}|\Lambda\rangle, \quad X^+_{25}X^+_{46}|\Lambda\rangle, \\ &X^+_{26}X^+_{46}|\Lambda\rangle, \quad X^+_{45}X^+_{25}|\Lambda\rangle, \quad X^+_{26}X^+_{25}X^+_{46}|\Lambda\rangle, \\ &X^+_{45}X^+_{25}X^+_{46}|\Lambda\rangle, \quad X^+_{26}X^+_{45}X^+_{25}X^+_{46}|\Lambda\rangle. \end{aligned} \tag{2.202}$$

The corresponding signatures – conformal and su(2) – in format $[d, j_1, j_2; r]$ are

$$[d \equiv d^{ac}, j_1, j_2; 0], \quad [d + \tfrac{p+q-2}{2}, j_1 + \tfrac{p+q-2}{2}, j_2; 1],$$

$$\begin{aligned}
&[d + \tfrac{p+q-2}{2}, j_1, j_2 + \tfrac{p+q-2}{2}; 1], \quad [d+1, j_1+1, j_2; 0],\\
&[d+1, j_1, j_2+1; 0], \quad [d+1, j_1 + \tfrac{p+q-2}{2}, j_2 + \tfrac{p+q-2}{2}; 2],\\
&[d+1, j_1 + \tfrac{p+q-2}{2}, j_2 + \tfrac{p+q-2}{2}; 0], \quad [d+1, j_1 + \tfrac{p+q-2}{2}, j_2 + \tfrac{p+q-2}{2}; 0],\\
&[d + \tfrac{3}{2}, j_1+1, j_2 + \tfrac{p+q-2}{2}; 1], \quad [d + \tfrac{3}{2}, j_1 + \tfrac{p+q-2}{2}, j_2+1; 1],\\
&[d+2, j_1+1, j_2+1; 0]. \qquad (2.203)
\end{aligned}$$

Note that all conformal entries are on the conformal unitarity threshold $d = 2 + j_1 + j_2$, $(j_1 j_2 > 0)$, i. e., shall use as input equation (2.140). Thus, the character formula is

$$\begin{aligned}
\operatorname{ch} L_\Lambda = {} & \frac{e(\Lambda)}{(1-t_2)(1-t_1t_2)(1-t_2t_3)(1-t_1t_2t_3)}\\
& \times \{\mathcal{P}_{n_1,n_2} + e(\alpha_{25})\mathcal{P}_{n_1+1,n_2}\mathcal{S}_1 + e(\alpha_{46})\mathcal{P}_{n_1,n_2+1}\mathcal{S}_1\\
& + e(\alpha_{26})e(\alpha_{25})\mathcal{P}_{n_1+2,n_2} + e(\alpha_{45})e(\alpha_{46})\mathcal{P}_{n_1,n_2+2}\\
& + e(\alpha_{25})e(\alpha_{46})\mathcal{P}_{n_1+1,n_2+1}\mathcal{S}_2\\
& + (e(\alpha_{26})e(\alpha_{46}) + e(\alpha_{45})e(\alpha_{25}))\mathcal{P}_{n_1+1,n_2+1}\\
& + (e(\alpha_{26})\mathcal{P}_{n_1+2,n_2+1} + e(\alpha_{45})\mathcal{P}_{n_1+1,n_2+2})e(\alpha_{25})e(\alpha_{46})\mathcal{S}_1\\
& + e(\alpha_{26})e(\alpha_{45})e(\alpha_{25})e(\alpha_{46})\mathcal{P}_{n_1+2,n_2+2}\} \qquad (2.204)\\
= {} & \frac{e(\Lambda)}{(1-t_2)(1-t_1t_2)(1-t_2t_3)(1-t_1t_2t_3)}\\
& \times \{\mathcal{P}_{n_1,n_2} + \theta_2\mathcal{P}_{n_1+1,n_2}(1+t_5) + \bar\theta_2\mathcal{P}_{n_1,n_2+1}(1+t_5)\\
& + \theta_1\theta_2\mathcal{P}_{n_1+2,n_2} + \bar\theta_1\bar\theta_2\mathcal{P}_{n_1,n_2+2}\\
& + \theta_2\bar\theta_2\mathcal{P}_{n_1+1,n_2+1}(1+t_5+t_5^2)\\
& + (\theta_1\bar\theta_2 + \bar\theta_1\theta_2)\mathcal{P}_{n_1+1,n_2+1}\\
& + (\theta_1\mathcal{P}_{n_1+2,n_2+1} + \bar\theta_1\mathcal{P}_{n_1+1,n_2+2})\theta_2\bar\theta_2(1+t_5)\\
& + \theta_1\bar\theta_1\theta_2\bar\theta_2\mathcal{P}_{n_1+2,n_2+2}\}
\end{aligned}$$

where we have used the characters of su(2): $\mathcal{S}_1 = 1 + t_5$, $\mathcal{S}_2 = 1 + t_5 + t_5^2$, $(t_5 = e(\alpha_5))$.

- $j_1 > 0, j_2 = 0$. Here we have the bare character formulas of equations (2.124) (without counterterms for $r \geq 4$). The states $X_{36}^+X_{46}^+|\Lambda\rangle$, $X_{15}^+|\Lambda\rangle$, and their descendants are eliminated. We recall from [123] that corresponding to the values of r, $r \geq 4$, $r = 3$, $r = 2$, $r = 1$, $r = 0$, there are, respectively, 64, 63, 58, 45, 16 states. In the last case, where $r = 0$, we eliminate the generator X_{16}^+ and exclude the generators $X_{3,4+k}^+$ from the anti-chiral sector. Then for $z = j_1$ this is a $\frac{1}{4}$-BPS case.

We shall present the character only for the last case. The 16 states of the superfield are

$$\begin{aligned}
&|\Lambda\rangle, \quad X_{25}^+|\Lambda\rangle, \quad X_{46}^+|\Lambda\rangle, \quad X_{26}^+X_{25}^+|\Lambda\rangle, \quad X_{45}^+X_{46}^+|\Lambda\rangle, \quad X_{25}^+X_{46}^+|\Lambda\rangle,\\
&X_{26}^+X_{46}^+|\Lambda\rangle, \quad X_{45}^+X_{25}^+|\Lambda\rangle, \quad X_{26}^+X_{25}^+X_{46}^+|\Lambda\rangle, \quad X_{45}^+X_{25}^+X_{46}^+|\Lambda\rangle,
\end{aligned}$$

$$X^+_{26}X^+_{45}X^+_{25}X^+_{46}|\Lambda\rangle, \quad X^+_{35}X^+_{25}X^+_{46}|\Lambda\rangle, \quad X^+_{36}X^+_{45}X^+_{25}|\Lambda\rangle,$$
$$X^+_{35}X^+_{45}X^+_{25}X^+_{46}|\Lambda\rangle, \quad X^+_{26}X^+_{35}X^+_{25}X^+_{46}|\Lambda\rangle, \quad X^+_{26}X^+_{36}X^+_{45}X^+_{25}|\Lambda\rangle. \tag{2.205}$$

The states of (2.202) appear as the first 11 of (2.205), though the content is different:

$$\begin{aligned}
&[d \equiv 2+j_1, j_1, 0;\, 0], \quad [d+\tfrac{p+q-2}{2}, j_1+\tfrac{p+q-2}{2}, 0;\, 1],\\
&[d+\tfrac{p+q-2}{2}, j_1, \tfrac{p+q-2}{2};\, 1], \quad [d+1, j_1+1, 0;\, 0], \quad [d+1, j_1, 1;\, 0],\\
&[d+1, j_1+\tfrac{p+q-2}{2}, \tfrac{p+q-2}{2};\, 2], \quad [d+1, j_1+\tfrac{p+q-2}{2}, \tfrac{p+q-2}{2};\, 0],\\
&[d+1, j_1+\tfrac{p+q-2}{2}, \tfrac{p+q-2}{2};\, 0], \quad [d+\tfrac{3}{2}, j_1+1, \tfrac{p+q-2}{2};\, 1],\\
&[d+\tfrac{3}{2}, j_1+\tfrac{p+q-2}{2}, 1;\, 1], \quad [d+2, j_1+1, 1;\, 0],\\
&[d+\tfrac{3}{2}, j_1+\tfrac{p+q-2}{2}, 0;\, 1], \quad [d+\tfrac{3}{2}, j_1+\tfrac{p+q-2}{2}, 0;\, 1],\\
&[d+2, j_1+\tfrac{p+q-2}{2}, \tfrac{p+q-2}{2};\, 0], \quad [d+2, j_1+1, 0;\, 0], \quad [d+2, j_1+1, 0;\, 0]
\end{aligned} \tag{2.206}$$

Since in all entries the value of d is above the conformal unitarity threshold, then we use equation (2.138) for the conformal part of the character formula:

$$\begin{aligned}
\operatorname{ch} L_\Lambda = {} & \frac{e(\Lambda)}{(1-t_2)(1-t_1t_2)(1-t_2t_3)(1-t_1t_2t_3)}\\
& \times \{\mathcal{Q}'_{n_1} + \theta_2\mathcal{Q}'_{n_1+1}(1+t_5) + \bar\theta_2\mathcal{Q}'_{n_1}(1+t_3)(1+t_5)\\
& + \theta_1\theta_2\mathcal{Q}'_{n_1+2} + \bar\theta_1\bar\theta_2\mathcal{Q}'_{n_1}(1+t_3+t_3^2)\\
& + \theta_2\bar\theta_2\mathcal{Q}'_{n_1+1}(1+t_3)(1+t_5+t_5^2)\\
& + (\theta_1\bar\theta_2+\bar\theta_1\theta_2)\mathcal{Q}'_{n_1+1}(1+t_3)\\
& + (\theta_1\mathcal{Q}'_{n_1+2}(1+t_3)+\bar\theta_1\mathcal{Q}'_{n_1+1}(1+t_3+t_3^2))\theta_2\bar\theta_2(1+t_5)\\
& + \theta_1\bar\theta_1\theta_2\bar\theta_2\mathcal{Q}'_{n_1+2}(1+t_3+t_3^2)\\
& + (\bar\vartheta_1\bar\theta_2+\bar\vartheta_2\bar\theta_1)\theta_2\mathcal{Q}'_{n_1+1}(1+t_5)\\
& + (\bar\vartheta_1\bar\theta_2+\bar\vartheta_2\bar\theta_1)\theta_1\theta_2\mathcal{Q}'_{n_1+2} + \bar\vartheta_1\bar\theta_1\theta_2\bar\theta_2\mathcal{Q}'_{n_1+1}(1+t_3)\}.
\end{aligned} \tag{2.207}$$

where we have used $\mathcal{Q}_{n,1} = \mathcal{Q}'_n$, $\mathcal{Q}_{n,2} = (1+t_3)\mathcal{Q}'_n$, $\mathcal{Q}_{n,3} = (1+t_3+t_3^2)\mathcal{Q}'_n$.

The next case is conjugate to the preceding.

- $j_1 = 0$, $j_2 > 0$. Here we have the bare character formulas of equations (3.89) from [123] (without counterterms for $r \geq 4$). The states $X^+_{15}X^+_{25}|\Lambda\rangle$, $X^+_{36}|\Lambda\rangle$ and their descendants are eliminated. When $r = 0$, we eliminate the generator X^+_{35} and exclude the generators $X^+_{1,4+k}$ from the chiral sector. Then for $z = -j_2$ this is a $\frac{1}{4}$-BPS case. We consider only the latter case. The 16 states of the superfield are

$$\begin{aligned}
&|\Lambda\rangle, \quad X^+_{25}|\Lambda\rangle, \quad X^+_{46}|\Lambda\rangle, \quad X^+_{26}X^+_{25}|\Lambda\rangle, \quad X^+_{45}X^+_{46}|\Lambda\rangle, \quad X^+_{25}X^+_{46}|\Lambda\rangle,\\
&X^+_{26}X^+_{46}|\Lambda\rangle, \quad X^+_{45}X^+_{25}|\Lambda\rangle, \quad X^+_{26}X^+_{25}X^+_{46}|\Lambda\rangle, \quad X^+_{45}X^+_{25}X^+_{46}|\Lambda\rangle,\\
&X^+_{26}X^+_{45}X^+_{25}X^+_{46}|\Lambda\rangle, \quad X^+_{16}X^+_{25}X^+_{46}|\Lambda\rangle, \quad X^+_{15}X^+_{26}X^+_{46}|\Lambda\rangle,\\
&X^+_{16}X^+_{45}X^+_{25}X^+_{46}|\Lambda\rangle, \quad X^+_{26}X^+_{15}X^+_{45}X^+_{46}|\Lambda\rangle, \quad X^+_{26}X^+_{16}X^+_{25}X^+_{46}|\Lambda\rangle.
\end{aligned} \tag{2.208}$$

The states of (2.202) appear as the first 11 of (2.208), the same states as in (2.205), though the content is different:

$$
\begin{aligned}
&[d \equiv 2 + j_2, 0, j_2;\, 0], \quad [d + \tfrac{p+q-2}{2}, \tfrac{p+q-2}{2}, j_2;\, 1],\\
&[d + \tfrac{p+q-2}{2}, 0, j_2 + \tfrac{p+q-2}{2};\, 1], \quad [d+1, 1, j_2;\, 0],\\
&[d+1, 0, j_2+1;\, 0], \quad [d+1, \tfrac{p+q-2}{2}, j_2 + \tfrac{p+q-2}{2};\, 2],\\
&[d+1, \tfrac{p+q-2}{2}, j_2 + \tfrac{p+q-2}{2};\, 0], \quad [d+1, \tfrac{p+q-2}{2}, j_2 + \tfrac{p+q-2}{2};\, 0],\\
&[d + \tfrac{3}{2}, 1, j_2 + \tfrac{p+q-2}{2};\, 1], \quad [d + \tfrac{3}{2}, \tfrac{p+q-2}{2}, j_2 + 1;\, 1],\\
&[d+2, 1, j_2+1;\, 0], \quad [d + \tfrac{3}{2}, 0, j_2 + \tfrac{p+q-2}{2};\, 1],\\
&[d + \tfrac{3}{2}, 0, j_2 + \tfrac{p+q-2}{2};\, 1], \quad [d+2, 0, j_2+1;\, 0],\\
&[d+2, 0, j_2+1;\, 0], \quad [d+2, \tfrac{p+q-2}{2}, j_2 + \tfrac{p+q-2}{2};\, 0].
\end{aligned}
\tag{2.209}
$$

The character formula is

$$
\begin{aligned}
\operatorname{ch} L_\Lambda = {}& \frac{e(\Lambda)}{(1-t_2)(1-t_1t_2)(1-t_2t_3)(1-t_1t_2t_3)} \\
&\times \{\mathcal{Q}_{n_3} + \theta_2\mathcal{Q}_{n_3}(1+t_1)(1+t_5) + \bar\theta_2\mathcal{Q}_{n_3+1}(1+t_5)\\
&+ \theta_1\theta_2\mathcal{Q}_{n_3}(1+t_1+t_1^2) + \bar\theta_1\bar\theta_2\mathcal{Q}_{n_3+2}\\
&+ \theta_2\bar\theta_2\mathcal{Q}_{n_3+1}(1+t_1)(1+t_5+t_5^2)\\
&+ (\theta_1\bar\theta_2 + \bar\theta_1\theta_2)\mathcal{Q}_{n_3+1}(1+t_1)\\
&+ (\theta_1\mathcal{Q}_{n_3+1}(1+t_1+t_1^2) + \bar\theta_1\mathcal{Q}_{n_3+2}(1+t_1))\theta_2\bar\theta_2(1+t_5)\\
&+ \theta_1\bar\theta_1\theta_2\bar\theta_2\mathcal{Q}_{n_3+2}(1+t_1+t_1^2)\\
&+ (\vartheta_1\theta_2 + \vartheta_2\theta_1)\bar\theta_2\mathcal{Q}_{n_3+1}(1+t_5)\\
&+ (\vartheta_1\theta_2 + \vartheta_2\theta_1)\bar\theta_1\bar\theta_2\mathcal{Q}_{n_3+2} + \theta_1\theta_2\bar\theta_1\bar\theta_2\mathcal{Q}_{n_3+1}(1+t_1)\}.
\end{aligned}
\tag{2.210}
$$

- $j_1 = j_2 = 0$. Here we have bare character formulas of equations (2.128) (without counterterms for $r \geq 4$). The states $X_{15}^+X_{25}^+|\Lambda\rangle$, $X_{36}^+X_{46}^+|\Lambda\rangle$, and their descendants are eliminated. We recall from [123] that corresponding to the values of r: $r \geq 4$, $r = 3$, $r = 2$, $r = 1$, $r = 0$, there are, respectively, 64, 63, 59, 47, 24 states. In the last case, when $r = 0$, we exclude the generators $X_{3,4+k}^+$ from the anti-chiral sector and the generators $X_{1,4+k}^+$ from the chiral sector and also the combination of impossible states:

$$X_{15}^+X_{26}^+X_{36}^+X_{45}^+|\Lambda\rangle. \tag{2.211}$$

We shall consider only the 24 states of the UIR at $r = 0$:

$$
\begin{aligned}
&|\Lambda\rangle, \quad X_{25}^+|\Lambda\rangle, \quad X_{46}^+|\Lambda\rangle, \quad X_{26}^+X_{25}^+|\Lambda\rangle, \quad X_{45}^+X_{46}^+|\Lambda\rangle, \quad X_{25}^+X_{46}^+|\Lambda\rangle,\\
&X_{26}^+X_{46}^+|\Lambda\rangle, \quad X_{45}^+X_{25}^+|\Lambda\rangle, \quad X_{26}^+X_{25}^+X_{46}^+|\Lambda\rangle, \quad X_{45}^+X_{25}^+X_{46}^+|\Lambda\rangle,\\
&X_{26}^+X_{45}^+X_{25}^+X_{46}^+|\Lambda\rangle, \quad X_{35}^+X_{25}^+X_{46}^+|\Lambda\rangle, \quad X_{36}^+X_{45}^+X_{25}^+|\Lambda\rangle,
\end{aligned}
$$

$$X_{35}^+X_{45}^+X_{25}^+X_{46}^+|\Lambda\rangle, \quad X_{26}^+X_{35}^+X_{25}^+X_{46}^+|\Lambda\rangle, \quad X_{26}^+X_{36}^+X_{45}^+X_{25}^+|\Lambda\rangle,$$
$$X_{16}^+X_{25}^+X_{46}^+|\Lambda\rangle, \quad X_{15}^+X_{26}^+X_{46}^+|\Lambda\rangle, \quad X_{16}^+X_{45}^+X_{25}^+X_{46}^+|\Lambda\rangle,$$
$$X_{26}^+X_{15}^+X_{45}^+X_{46}^+|\Lambda\rangle, \quad X_{26}^+X_{16}^+X_{25}^+X_{46}^+|\Lambda\rangle, \quad X_{16}^+X_{25}^+X_{35}^+X_{46}^+|\Lambda\rangle,$$
$$X_{16}^+X_{25}^+X_{36}^+X_{45}^+|\Lambda\rangle, \quad X_{15}^+X_{26}^+X_{35}^+X_{46}^+|\Lambda\rangle. \tag{2.212}$$

The states of (2.202) appear as the first 11 of (2.212), the states of (2.205) as the first 16 of (2.212), the states of (2.208) as the first 11 plus states 17–21 of (2.212), though of course the contents is different:

$$\begin{gathered}
[2,0,0;0], \quad [\tfrac{5}{2},\tfrac{p+q-2}{2},0;1], \quad [\tfrac{5}{2},0,\tfrac{p+q-2}{2};1], \quad [\tfrac{3}{2},1,0;0], \quad [\tfrac{3}{2},0,1;0],\\
[\tfrac{3}{2},\tfrac{p+q-2}{2},\tfrac{p+q-2}{2};2], \quad [\tfrac{3}{2},\tfrac{p+q-2}{2},\tfrac{p+q-2}{2};0], \quad [\tfrac{3}{2},\tfrac{p+q-2}{2},\tfrac{p+q-2}{2};0],\\
[\tfrac{7}{2},1,\tfrac{p+q-2}{2};1], \quad [\tfrac{7}{2},\tfrac{p+q-2}{2},1;1], \quad [4,1,1;0], \quad [\tfrac{7}{2},\tfrac{p+q-2}{2},0;1],\\
[\tfrac{7}{2},\tfrac{p+q-2}{2},0;1], \quad [4,\tfrac{p+q-2}{2},\tfrac{p+q-2}{2};0], \quad [4,1,0;0], \quad [4,1,0;0],\\
[\tfrac{7}{2},0,\tfrac{p+q-2}{2};1], \quad [\tfrac{7}{2},0,\tfrac{p+q-2}{2};1], \quad [4,0,1;0], \quad [4,0,1;0],\\
[4,\tfrac{p+q-2}{2},\tfrac{p+q-2}{2};0], \quad [4,0,0;0], \quad [4,0,0;0], \quad [4,0,0;0]
\end{gathered} \tag{2.213}$$

The character formula is

$$\begin{aligned}
\operatorname{ch} L_\Lambda = {}& \frac{e(\Lambda)}{(1-t_2)(1-t_1t_2)(1-t_2t_3)(1-t_1t_2t_3)}\\
&\times \{1+\theta_2(1+t_1)(1+t_5)+\bar\theta_2(1+t_3)(1+t_5)\\
&+\theta_1\theta_2(1+t_1+t_1^2)+\bar\theta_1\bar\theta_2(1+t_3+t_3^2)\\
&+\theta_2\bar\theta_2(1+t_1)(1+t_3)(1+t_5+t_5^2)+(\theta_1\bar\theta_2+\bar\theta_1\theta_2)(1+t_1)(1+t_3)\\
&+(\theta_1(1+t_1+t_1^2)(1+t_3)+\bar\theta_1(1+t_1)(1+t_3+t_3^2))\theta_2\bar\theta_2(1+t_5)\\
&+\theta_1\bar\theta_1\theta_2\bar\theta_2(1+t_1+t_1^2)(1+t_3+t_3^2)\\
&+(\bar\vartheta_1\bar\theta_2+\bar\vartheta_2\bar\theta_1)\theta_2(1+t_1)(1+t_5)\\
&+(\bar\vartheta_1\bar\theta_2+\bar\vartheta_2\bar\theta_1)\theta_1\theta_2(1+t_1+t_1^2)+\bar\vartheta_1\bar\theta_1\theta_2\bar\theta_2(1+t_1)(1+t_3)\\
&+(\vartheta_1\theta_2+\vartheta_2\theta_1)\bar\theta_2(1+t_3)(1+t_5)\\
&+(\vartheta_1\theta_2+\vartheta_2\theta_1)\bar\theta_1\bar\theta_2(1+t_3+t_3^2)+\vartheta_1\theta_1\theta_2\bar\theta_2(1+t_3)(1+t_1)\\
&+\vartheta_1\theta_2\bar\vartheta_1\bar\theta_2+\vartheta_1\theta_2\bar\vartheta_2\bar\theta_1+\vartheta_2\theta_1\bar\vartheta_1\bar\theta_2\}.
\end{aligned} \tag{2.214}$$

• **ad** $d = d_{21}^1 = d_{22}^4 = 1 + j_2 + r, j_1 = 0, z = -1 - j_2$.

Here we have the bare character formulas (3.95) from [123] when $j_2 r > 0$, (3.96) when $j_2 = 0, r > 0$ (both these cases without counterterms for $r \geq 4$), and, finally, when $r = 0$ we have (3.97) independently of the value of j_2—these are the anti-chiral massless UIRs.

The generators X_{15}^+, X_{25}^+, and in addition X_{36}^+ for $j_2 > 0$ (respectively, the state $X_{36}^+X_{46}^+|\Lambda\rangle$, and its descendants for $j_2 = 0$) are eliminated. The maximal number of states is $24 = 3(\text{chiral}) \times 8(\text{anti-chiral})$, achieved for $r \geq 4$. The chiral sector for $r > 0$ consists of the states

$$X_{26}^+|\Lambda\rangle, \quad r \geq 1, X_{16}^+X_{26}^+|\Lambda\rangle, \quad r \geq 2, \tag{2.215}$$

and the vacuum, while the anti-chiral sector is given by

$$\begin{aligned}
&X_{46}^+, \quad \varepsilon_r^a = 1,\ \varepsilon_j^a = 1,\\
&X_{45}^+X_{46}^+, \quad \varepsilon_r^a = 0,\ \varepsilon_j^a = 2,\\
&\mathbf{1}, \quad X_{35}^+X_{46}^+, \quad \varepsilon_r^a = 0,\ \varepsilon_j^a = 0,\\
&X_{45}^+, \quad X_{45}^+X_{35}^+X_{46}^+, \quad \varepsilon_r^a = -1,\ \varepsilon_j^a = 1,\\
&X_{35}^+, \quad \varepsilon_r^a = -1,\ \varepsilon_j^a = -1,\\
&X_{35}^+X_{45}^+, \quad \varepsilon_r^a = -2,\ \varepsilon_j^a = 0
\end{aligned} \tag{2.216}$$

for $j_2 > 0$ and by

$$\begin{aligned}
&X_{46}^+, \quad \varepsilon_r^a = 1,\ \varepsilon_j^a = 1,\\
&X_{45}^+X_{46}^+, \quad \varepsilon_r^a = 0,\ \varepsilon_j^a = 2,\\
&\mathbf{1}, \quad X_{35}^+X_{46}^+, \quad X_{45}^+X_{36}^+, \quad \varepsilon_r^a = 0,\ \varepsilon_j^a = 0,\\
&X_{45}^+, \quad X_{45}^+X_{35}^+X_{46}^+, \quad \varepsilon_r^a = -1,\ \varepsilon_j^a = 1,\\
&X_{35}^+X_{45}^+, \quad \varepsilon_r^a = -2,\ \varepsilon_j^a = 0
\end{aligned} \tag{2.217}$$

for $j_2 = 0$. Finally, for $r = 0$ also the generators $X_{16}^+, X_{26}^+, X_{35}^+$ are eliminated.

The possible 24 states for $j_2 > 0$ are given explicitly by

$$\begin{aligned}
&|\Lambda\rangle, \quad X_{46}^+|\Lambda\rangle, \quad X_{45}^+X_{46}^+|\Lambda\rangle, \quad r \geq 0,\\
&X_{26}^+|\Lambda\rangle, \quad X_{45}^+|\Lambda\rangle, \quad X_{35}^+X_{46}^+|\Lambda\rangle, \quad X_{26}^+X_{46}^+|\Lambda\rangle, \quad X_{26}^+X_{45}^+X_{46}^+|\Lambda\rangle, \quad r \geq 1,\\
&X_{26}^+X_{35}^+X_{46}^+|\Lambda\rangle, \quad X_{16}^+X_{26}^+X_{46}^+|\Lambda\rangle, \quad X_{45}^+X_{35}^+X_{46}^+|\Lambda\rangle, \quad r \geq 1,\\
&X_{35}^+|\Lambda\rangle, \quad r \geq 1, \qquad (*)\\
&X_{26}^+X_{45}^+|\Lambda\rangle, \quad X_{16}^+X_{26}^+|\Lambda\rangle, \quad X_{35}^+X_{45}^+|\Lambda\rangle, \quad X_{26}^+X_{45}^+X_{35}^+X_{46}^+|\Lambda\rangle, \quad r \geq 2,\\
&X_{16}^+X_{26}^+X_{45}^+X_{46}^+|\Lambda\rangle, \quad X_{16}^+X_{26}^+X_{35}^+X_{46}^+|\Lambda\rangle, \quad r \geq 2,\\
&X_{26}^+X_{35}^+|\Lambda\rangle, \quad r \geq 2, \qquad (*)\\
&X_{26}^+X_{35}^+X_{45}^+|\Lambda\rangle, \quad X_{16}^+X_{26}^+X_{45}^+|\Lambda\rangle, \quad X_{16}^+X_{26}^+X_{45}^+X_{35}^+X_{46}^+|\Lambda\rangle, \quad r \geq 3,\\
&X_{16}^+X_{26}^+X_{35}^+|\Lambda\rangle, \quad r \geq 3, \qquad (*)\\
&X_{16}^+X_{26}^+X_{35}^+X_{45}^+|\Lambda\rangle, \quad r \geq 4.
\end{aligned} \tag{2.218}$$

Corresponding to the values of r: $r \geq 4$, $r = 3$, $r = 2$, $r = 1$, $r = 0$, there are, respectively, 24, 23, 19, 12, 3 states. Three states are marked with $(*)$—these states are not present when $j_2 = 0$. Thus, the 24 states for $j_2 = 0$ are given as the 21 from (2.218) without $(*)$ and the following three states:

$$X^+_{36}X^+_{45}|\Lambda\rangle, \quad X^+_{26}X^+_{36}X^+_{45}|\Lambda\rangle, \quad r \geq 1,$$
$$X^+_{16}X^+_{26}X^+_{36}X^+_{45}|\Lambda\rangle, \quad r \geq 2. \tag{2.219}$$

Thus, for $j_2 = 0$ corresponding to the values of r there are, respectively, 24, 23, 20, 13, 3 states. The content of the states in (2.218) is as follows:

$$\begin{aligned}
&[d \equiv 1 + j_2 + r, 0, j_2; r], \quad [d + \tfrac{p+q-2}{2}, 0, j_2 + \tfrac{p+q-2}{2}; r+1], \quad r \geq 0\\
&[d+1, 0, j_2+1; r], \quad r \geq 0\\
&[d + \tfrac{p+q-2}{2}, \tfrac{p+q-2}{2}, j_2; r-1], \quad [d + \tfrac{p+q-2}{2}, 0, j_2 + \tfrac{p+q-2}{2}; r-1], \quad r \geq 1\\
&[d+1, 0, j_2; r], \quad [d+1, \tfrac{p+q-2}{2}, j_2 + \tfrac{p+q-2}{2}; r], \quad r \geq 1\\
&[d + \tfrac{3}{2}, \tfrac{p+q-2}{2}, j_2+1; r-1], \quad [d + \tfrac{3}{2}, \tfrac{p+q-2}{2}, j_2; r-1], \quad r \geq 1\\
&[d + \tfrac{3}{2}, 0, j_2 + \tfrac{p+q-2}{2}; r-1], \quad [d + \tfrac{3}{2}, 0, j_2 + \tfrac{p+q-2}{2}; r-1], \quad r \geq 1\\
&[d + \tfrac{p+q-2}{2}, 0, j_2 - \tfrac{p+q-2}{2}; r-1], \quad r \geq 1 \quad (*)\\
&[d+1, \tfrac{p+q-2}{2}, j_2 + \tfrac{p+q-2}{2}; r-2], \quad [d+1, 0, j_2; r-2], \quad r \geq 2\\
&[d+1, 0, j_2; r-2], \quad [d+2, \tfrac{p+q-2}{2}, j_2 + \tfrac{p+q-2}{2}; r-2], \quad r \geq 2\\
&[d+2, 0, j_2+1; r-2], \quad [d+2, 0, j_2; r-2], \quad r \geq 2\\
&[d+1, \tfrac{p+q-2}{2}, j_2 - \tfrac{p+q-2}{2}; r-2], \quad r \geq 2, \quad (*)\\
&[d + \tfrac{3}{2}, \tfrac{p+q-2}{2}, j_2; r-3], \quad [d + \tfrac{3}{2}, 0, j_2 + \tfrac{p+q-2}{2}; r-3], \quad r \geq 3,\\
&[d + \tfrac{5}{2}, 0, j_2 + \tfrac{p+q-2}{2}; r-3], \quad r \geq 3,\\
&[d + \tfrac{3}{2}, 0, j_2 - \tfrac{p+q-2}{2}; r-3], \quad r \geq 3, \quad (*)\\
&[d+2, 0, j_2; r-4], \quad r \geq 4.
\end{aligned} \tag{2.220}$$

and for the states in (2.219):

$$\begin{aligned}
&[d+1, 0, j_2; r], \quad [d+1, \tfrac{p+q-2}{2}, j_2; r-1], \quad r \geq 1\\
&[d+2, 0, j_2; r-2], \quad r \geq 2.
\end{aligned} \tag{2.221}$$

The character formula for $r > 0$ is

$$\begin{aligned}
\operatorname{ch} L_\Lambda = {} & \frac{e(\Lambda)}{(1-t_2)(1-t_1t_2)(1-t_2t_3)(1-t_1t_2t_3)}\\
& \times \{\mathcal{Q}_{n_3}\mathcal{S}_r + \bar{\theta}_2\mathcal{Q}_{n_3+1}\mathcal{S}_{r+1} + \bar{\theta}_1\bar{\theta}_2\mathcal{Q}_{n_3+2}\mathcal{S}_r\\
& + \theta_1(1+t_1)\mathcal{Q}_{n_3}\mathcal{S}_{r-1} + \bar{\theta}_1\mathcal{Q}_{n_3+1}\mathcal{S}_{r-1}\\
& + \bar{\vartheta}_1\bar{\theta}_2\mathcal{Q}_{n_3}\mathcal{S}_r + \theta_1\bar{\theta}_2(1+t_1)\mathcal{Q}_{n_3+1}\mathcal{S}_r\\
& + \theta_1\bar{\theta}_1\bar{\theta}_2(1+t_1)\mathcal{Q}_{n_3+2}\mathcal{S}_{r-1} + \theta_1\bar{\vartheta}_1\bar{\theta}_2(1+t_1)\mathcal{Q}_{n_3}\mathcal{S}_{r-1}\\
& + \vartheta_1\theta_1\bar{\theta}_2\mathcal{Q}_{n_3+1}\mathcal{S}_{r-1} + \bar{\theta}_1\bar{\vartheta}_1\bar{\theta}_2\mathcal{Q}_{n_3+1}\mathcal{S}_{r-1}\\
& + \bar{\vartheta}_1\mathcal{Q}_{n_3-1}\mathcal{S}_{r-1} + \bar{\vartheta}_1\bar{\theta}_1\mathcal{Q}_{n_3}\mathcal{S}_{r-2}\\
& + \theta_1\bar{\theta}_1(1+t_1)\mathcal{Q}_{n_3+1}\mathcal{S}_{r-2} + \vartheta_1\theta_1\mathcal{Q}_{n_3}\mathcal{S}_{r-2}
\end{aligned}$$

$$
\begin{aligned}
&+ \theta_1\bar{\theta}_1\bar{\vartheta}_1\bar{\theta}_2(1 + t_1)\mathcal{Q}_{n_3+1}\mathcal{S}_{r-2} + \vartheta_1\theta_1\bar{\theta}_1\bar{\theta}_2\mathcal{Q}_{n_3+2}\mathcal{S}_{r-2} \\
&+ \vartheta_1\theta_1\bar{\vartheta}_1\bar{\theta}_2\mathcal{Q}_{n_3}\mathcal{S}_{r-2} + \theta_1\bar{\vartheta}_1(1 + t_1)\mathcal{Q}_{n_3-1}\mathcal{S}_{r-2} \\
&+ \theta_1\bar{\vartheta}_1\bar{\theta}_1(1 + t_1)\mathcal{Q}_{n_3}\mathcal{S}_{r-3} + \vartheta_1\theta_1\bar{\theta}_1\mathcal{Q}_{n_3+1}\mathcal{S}_{r-3} \\
&+ \vartheta_1\theta_1\bar{\theta}_1\bar{\vartheta}_1\bar{\theta}_2\mathcal{Q}_{n_3+1}\mathcal{S}_{r-3} + \vartheta_1\theta_1\bar{\vartheta}_1\mathcal{Q}_{n_3-1}\mathcal{S}_{r-3} \\
&+ \vartheta_1\theta_1\bar{\vartheta}_1\bar{\theta}_1\mathcal{Q}_{n_3}\mathcal{S}_{r-4} + \delta_{j_2,0}\vartheta_1\theta_1\bar{\vartheta}_2\bar{\theta}_1\mathcal{Q}_{n_3}\mathcal{S}_{r-2} \\
&+ \delta_{j_2,0}\bar{\vartheta}_2\bar{\theta}_1\mathcal{Q}_{n_3}\mathcal{S}_r + \delta_{j_2,0}\theta_1\bar{\vartheta}_2\bar{\theta}_1(1 + t_1)\mathcal{Q}_{n_3}\mathcal{S}_{r-1}\} \qquad (2.222)
\end{aligned}
$$

where we use the su(2) character factors $\mathcal{S}_p = \sum_{s=0}^{p} t_5^s$ for $p \in \mathbb{Z}_+$, and for continuity we use: $\mathcal{S}_p = 0$ for $p \in -\mathbb{N}$, we also use the convention $\mathcal{Q}_p = 0$ for $p \le 0$. Thus, the formula is valid for all $r > 0$ and for all j_2.

For $r = 0$ we have the character formula for the anti-chiral massless UIRs and we have to use the massless conformal characters (2.142), (2.143), (2.144):

$$
\begin{aligned}
\operatorname{ch} L_\Lambda = &\frac{e(\Lambda)}{(1 - t_2)(1 - t_1t_2)(1 - t_2t_3)(1 - t_1t_2t_3)} \\
&\times \{\mathcal{P}_{n_3} + \bar{\theta}_2\mathcal{P}_{n_3+1}(1 + t_5) + \bar{\theta}_1\bar{\theta}_2\mathcal{P}_{n_3+2}\}. \qquad (2.223)
\end{aligned}
$$

• **bc** $d = d_{21}^2 = d_{22}^3 = 1 + j_1 + r, j_2 = 0, z = 1 + j_1$.

Here we have the bare character formulas (3.98) from [123] when $j_1r > 0$, (3.99) when $j_1 = 0, r > 0$ (both these cases without counterterms for $r \ge 4$), and, finally, when $r = 0$ we have (3.100) independently of the value of j_1—these are the chiral massless UIRs.

This case is conjugate to the previous one **ad** and everything may be obtained from it by the mirror symmetry. We give only the character of the chiral massless case:

$$
\begin{aligned}
\operatorname{ch} L_\Lambda = &\frac{e(\Lambda)}{(1 - t_2)(1 - t_1t_2)(1 - t_2t_3)(1 - t_1t_2t_3)} \\
&\times \{\mathcal{P}'_{n_1} + \theta_2\mathcal{P}'_{n_1+1}(1 + t_5) + \theta_1\theta_2\mathcal{P}'_{n_1+2}\}. \qquad (2.224)
\end{aligned}
$$

• **bd** $d = d_{21}^2 = d_{22}^4 = r, j_1 = j_2 = 0 = z$.

The generators $X_{15}^+, X_{25}^+, X_{36}^+, X_{46}^+$ are eliminated. Thus, for $r > 1$ these are $\frac{1}{2}$-BPS cases. For $r = 1$ also the generators X_{16}^+, X_{35}^+ are eliminated. Thus, the latter is a $\frac{3}{4}$-BPS case, and it is also the $N = 2$ mixed massless irrep. For $r = 0$ the remaining two generators X_{26}^+, X_{45}^+ are eliminated and we have the trivial irrep as explained in general.

For $r > 0$ the bare character formula is (3.101) from [123] with $i_0 = i'_0 = 0$. The maximal number of states is nine and the list of states together with the conditions when they exist are

$$
\begin{aligned}
&|\Lambda\rangle, \quad r \ge 0, \\
&X_{26}^+|\Lambda\rangle, \quad X_{45}^+|\Lambda\rangle, \quad r \ge 1, \\
&X_{16}^+X_{26}^+|\Lambda\rangle, \quad X_{35}^+X_{45}^+|\Lambda\rangle, \quad X_{26}^+X_{45}^+|\Lambda\rangle, \quad r \ge 2,
\end{aligned}
$$

$$X^+_{16}X^+_{26}X^+_{45}|\Lambda\rangle, \quad X^+_{26}X^+_{35}X^+_{45}|\Lambda\rangle, \quad r \geq 3,$$
$$X^+_{16}X^+_{26}X^+_{35}X^+_{45}|\Lambda\rangle, \quad r \geq 4. \tag{2.225}$$

Thus, corresponding to the values of r we have 9, 8, 6, 3, 1 states in the superfield decomposition. The mixed massless $\frac{3}{4}$-BPS irrep is obtained for $d = r = 1$ and consists of the first three states above (as was shown in general).

The explicit character formula for $r > 1$ is

$$\begin{aligned} \operatorname{ch} L_\Lambda = {} & \frac{e(\Lambda)}{(1-t_2)(1-t_1t_2)(1-t_2t_3)(1-t_1t_2t_3)} \\ & \times \{\mathcal{S}_r + \theta_1(1+t_1)\mathcal{S}_{r-1} + \bar{\theta}_1(1+t_3)\mathcal{S}_{r-1} \\ & + \vartheta_1\theta_1\mathcal{S}_{r-2} + \bar{\vartheta}_1\bar{\theta}_1\mathcal{S}_{r-2} + \theta_1\bar{\theta}_1(1+t_1)(1+t_3)\mathcal{S}_{r-2} \\ & + \vartheta_1\theta_1\bar{\theta}_1(1+t_3)\mathcal{S}_{r-3} + \theta_1\bar{\vartheta}_1\bar{\theta}_1(1+t_1)\mathcal{S}_{r-3} + \vartheta_1\theta_1\bar{\vartheta}_1\bar{\theta}_1\mathcal{S}_{r-4}\} \end{aligned} \tag{2.226}$$

where we have used the conformal factor $\mathcal{Q}_{n_1,n_3}$ from (2.138) (only for $n_1, n_3 \leq 2$).

For the $\frac{3}{4}$-BPS mixed massless irrep, $r = 1$, we have to use the massless conformal characters from (2.143), (2.144), (2.147), and we have

$$\begin{aligned} \operatorname{ch} L_\Lambda = {} & \frac{e(\Lambda)}{(1-t_2)(1-t_1t_2)(1-t_2t_3)(1-t_1t_2t_3)} \\ & \times \{(1-t_2t_{13})(1+t_5) + \theta_1(1+t_1-t_{12}-t_{13}) \\ & + \bar{\theta}_1(1+t_3-t_{23}-t_{13})\} \end{aligned} \tag{2.227}$$

2.3.4 $N = 4$

For $N = 4$ we consider only some important examples.

First we introduce notation for the odd roots when $N = 4$ using (2.13):

$$\begin{aligned} & \alpha_{25} = \gamma_3, \alpha_{26} = \gamma_3 + \alpha_5, \alpha_{27} = \gamma_3 + \alpha_5 + \alpha_6, \\ & \alpha_{28} = \gamma_3 + \alpha_5 + \alpha_6 + \alpha_7, \alpha_{45} = \gamma_4 + \alpha_5 + \alpha_6 + \alpha_7, \\ & \alpha_{46} = \gamma_4 + \alpha_5 + \alpha_6, \alpha_{47} = \gamma_4 + \alpha_5, \alpha_{48} = \gamma_4, \\ & \alpha_{1k} = \alpha_1 + \alpha_{2k}, \quad k = 5,6,7,8; \\ & \alpha_{3k} = \alpha_3 + \alpha_{4k}, \quad k = 5,6,7,8. \end{aligned} \tag{2.228}$$

Note that as a consequence of our mirror symmetry (2.15) we have

$$\begin{aligned} & \alpha_{15} \longleftrightarrow \alpha_{38}, \quad \alpha_{16} \longleftrightarrow \alpha_{37}, \quad \alpha_{17} \longleftrightarrow \alpha_{36}, \quad \alpha_{18} \longleftrightarrow \alpha_{35}, \\ & \alpha_{25} \longleftrightarrow \alpha_{48}, \quad \alpha_{26} \longleftrightarrow \alpha_{47}, \quad \alpha_{27} \longleftrightarrow \alpha_{46}, \quad \alpha_{28} \longleftrightarrow \alpha_{45}. \end{aligned} \tag{2.229}$$

For the characters we shall need also the following notation:

$$\begin{aligned} & \vartheta_k \equiv e(\alpha_{1,9-k}), \quad \theta_k \equiv e(\alpha_{2,9-k}), \quad k = 1,2,3,4 \\ & \bar{\vartheta}_k \equiv e(\alpha_{3,4+k}), \quad \bar{\theta}_k \equiv e(\alpha_{4,4+k}), \quad k = 1,2,3,4. \end{aligned} \tag{2.230}$$

- First we consider the *massless UIRs* (multiplets). As in all cases when $N > 1$ there are *three* cases of massless UIRs. In our classification they are DRC cases **ad, bc, bd.**

• **ad** $d = d^1_{41} = d^4_{44} = d^{ad} = 1 + j_2 = -z$, $j_1 = 0$, $r_i = 0, \forall i$, and all generators $X^+_{1,4+k}, X^+_{2,4+k}, X^+_{3,4+k}$ are eliminated. These anti-chiral irreps were denoted χ^+_s, $s = j_2 = 0, \frac{1}{2}, 1, \ldots$, in Section 3 of [140]. Besides the vacuum they contain only N states in $\hat{\Lambda}_\Lambda$ and should be called ultrashort UIRs. The bare character formula can be written in the most explicit way [123]:

$$\begin{aligned}\operatorname{ch}\hat{\Lambda}_\Lambda &= 1 + e(\alpha_{48}) + e(\alpha_{47})e(\alpha_{48}) + e(\alpha_{46})e(\alpha_{47})e(\alpha_{48})\\ &\quad + e(\alpha_{45})e(\alpha_{46})e(\alpha_{47})e(\alpha_{48}).\end{aligned} \tag{2.231}$$

Their signatures are

$$\begin{aligned}&[d \equiv 1 + j_2;\ 0, j_2;\ 0,0,0], \quad [d + \tfrac{1}{2};\ 0, j_2 + \tfrac{1}{2};\ 1,0,0],\\ &[d + 1;\ 0, j_2 + 1;\ 0,1,0],\\ &[d + \tfrac{3}{2};\ 0, j_2 + \tfrac{3}{2};\ 0,0,1], \quad [d + 2;\ 0, j_2 + 2;\ 0,0,0],\end{aligned} \tag{2.232}$$

All conformal entries are on the massless unitarity threshold, thus for the conformal entries we must use equation (2.142). For the su(4) entries we use formulas from (2.157). Then we have the explicit character formula for the anti-chiral massless case:

$$\begin{aligned}\operatorname{ch} L_\Lambda &= \frac{e(\Lambda)}{(1-t_2)(1-t_1t_2)(1-t_2t_3)(1-t_1t_2t_3)}\\ &\quad \times \{\mathcal{P}_{n_3} + \bar{\theta}_4\mathcal{P}_{n_3+1}\mathcal{S}_{100} + \bar{\theta}_3\bar{\theta}_4\mathcal{P}_{n_3+2}\mathcal{S}_{010}\\ &\quad + \bar{\theta}_2\bar{\theta}_3\bar{\theta}_4\mathcal{P}_{n_3+3}\mathcal{S}_{001} + \bar{\theta}_1\bar{\theta}_2\bar{\theta}_3\bar{\theta}_4\mathcal{P}_{n_3+4}\}\end{aligned} \tag{2.233}$$

• **bc** $d = d^2_{41} = d^3_{44} = d^{bc} = 1 + j_1 = z$, $j_2 = 0$, $r_i = 0, \forall i$, and all generators $X^+_{1,4+k}$, $X^+_{3,4+k}, X^+_{4,4+k}$ are eliminated. These chiral irreps were denoted χ_s, $s = j_1 = 0, \frac{1}{2}, 1, \ldots$, in Section 3 of [140]. Besides the vacuum they contain only N states in $\hat{\Lambda}_\Lambda$ and should be called ultrashort UIRs. The bare character formula is [123]:

$$\begin{aligned}\operatorname{ch}\hat{\Lambda}_\Lambda &= 1 + e(\alpha_{25}) + e(\alpha_{26})e(\alpha_{25}) + e(\alpha_{27})e(\alpha_{26})e(\alpha_{25})\\ &\quad + e(\alpha_{28})e(\alpha_{27})e(\alpha_{26})e(\alpha_{25}).\end{aligned} \tag{2.234}$$

Their signatures are

$$\begin{aligned}&[d \equiv 1 + j_1;\ j_1, 0;\ 0,0,0], \quad [d + \tfrac{1}{2};\ j_1 + \tfrac{1}{2}, 0;\ 0,0,1],\\ &[d + 1;\ j_1 + 1, 0;\ 0,1,0],\\ &[d + \tfrac{3}{2};\ j_1 + \tfrac{3}{2}, 0;\ 1,0,0], \quad [d + 2;\ j_1 + 2, 0;\ 0,0,0].\end{aligned} \tag{2.235}$$

All conformal entries are on the massless unitarity threshold, thus we use equation (2.145). Then we have the explicit character formula for the chiral massless case:

$$\begin{aligned}\operatorname{ch} L_\Lambda = &\frac{e(\Lambda)}{(1-t_2)(1-t_1t_2)(1-t_2t_3)(1-t_1t_2t_3)}\\ &\times \{\mathcal{P}'_{n_1} + \theta_4\mathcal{P}'_{n_1+1}\mathcal{S}_{001} + \theta_3\theta_4\mathcal{P}'_{n_1+2}\mathcal{S}_{010}\\ &+ \theta_2\theta_3\theta_4\mathcal{P}'_{n_1+3}\mathcal{S}_{100} + \theta_1\theta_2\theta_3\theta_4\mathcal{P}'_{n_1+4}\}\end{aligned} \tag{2.236}$$

We note the mirror symmetry between character formulas (2.233) and (2.236).

• **bd** $d = d^2_{41} = d^4_{44} = m_1 = 1$, $i_0 = 0, 1, 2$, $z = (i_0 - 1)/2$, $j_1 = j_2 = 0$, $r_i = \delta_{i,i_0+1}$.

In Section 3 of [140] they are parametrized by $n = 2, 3$, and denoted by χ'_n, $n = m$ ($z = n/2 - 1$), χ'^{+}_n, $n = N - m$ ($z = 1 - n/2$), but there is the coincidence for $n = 2$: $\chi'_2 = \chi'^{+}_2$. Here they are enumerated by the parameter i_0. The self-conjugate case ($i_0 = 1$, $z = 0$) is a $\frac{3}{4}$-BPS state. The following generators are eliminated: the chiral X^+_{1k}, $\forall k$, X^+_{2k}, $k = 5, \ldots, 5 + i_0$, and the anti-chiral X^+_{3k}, $\forall k$, $X^+_{4,N+4-k}$, $k = 0, \ldots, i_0$.

The bare character formula is (2.136) for $N = 4$:

$$\operatorname{ch} \hat{\Lambda}_\Lambda = 1 + \sum_{j=0}^{i_0} \prod_{i=j}^{i_0} e(\alpha_{2,8-i}) + \sum_{k=1}^{3-i_0} \prod_{i=k}^{3-i_0} e(\alpha_{4,4+i}) \tag{2.237}$$

We write out the signatures for the three subcases separately:

$$\begin{aligned}&[1; 0, 0; 1, 0, 0], \quad [\tfrac{3}{2}; \tfrac{1}{2}, 0; 0, 0, 0], \quad [\tfrac{3}{2}; 0, \tfrac{1}{2}; 0, 1, 0],\\ &[2; 0, 1; 0, 0, 1], \quad [\tfrac{5}{2}; 0, 1; 0, 0, 0], \quad i_0 = 0\\ &[1; 0, 0; 0, 1, 0], \quad [\tfrac{3}{2}; \tfrac{1}{2}, 0; 1, 0, 0], \quad [2; 1, 0; 0, 0, 0],\\ &[\tfrac{3}{2}; 0, \tfrac{1}{2}; 0, 0, 1], \quad [2; 0, 1; 0, 0, 0], \quad i_0 = 1\\ &[1; 0, 0; 0, 0, 1], \quad [\tfrac{3}{2}; \tfrac{1}{2}, 0; 0, 1, 0], \quad [2; 1, 0; 1, 0, 0],\\ &[\tfrac{5}{2}; \tfrac{3}{2}, 0; 0, 0, 0], \quad [\tfrac{3}{2}; 0, \tfrac{1}{2}; 0, 0, 0], \quad i_0 = 2\end{aligned} \tag{2.238}$$

All conformal entries are on the massless unitarity threshold, thus we use formulas (2.142) and (2.145). Then we have the explicit character formula for the mixed chiral–anti-chiral massless cases:

$$\begin{aligned}\operatorname{ch} L^0_\Lambda = &\frac{e(\Lambda)}{(1-t_2)(1-t_1t_2)(1-t_2t_3)(1-t_1t_2t_3)}\\ &\times \{(1 - t_2t_{13})\mathcal{S}_{100} + \theta_2(1 + t_1 - t_{12} - t_{13})\\ &+ \bar{\theta}_3(1 + t_3 - t_{23} - t_{13})\mathcal{S}_{010} + \bar{\theta}_2\bar{\theta}_3\mathcal{P}_3(t_3)\mathcal{S}_{001} + \bar{\theta}_1\bar{\theta}_2\bar{\theta}_3\mathcal{P}_4(t_3)\}\end{aligned} \tag{2.239a}$$

$$\begin{aligned}\operatorname{ch} L^1_\Lambda = &\frac{e(\Lambda)}{(1-t_2)(1-t_1t_2)(1-t_2t_3)(1-t_1t_2t_3)}\\ &\times \{(1 - t_2t_{13})\mathcal{S}_{010} + \theta_2(1 + t_1 - t_{12} - t_{13})\mathcal{S}_{100} + \theta_1\theta_2\mathcal{P}'_3(t_1)\\ &+ \bar{\theta}_2(1 + t_3 - t_{23} - t_{13})\mathcal{S}_{001} + \bar{\theta}_1\bar{\theta}_2\mathcal{P}_3(t_3)\}\end{aligned} \tag{2.239b}$$

$$\operatorname{ch} L^2_\Lambda = \frac{e(\Lambda)}{(1-t_2)(1-t_1t_2)(1-t_2t_3)(1-t_1t_2t_3)} \times \{(1-t_2t_{13})\mathcal{S}_{001} + \theta_3(1+t_1-t_{12}-t_{13})\mathcal{S}_{010} + \theta_2\theta_3\mathcal{P}'_3(t_1)\mathcal{S}_{100} + \theta_1\theta_2\theta_3\mathcal{P}'_4(t_1) + \bar\theta_2(1+t_3-t_{23}-t_{13})\} \tag{2.239c}$$

We note the mirror symmetry between character formulas (2.239a) and (2.239c), while (2.239b) is self-conjugate (as expected).

- Next we consider the *graviton supermultiplet* [217]. In our classification this is a DRC case:

- **bd** $d = d^2_{41} = d^4_{44} = d^{bd} = m_1 = 2, j_1 = j_2 = 0, r_2 = 2, r_1 = r_3 = 0, z = 0.$

Here the generators X^+_{1k}, X^+_{2k} for $k = 5, 6$ are eliminated and the generators X^+_{3k}, X^+_{4k} for $k = 7, 8$ and this is a $\frac{1}{2}$-BPS case.

The bare character formula is (2.134) taken for the case $i_0 = i'_0 = 1$:

$$\operatorname{ch} \hat\Lambda_\Lambda = \prod_{\substack{\alpha\in\Delta^+_{\bar 1} \\ \alpha\neq\alpha_{j,5+N-k} \\ j=3,4,k=1,2 \\ \alpha\neq\alpha_{j',4+k'} \\ j'=1,2,k'=1,2}} (1+e(\alpha)) - \mathcal{R} \tag{2.240}$$

Explicitly, the states (fields) are

$$\begin{aligned}
&|\Lambda\rangle, \quad [2; 0,0; 0,2,0], \quad \varphi_{(1)} \\
&X^+_{27}|\Lambda\rangle, \quad [\tfrac{5}{2}; \tfrac{1}{2}, 0; 1,1,0], \quad \lambda^+_{(1)} \\
&X^+_{46}|\Lambda\rangle, \quad [\tfrac{5}{2}; 0, \tfrac{1}{2}; 0,1,1] \quad \lambda^-_{(1)} \\
&X^+_{17}X^+_{27}|\Lambda\rangle, \quad [3; 0,0; 2,0,0], \quad \varphi_{(2)} \\
&X^+_{36}X^+_{46}|\Lambda\rangle, \quad [3; 0,0; 0,0,2], \quad \bar\varphi_{(2)} \\
&X^+_{18}X^+_{27}|\Lambda\rangle, \quad [3; 0,0; 0,1,0], \quad *\varphi'_{(2)} \\
&X^+_{35}X^+_{46}|\Lambda\rangle, \quad [3; 0,0; 0,1,0], \quad *\varphi''_{(2)} \\
&X^+_{28}X^+_{27}|\Lambda\rangle, \quad [3; 1,0; 0,1,0], \quad A^+_{\mu\nu} \\
&X^+_{45}X^+_{46}|\Lambda\rangle, \quad [3; 0,1; 0,1,0], \quad A^-_{\mu\nu} \\
&X^+_{46}X^+_{27}|\Lambda\rangle, \quad [3; \tfrac{1}{2}, \tfrac{p+q-2}{2}; 1,0,1], \quad A_\mu \\
&X^+_{28}X^+_{17}X^+_{27}|\Lambda\rangle, \quad [\tfrac{7}{2}; \tfrac{1}{2}, 0; 1,0,0], \quad \lambda^+_2 \\
&X^+_{35}X^+_{27}X^+_{46}|\Lambda\rangle, \quad [\tfrac{7}{2}; \tfrac{1}{2}, 0; 1,0,0], \quad *\lambda'^+_2 \\
&X^+_{45}X^+_{36}X^+_{46}|\Lambda\rangle, \quad [\tfrac{7}{2}; 0, \tfrac{1}{2}; 0,0,1], \quad \lambda^-_2 \\
&X^+_{18}X^+_{46}X^+_{27}|\Lambda\rangle, \quad [\tfrac{7}{2}; 0, \tfrac{1}{2}; 0,0,1], \quad *\lambda'^-_2 \\
&X^+_{28}X^+_{46}X^+_{27}|\Lambda\rangle, \quad [\tfrac{7}{2}; 1, \tfrac{1}{2}; 0,0,1], \quad \psi^+_\mu \\
&X^+_{45}X^+_{27}X^+_{46}|\Lambda\rangle, \quad [\tfrac{7}{2}; \tfrac{1}{2}, 1; 1,0,0], \quad \psi^-_\mu \\
&X^+_{18}X^+_{28}X^+_{17}X^+_{27}|\Lambda\rangle, \quad [4; 0,0; 0,0,0], \quad \varphi_{(3)} \\
&X^+_{35}X^+_{45}X^+_{36}X^+_{46}|\Lambda\rangle, \quad [4; 0,0; 0,0,0], \quad \varphi_{(3)}
\end{aligned}$$

$$\begin{aligned}
&X^+_{18}X^+_{35}X^+_{46}X^+_{27}|\Lambda\rangle, && [4;0,0;0,0,0], && *\varphi_{(3)}\\
&X^+_{28}X^+_{35}X^+_{46}X^+_{27}|\Lambda\rangle, && [4;1,0;0,0,0], && *B^+_{\mu\nu}\\
&X^+_{18}X^+_{45}X^+_{27}X^+_{46}|\Lambda\rangle, && [4;0,1;0,0,0], && *B^-_{\mu\nu}\\
&X^+_{28}X^+_{45}X^+_{27}X^+_{46}|\Lambda\rangle, && [4;1,1;0,0,0], && h_{\mu\nu}.
\end{aligned} \tag{2.241}$$

This multiplet appeared first in [217] and we use this notation for the states, though we should note that the fields marked with $*$ are missing there (however, for some it is just a question of multiplicity).

For the conformal entries in the explicit character formula we shall use equation (2.140) for the fields on the unitarity threshold: graviton $h_{\mu\nu}$, gravitini $\psi^\pm_\mu$, vector A_μ, and equation (2.138) for the rest. Thus, we have

$$\begin{aligned}
\operatorname{ch} L_\Lambda = {} & \frac{e(\Lambda)}{(1-t_2)(1-t_1t_2)(1-t_2t_3)(1-t_1t_2t_3)} \\
& \times \{\mathcal{S}_{020} + \theta_2\mathcal{S}_{110}(1+t_1) + \bar\theta_2\mathcal{S}_{011}(1+t_3) \\
& + \theta_2\bar\theta_2\mathcal{S}_{101}(1+t_1)(1+t_3) + \theta_2\vartheta_2\mathcal{S}_{200} + \bar\theta_2\bar\vartheta_2\mathcal{S}_{002} \\
& + (\theta_1\theta_2(1+t_1+t_1^2) + \bar\theta_1\bar\theta_2(1+t_3+t_3^2))\mathcal{S}_{010} \\
& + (\theta_2\vartheta_1 + \bar\theta_2\bar\vartheta_1)\mathcal{S}_{010} \qquad * \\
& + \theta_2(\theta_1\vartheta_2 + \bar\theta_2\bar\vartheta_1)(1+t_1)\mathcal{S}_{100} \qquad *' \\
& + \bar\theta_2(\bar\theta_1\bar\vartheta_2 + \theta_2\vartheta_1)(1+t_3)\mathcal{S}_{001} \qquad *' \\
& + \theta_2\bar\theta_2(\theta_1(1+t_1+t_1^2)(1+t_3)\mathcal{S}_{001} + \bar\theta_1(1+t_1)(1+t_3+t_3^2)\mathcal{S}_{100}) \\
& + \theta_1\theta_2\vartheta_1\vartheta_2 + \bar\theta_1\bar\theta_2\bar\vartheta_1\bar\vartheta_2 + \theta_2\vartheta_1\bar\theta_2\bar\vartheta_1 \qquad *' \\
& + \theta_2\bar\theta_2(\theta_1\bar\vartheta_1(1+t_1+t_1^2) + \vartheta_1\bar\theta_1(1+t_3+t_3^2)) \qquad * \\
& + \theta_1\theta_2\bar\theta_1\bar\theta_2(1+t_1+t_1^2)(1+t_3+t_3^2)\}
\end{aligned} \tag{2.242}$$

where $*$ designates a line that would be missing if we use [217], while $*'$ designates a line from which the last term would be missing.

- Next we consider the *Konishi supermultiplet*, [264], this the spinless R-symmetry scalar on the unitarity threshold, thus, $d = 2$. In our classification this is a DRC case **ac**:

  - **ac** $d = d_{\max} = d^1_{N1} = d^3_{NN} = d^{ac} = 2, j_1 = j_2 = 0, r_i = 0, \forall i, z = 0.$

  Here all generators X^+_{1k} and X^+_{3k} are eliminated and this is $\frac{1}{2}$-BPS case.

  The bare character formula is (3.84) from [123]:

$$\begin{aligned}
\operatorname{ch}\hat\Lambda_\Lambda = {} & \sum_{k=1}^{N}\prod_{i=1}^{k} e(\alpha_{2,4+i}) + \sum_{k=1}^{N}\prod_{i=1}^{k} e(\alpha_{4,5+N-i}) \\
& + \prod_{\substack{\alpha\in\Delta_{\bar 1}^+ \\ \varepsilon_2>0,\varepsilon_4>0}} (1+e(\alpha)) - \mathcal{R}.
\end{aligned} \tag{2.243}$$

The states of the first line of (2.243) are the four chiral and anti-chiral states of this case:

$$\begin{aligned}&X^+_{25}|\Lambda\rangle, \quad X^+_{26}X^+_{25}|\Lambda\rangle, \quad X^+_{27}X^+_{26}X^+_{25}|\Lambda\rangle, \quad X^+_{28}X^+_{27}X^+_{26}X^+_{25}|\Lambda\rangle,\\ &X^+_{48}|\Lambda\rangle, \quad X^+_{47}X^+_{48}|\Lambda\rangle, \quad X^+_{46}X^+_{47}X^+_{48}|\Lambda\rangle, \quad X^+_{45}X^+_{46}X^+_{47}X^+_{48}|\Lambda\rangle\end{aligned} \tag{2.244}$$

All other states (besides the vacuum) are of mixed chirality, and our first task is to make explicit the second line of the above equation (2.243). It turns out that the counterterm has 22 states and there remain 42 states contributing to this second line, which explicitly are

$$\begin{aligned}
&|\Lambda\rangle, \quad X^+_{25}X^+_{48}|\Lambda\rangle, \qquad * \\
&X^+_{45}X^+_{25}|\Lambda\rangle, \quad X^+_{28}X^+_{48}|\Lambda\rangle,\\
&X^+_{28}X^+_{25}X^+_{48}|\Lambda\rangle, \quad X^+_{45}X^+_{25}X^+_{48}|\Lambda\rangle,\\
&X^+_{26}X^+_{25}X^+_{48}|\Lambda\rangle, \quad X^+_{47}X^+_{25}X^+_{48}|\Lambda\rangle,\\
&X^+_{46}X^+_{26}X^+_{25}|\Lambda\rangle, \quad X^+_{27}X^+_{47}X^+_{48}|\Lambda\rangle,\\
&X^+_{46}X^+_{26}X^+_{25}X^+_{48}|\Lambda\rangle, \quad X^+_{27}X^+_{47}X^+_{25}X^+_{48}|\Lambda\rangle,\\
&X^+_{47}X^+_{26}X^+_{25}X^+_{48}|\Lambda\rangle, \quad X^+_{28}X^+_{45}X^+_{25}X^+_{48}|\Lambda\rangle, \qquad *\\
&X^+_{45}X^+_{46}X^+_{26}X^+_{25}|\Lambda\rangle, \quad X^+_{28}X^+_{27}X^+_{47}X^+_{48}|\Lambda\rangle,\\
&X^+_{27}X^+_{26}X^+_{25}X^+_{48}|\Lambda\rangle, \quad X^+_{46}X^+_{47}X^+_{25}X^+_{48}|\Lambda\rangle,\\
&X^+_{26}X^+_{28}X^+_{25}X^+_{48}|\Lambda\rangle, \quad X^+_{45}X^+_{47}X^+_{25}X^+_{48}|\Lambda\rangle,\\
&X^+_{47}X^+_{27}X^+_{26}X^+_{25}|\Lambda\rangle, \quad X^+_{26}X^+_{46}X^+_{47}X^+_{48}|\Lambda\rangle,\\
&X^+_{46}X^+_{47}X^+_{27}X^+_{26}X^+_{25}|\Lambda\rangle, \quad X^+_{27}X^+_{26}X^+_{46}X^+_{47}X^+_{48}|\Lambda\rangle,\\
&X^+_{28}X^+_{46}X^+_{26}X^+_{25}X^+_{48}|\Lambda\rangle, \quad X^+_{45}X^+_{27}X^+_{47}X^+_{25}X^+_{48}|\Lambda\rangle,\\
&X^+_{28}X^+_{27}X^+_{47}X^+_{25}X^+_{48}|\Lambda\rangle, \quad X^+_{45}X^+_{46}X^+_{26}X^+_{25}X^+_{48}|\Lambda\rangle,\\
&X^+_{26}X^+_{27}X^+_{47}X^+_{25}X^+_{48}|\Lambda\rangle, \quad X^+_{47}X^+_{46}X^+_{26}X^+_{25}X^+_{48}|\Lambda\rangle,\\
&X^+_{28}X^+_{27}X^+_{26}X^+_{25}X^+_{48}|\Lambda\rangle, \quad X^+_{45}X^+_{46}X^+_{47}X^+_{25}X^+_{48}|\Lambda\rangle,\\
&X^+_{28}X^+_{45}X^+_{46}X^+_{26}X^+_{25}X^+_{48}|\Lambda\rangle, \quad X^+_{45}X^+_{28}X^+_{27}X^+_{47}X^+_{25}X^+_{48}|\Lambda\rangle,\\
&X^+_{45}X^+_{46}X^+_{47}X^+_{27}X^+_{26}X^+_{25}|\Lambda\rangle, \quad X^+_{28}X^+_{27}X^+_{26}X^+_{46}X^+_{47}X^+_{48}|\Lambda\rangle,\\
&X^+_{47}X^+_{28}X^+_{27}X^+_{26}X^+_{25}X^+_{48}|\Lambda\rangle, \quad X^+_{26}X^+_{45}X^+_{46}X^+_{47}X^+_{25}X^+_{48}|\Lambda\rangle,\\
&X^+_{46}X^+_{47}X^+_{28}X^+_{27}X^+_{26}X^+_{25}X^+_{48}|\Lambda\rangle, \quad X^+_{27}X^+_{26}X^+_{45}X^+_{46}X^+_{47}X^+_{25}X^+_{48}|\Lambda\rangle,\\
&X^+_{46}X^+_{27}X^+_{47}X^+_{26}X^+_{25}X^+_{48}|\Lambda\rangle, \qquad *\\
&X^+_{28}X^+_{27}X^+_{26}X^+_{45}X^+_{46}X^+_{47}X^+_{25}X^+_{48}|\Lambda\rangle, \qquad *
\end{aligned} \tag{2.245}$$

where $*$ denotes lines in which the states are self-conjugate, while in all other cases the two states in each line are conjugate to each other. The corresponding signatures of both (2.244) and (2.245) are

$$[\tfrac{5}{2};\ \tfrac{p+q-2}{2}, 0;\ 0, 0, 1], \quad [3;\ 1, 0;\ 0, 1, 0], \quad [\tfrac{7}{2};\ \tfrac{3}{2}, 0;\ 1, 0, 0], \quad [4;\ 2, 0;\ 0, 0, 0],$$

$$
\begin{aligned}
&[\tfrac{5}{2};\, 0, \tfrac{p+q-2}{2};\, 1,0,0], \quad [3;\, 0,1;\, 0,1,0], \quad [\tfrac{7}{2};\, 0, \tfrac{3}{2};\, 0,0,1], \quad [4;\, 0,2;\, 0,0,0],\\
&[2;\, 0,0;\, 0,0,0], \quad [3;\, \tfrac{p+q-2}{2}, \tfrac{p+q-2}{2};\, 1,0,1],\\
&[3;\, \tfrac{p+q-2}{2}, \tfrac{p+q-2}{2};\, 0,0,0], \quad [3;\, \tfrac{p+q-2}{2}, \tfrac{p+q-2}{2};\, 0,0,0],\\
&[\tfrac{7}{2};\, 1, \tfrac{p+q-2}{2};\, 0,0,1], \quad [\tfrac{7}{2};\, \tfrac{p+q-2}{2}, 1;\, 1,0,0],\\
&[\tfrac{7}{2};\, 1, \tfrac{p+q-2}{2};\, 1,1,0], \quad [\tfrac{7}{2};\, \tfrac{p+q-2}{2}, 1;\, 0,1,1],\\
&[\tfrac{7}{2};\, 1, \tfrac{p+q-2}{2};\, 0,0,1], \quad [\tfrac{7}{2};\, \tfrac{p+q-2}{2}, 1;\, 1,0,0],\\
&[4;\, 1,1;\, 1,0,1], \quad [4;\, 1,1;\, 1,0,1],\\
&[4;\, 1,1;\, 0,2,0], \quad [4;\, 1,1;\, 0,0,0],\\
&[4;\, 1,1;\, 0,0,0], \quad [4;\, 1,1;\, 0,0,0],\\
&[4;\, \tfrac{3}{2}, \tfrac{p+q-2}{2};\, 2,0,0], \quad [4;\, \tfrac{p+q-2}{2}, \tfrac{3}{2};\, 0,0,2],\\
&[4;\, \tfrac{3}{2}, \tfrac{p+q-2}{2};\, 0,1,0], \quad [4;\, \tfrac{p+q-2}{2}, \tfrac{3}{2};\, 0,1,0],\\
&[4;\, \tfrac{3}{2}, \tfrac{p+q-2}{2};\, 0,1,0], \quad [4;\, \tfrac{p+q-2}{2}, \tfrac{3}{2};\, 0,1,0],\\
&[\tfrac{9}{2};\, \tfrac{3}{2}, 1;\, 0,0,1], \quad [\tfrac{9}{2};\, 1, \tfrac{3}{2};\, 1,0,0],\\
&[\tfrac{9}{2};\, \tfrac{3}{2}, 1;\, 0,0,1], \quad [\tfrac{9}{2};\, 1, \tfrac{3}{2};\, 1,0,0],\\
&[\tfrac{9}{2};\, \tfrac{3}{2}, 1;\, 0,0,1], \quad [\tfrac{9}{2};\, 1, \tfrac{3}{2};\, 1,0,0],\\
&[\tfrac{9}{2};\, \tfrac{3}{2}, 1;\, 1,1,0], \quad [\tfrac{9}{2};\, 1, \tfrac{3}{2};\, 0,1,1],\\
&[\tfrac{9}{2};\, 2, \tfrac{p+q-2}{2};\, 1,0,0], \quad [\tfrac{9}{2};\, \tfrac{p+q-2}{2}, 2;\, 0,0,1],\\
&[5;\, \tfrac{3}{2}, \tfrac{3}{2};\, 0,0,0], \quad [5;\, \tfrac{3}{2}, \tfrac{3}{2};\, 0,0,0],\\
&[5;\, \tfrac{3}{2}, \tfrac{3}{2};\, 0,0,0], \quad [5;\, \tfrac{3}{2}, \tfrac{3}{2};\, 0,0,0],\\
&[5;\, 2,1;\, 0,1,0], \quad [5;\, 1,2;\, 0,1,0],\\
&[\tfrac{11}{2};\, 2, \tfrac{3}{2};\, 0,0,1], \quad [\tfrac{11}{2};\, \tfrac{3}{2}, 2;\, 1,0,0],\\
&[5;\, \tfrac{3}{2}, \tfrac{3}{2};\, 1,0,1], \quad [6;\, 2,2;\, 0,0,0],
\end{aligned} \tag{2.246}
$$

where the first two lines are the eight states in (2.244) and the rest are from (2.245).

Note that for all conformal signatures we have $d-j_1-j_2 = 2$. For $j_1 j_2 \neq 0$ this is *on* the conformal unitarity threshold and then we shall use equation (2.140). For $j_1 j_2 = 0$ this is *above* the conformal unitarity threshold and then we shall use the generic equation (2.138):

$$
\begin{aligned}
\operatorname{ch} L_\Lambda = {}& \frac{e(\Lambda)}{(1-t_2)(1-t_1t_2)(1-t_2t_3)(1-t_1t_2t_3)}\\
&\times \{\mathbf{1} + \theta_4(1+t_1)\mathcal{S}_{001} + \theta_3\theta_4(1+t_1+t_1^2)\mathcal{S}_{010}\\
&+ \theta_2\theta_3\theta_4(1+t_1+t_1^2+t_1^3)\mathcal{S}_{100} + \bar\theta_2\bar\theta_3\bar\theta_4(1+t_3+t_3^2+t_3^3)\mathcal{S}_{001}\\
&+ \theta_1\theta_2\theta_3\theta_4(1+t_1+t_1^2+t_1^3+t_1^4) + \bar\theta_1\bar\theta_2\bar\theta_3\bar\theta_4(1+t_3+t_3^2+t_3^3+t_3^4)\\
&+ \bar\theta_4(1+t_3)\mathcal{S}_{100} + \bar\theta_3\bar\theta_4(1+t_3+t_3^2)\mathcal{S}_{010}\\
&+ (\theta_4\bar\theta_4\mathcal{S}_{101} + \bar\theta_1\theta_4 + \theta_1\bar\theta_4)\mathcal{P}_{22}\\
&+ ((\theta_1\theta_4\bar\theta_4 + \bar\theta_2\theta_3\theta_4)\mathcal{S}_{001} + \theta_3\theta_4\bar\theta_4\mathcal{S}_{110})\mathcal{P}_{3,2}
\end{aligned}
$$

$$
\begin{aligned}
&+ ((\bar\theta_1\theta_4\bar\theta_4 + \theta_2\bar\theta_3\bar\theta_4)\mathcal{S}_{100} + \bar\theta_3\theta_4\bar\theta_4\mathcal{S}_{011})\mathcal{P}_{2,3} \\
&+ ((\bar\theta_2\theta_3\theta_4\bar\theta_4 + \theta_2\bar\theta_3\theta_4\bar\theta_4)\mathcal{S}_{101} + \bar\theta_3\theta_3\theta_4\bar\theta_4\mathcal{S}_{020} \\
&+ \theta_1\bar\theta_1\theta_4\bar\theta_4 + \bar\theta_1\bar\theta_2\theta_3\theta_4 + \theta_1\theta_2\bar\theta_3\bar\theta_4)\mathcal{P}_{3,3} \\
&+ ((\theta_1\bar\theta_4 + \bar\theta_3\theta_2)\mathcal{S}_{010} + \theta_2\bar\theta_4\mathcal{S}_{200})\theta_3\theta_4\mathcal{P}_{4,2} \\
&+ ((\bar\theta_1\theta_4 + \theta_3\bar\theta_2)\mathcal{S}_{010} + \bar\theta_2\theta_4\mathcal{S}_{002})\bar\theta_3\bar\theta_4\mathcal{P}_{2,4} \\
&+ ((\theta_2\theta_3\bar\theta_2\bar\theta_3 + \theta_1\bar\theta_2\theta_3\bar\theta_4 + \theta_1\theta_2\bar\theta_3\bar\theta_4)\mathcal{S}_{001} \\
&+ \theta_3\theta_2\bar\theta_3\bar\theta_4\mathcal{S}_{110})\theta_4\mathcal{P}_{4,3} \\
&+ ((\theta_2\theta_3\bar\theta_2\bar\theta_3 + \bar\theta_1\theta_2\bar\theta_3\theta_4 + \bar\theta_1\bar\theta_2\theta_3\theta_4)\mathcal{S}_{100} \\
&+ \bar\theta_3\bar\theta_2\theta_3\theta_4\mathcal{S}_{011})\bar\theta_4\mathcal{P}_{3,4} \\
&+ \theta_1\theta_2\theta_3\theta_4\bar\theta_4\mathcal{S}_{100}\mathcal{P}_{5,2} + \bar\theta_1\bar\theta_2\bar\theta_3\bar\theta_4\theta_4\mathcal{S}_{001}\mathcal{P}_{2,5} \\
&+ (\theta_1\bar\theta_1\bar\theta_2\theta_3\theta_4\bar\theta_4 + \theta_1\bar\theta_1\theta_2\bar\theta_3\theta_4\bar\theta_4 \\
&+ \bar\theta_1\theta_2\bar\theta_2\theta_3\bar\theta_3\theta_4 + \theta_1\theta_2\bar\theta_2\theta_3\bar\theta_3\bar\theta_4 + \theta_2\bar\theta_2\theta_3\bar\theta_3\theta_4\bar\theta_4\mathcal{S}_{101})\mathcal{P}_{4,4} \\
&+ (\theta_1\theta_2\theta_3\bar\theta_3\theta_4\bar\theta_4\mathcal{P}_{5,3} + \bar\theta_1\bar\theta_2\theta_3\bar\theta_3\theta_4\bar\theta_4\mathcal{P}_{3,5})\mathcal{S}_{101} \\
&+ \theta_1\theta_2\theta_3\theta_4\bar\theta_2\bar\theta_3\bar\theta_4\mathcal{P}_{5,4}\mathcal{S}_{001} + \bar\theta_1\bar\theta_2\bar\theta_3\bar\theta_4\theta_2\theta_3\theta_4\mathcal{P}_{4,5}\mathcal{S}_{100} \\
&+ \theta_1\theta_2\theta_3\theta_4\bar\theta_1\bar\theta_2\bar\theta_3\bar\theta_4\mathcal{P}_{5,5}\}.
\end{aligned}
\tag{2.247}
$$

2.4 BPS states for $D = 4$

Our first task in this section is to present explicitly the reduction of the supersymmetries in the irreducible UIRs. This means that we are to give explicitly the number κ of odd generators which are eliminated from the corresponding lowest weight module (or equivalently, the number of super-derivatives that annihilate the corresponding superfield).

2.4.1 R-symmetry scalars

We start with the simpler cases of R-symmetry scalars when $r_i = 0$ for all i, which means also that $m_1 = m = m^* = 0$. These cases are valid also for $N = 1$. More explicitly

$$
\begin{aligned}
\bullet\mathbf{a}\quad & d = d^a_{|_{m=0}} = 2 + 2j_2 + z > 2 + 2j_1 - z, \quad j_1 \text{ arbitrary}, \\
& \kappa = N + (1-N)\delta_{j_2,0}
\end{aligned}
\tag{2.248}
$$

Here κ is the number of anti-chiral generators $X^+_{3,4+k}$, $k = 1, \dots, \kappa$, that are eliminated. Thus, in the cases when $\kappa = N$ the semi-short UIRs may be called semi-chiral since they lack half of the anti-chiral generators.

$$
\begin{aligned}
\bullet\mathbf{b}\quad & d = d^b_{|_{m=0}} = z > 2 + 2j_1 - z, \quad j_1 \text{ arbitrary}, \; j_2 = 0, \\
& \kappa = 2N
\end{aligned}
\tag{2.249}
$$

These short UIRs may be called chiral since they lack all anti-chiral generators $X^+_{3,4+k}$, $X^+_{4,4+k}$, $k = 1, \dots, N$.

$$\bullet\mathbf{c} \quad d = d^c_{|m=0} = 2 + 2j_1 - z > 2 + 2j_2 + z, \quad j_2 \text{ arbitrary}, \\ \kappa = N + (1 - N)\delta_{j_1,0} \tag{2.250}$$

Here, κ is the number of chiral generators $X^+_{1,4+k}$, $k = 1, \dots, \kappa$, that are eliminated. Thus, in the cases when $\kappa = N$ the semi-short UIRs may be called semi-anti-chiral since they lack half of the chiral generators.

$$\bullet\mathbf{d} \quad d = d^d_{|m=0} = -z > 2 + 2j_2 + z, \quad j_2 \text{ arbitrary}, \ j_1 = 0, \\ \kappa = 2N. \tag{2.251}$$

These short UIRs may be called anti-chiral since they lack all chiral generators $X^+_{1,4+k}$, $X^+_{2,4+k}$, $k = 1, \dots, N$.

$$\bullet\mathbf{ac} \quad d = d^{ac}_{|m=0} = 2 + j_1 + j_2, \quad z = j_1 - j_2, \\ \kappa = 2N + (1 - N)(\delta_{j_1,0} + \delta_{j_2,0}) \tag{2.252}$$

Here κ is the number of mixed elimination: chiral generators $X^+_{1,4+k}$, $(k = 1, \dots, N + (1 - N)\delta_{j_1,0})$, and anti-chiral generators $X^+_{3,4+k}$ $(k = 1, \dots, N + (1 - N)\delta_{j_2,0})$. Thus, in the cases when $\kappa = 2N$ the semi-short UIRs may be called semi-chiral–anti-chiral since they lack half of the chiral and half of the anti-chiral generators. (They may be called Grassmann analytic following [174].)

$$\bullet\mathbf{ad} \quad d = d^{ad}_{|m=0} = 1 + j_2 = -z, \quad j_1 = 0, \\ \kappa = 3N + (1 - N)\delta_{j_2,0} \tag{2.253}$$

Here κ is the number of mixed elimination: both types chiral generators $X^+_{1,4+k}$, $X^+_{2,4+k}$ $(k = 1, \dots, N)$, and anti-chiral generators $X^+_{3,4+k}$ $(k = 1, \dots, N + (1 - N)\delta_{j_2,0})$. Thus, in the cases when $\kappa = 3N$ the semi-short UIRs may be called semi-chiral and anti-chiral since they lack all the chiral and half of the anti-chiral generators.

$$\bullet\mathbf{bc} \quad d = d^{bc}_{|m=0} = 1 + j_1 = z, \quad j_2 = 0, \\ \kappa = 3N + (1 - N)\delta_{j_1,0} \tag{2.254}$$

Here κ is the number of mixed elimination: chiral generators $X^+_{1,4+k}$ $(k = 1, \dots, N + (1 - N)\delta_{j_1,0})$ and both types anti-chiral generators $X^+_{3,4+k}$, $X^+_{2,4+k}$ $(k = 1, \dots, N)$. Thus, in the cases when $\kappa = 3N$ the semi-short UIRs may be called chiral and semi-anti-chiral since they lack half of the chiral and all of the anti-chiral generators.

The last two cases (ad, bc) form two of the three series of massless states, holomorphic and antiholomorphic [140]; see also [139, 123].

The case • **bd** for R-symmetry scalars is trivial, since also all other quantum numbers are zero ($d = j_1 = j_2 = z = 0$).

2.4.2 R-symmetry non-scalars

Here we need some additional notation. Let $N > 1$ and let i_0 be an integer such that $0 \leq i_0 \leq N-1$, $r_i = 0$ for $i \leq i_0$, and if $i_0 < N-1$ then $r_{i_0+1} > 0$. Let now i'_0 be an integer such that $0 \leq i'_0 \leq N-1$, $r_{N-i} = 0$ for $i \leq i'_0$, and if $i'_0 < N-1$ then $r_{N-1-i'_0} > 0$.[13]

With this notation the cases of R-symmetry scalars occur when $i_0 + i'_0 = N-1$, thus, from now on we have the restriction

$$0 \leq i_0 + i'_0 \leq N-2. \tag{2.255}$$

Now we can make a list for the values of κ, with the same interpretation as in the previous subsection, only the last case is added here.

$$\begin{aligned} \bullet\mathbf{a}\quad & d = d^a = 2 + 2j_2 + z + 2m_1 - 2m/N > 2 + 2j_1 - z + 2m/N, \\ & j_1, j_2 \text{ arbitrary}, \\ & \kappa = 1 + i_0(1 - \delta_{j_2,0}) \leq N-1. \end{aligned} \tag{2.256}$$

Here are eliminated the anti-chiral generators $X^+_{3,4+k}$, $k \leq \kappa$.

$$\begin{aligned} \bullet\mathbf{b}\quad & d = d^b = z + 2m_1 - 2m/N > 2 + 2j_1 - z + 2m/N, \\ & j_2 = 0,\ j_1 \text{ arbitrary}, \\ & \kappa = 2 + 2i_0 \leq 2N-2. \end{aligned} \tag{2.257}$$

Here are eliminated the anti-chiral generators $X^+_{3,4+k}$, $X^+_{3,4+k}$, $k \leq 1 + i_0$.

$$\begin{aligned} \bullet\mathbf{c}\quad & d = d^c = 2 + 2j_1 - z + 2m/N > 2 + 2j_2 + z + 2m_1 - 2m/N, \\ & j_1, j_2 \text{ arbitrary}, \\ & \kappa = 1 + i'_0(1 - \delta_{j_1,0}) \leq N-1. \end{aligned} \tag{2.258}$$

Here are eliminated the chiral generators $X^+_{1,4+k}$, $k \leq \kappa$.

$$\begin{aligned} \bullet\mathbf{d}\quad & d = d^d = 2m/N - z > 2 + 2j_2 + z + 2m_1 - 2m/N, \\ & j_1 = 0,\ j_2 \text{ arbitrary}, \\ & \kappa = 2 + 2i'_0 \leq 2N-2. \end{aligned} \tag{2.259}$$

Here are eliminated the chiral generators $X^+_{1,4+k}$, $X^+_{2,4+k}$, $k \leq 1 + i'_0$.

$$\begin{aligned} \bullet\mathbf{ac}\quad & d = d^{ac}, \quad z = j_1 - j_2 + 2m/N - m_1, \quad j_1, j_2 \text{ arbitrary}, \\ & \kappa = 2 + i_0(1 - \delta_{j_2,0}) + i'_0(1 - \delta_{j_1,0}) \leq N. \end{aligned} \tag{2.260}$$

---

13 Both definitions are formally valid for $N = 1$ with $i_0 = 0$ since $r_0 \equiv 0$ by convention and with $i'_0 = 0$ since $r_N \equiv 0$ by convention.

Here the chiral generators $X^+_{1,4+k}$, $k \le 1 + i'_0(1 - \delta_{j_1,0})$, and anti-chiral generators $X^+_{3,4+k}$, $k \le 1 + i_0(1 - \delta_{j_2,0})$, are eliminated.

$$\bullet\mathbf{ad} \quad d = d^{ad}, \quad j_1 = 0, \quad z = 2m/N - m_1 - 1 - j_2, \quad j_2 \text{ arbitrary}, \\ \kappa = 3 + i_0(1 - \delta_{j_2,0}) + 2i'_0 \le 1 + N + i'_0 \le 2N - 1. \tag{2.261}$$

Here the chiral generators $X^+_{1,4+k}$, $X^+_{2,4+k}$, $k \le 1 + i'_0$, and anti-chiral generators $X^+_{3,4+k}$, $k \le 1 + i_0(1 - \delta_{j_2,0})$, are eliminated.

$$\bullet\mathbf{bc} \quad d = d^{bc}, \quad j_2 = 0, \quad z = 2m/N - m_1 + 1 + j_1, \quad j_1 \text{ arbitrary}, \\ \kappa = 3 + 2i_0 + i'_0(1 - \delta_{j_1,0}) \le 1 + N + i_0 \le 2N - 1. \tag{2.262}$$

Here are eliminated chiral generators $X^+_{1,4+k}$, $k \le 1 + i'_0(1 - \delta_{j_1,0})$, and anti-chiral generators $X^+_{3,4+k}$, $X^+_{4,4+k}$, $k \le 1 + i_0$.

$$\bullet\mathbf{bd} \quad d = d^{bd} = m_1, \quad j_1 = j_2 = 0, \quad z = 2m/N - m_1, \\ \kappa = 4 + 2i_0 + 2i'_0 \le 2N. \tag{2.263}$$

Here are eliminated chiral generators $X^+_{1,4+k}$, $X^+_{2,4+k}$, $k \le 1 + i'_0$, and anti-chiral generators $X^+_{3,4+k}$, $X^+_{3,4+k}$, $k \le 1 + i_0$.

Note that the case $\kappa = 2N$ is possible exactly when $i_0 + i'_0 = N - 2$, i. e., when there is only one non-zero r_i, namely, $r_{i_0+1} \ne 0$, $i_0 = 0, 1, \ldots, N - 2$:

$$\bullet\ \mathbf{bd} \quad \kappa = 2N : d = m_1 = r_{i_0+1}, \quad j_1 = j_2 = 0, \quad z = r_{i_0+1}\frac{2 + 2i_0 - N}{N}. \tag{2.264}$$

When $d = m_1 = 1$ these $\frac{1}{2}$-eliminated UIRs form the 'mixed' series of massless representations [140]; see also [139, 123].[14]

In the next sections we shall use the above classification to the so-called BPS states.

2.4.3 PSU(2,2/4)

The most interesting case is when $N = 4$. This is related to super-Yang–Mills theory and contains the so-called BPS states, cf., [9, 172, 173, 176, 174, 159, 338, 106, 22, 104, 149]. They are characterized by the number κ of odd generators which annihilate them—then the corresponding state is called a $\frac{\kappa}{4N}$-BPS state. Group-theoretically the case $N = 4$ is special since the $u(1)$ subalgebra carrying the quantum number z becomes central and one can invariantly set $z = 0$.

---

14 This series is absent for $N = 1$.

We give now the explicit list of these states:

• **a** $d = d^1_{41} = 2 + 2j_2 + 2m_1 - \frac{1}{2}m > d^3_{44}$. The last inequality leads to the restriction

$$2j_2 + r_1 > 2j_1 + r_3. \tag{2.265}$$

In the case of R-symmetry scalars, i. e., $m_1 = 0$, it follows that $j_2 > j_1$, i. e., $j_2 > 0$, and then we have

$$\kappa = 4, \quad m_1 = 0, \quad j_2 > 0. \tag{2.266}$$

In the case of R-symmetry non-scalars, i. e., $m_1 \neq 0$, we have the range: $i_0 + i'_0 \leq 2$, and thus

$$\kappa = 1 + i_0(1 - \delta_{j_2,0}) \leq 3. \tag{2.267}$$

• **b** $d = d^2_{41} = \frac{1}{2}m^* > d^3_{44}, j_2 = 0$. The last inequality leads to the restriction

$$r_1 > 2 + 2j_1 + r_3. \tag{2.268}$$

The latter means that $r_1 > 2$, i. e., $m_1 \neq 0$, $i_0 = 0$, and thus

$$\kappa = 2. \tag{2.269}$$

The next two cases are conjugate to the previous two, so we present them briefly:

$$\bullet\,\mathbf{c} \quad d = d^3_{44} = 2 + 2j_1 + \frac{1}{2}m > d^1_{41} \quad \Rightarrow \quad 2j_1 + r_3 > 2j_2 + r_1, \tag{2.270}$$

$$m_1 = 0 \quad \Rightarrow \quad j_1 > j_2 \quad \Rightarrow \quad j_1 > 0 \quad \Rightarrow \quad \kappa = 4, \quad m_1 = 0,\ j_1 > 0. \tag{2.271}$$

$$m_1 \neq 0 \quad \Rightarrow \quad i_0 + i'_0 \leq 2 \quad \Rightarrow \quad \kappa = 1 + i'_0(1 - \delta_{j_1,0}) \leq 3. \tag{2.272}$$

$$\bullet\,\mathbf{d} \quad d = d^4_{44} = \frac{1}{2}m > d^1_{41}, \quad j_1 = 0, \quad \Rightarrow \quad r_3 > 2 + 2j_2 + r_1, \tag{2.273}$$

$$\Rightarrow \quad r_3 > 2 \quad \Rightarrow \quad m_1 \neq 0, \quad i'_0 = 0 \quad \Rightarrow \quad \kappa = 2. \tag{2.274}$$

• **ac** $d = d^{ac} = 2 + j_1 + j_2 + m_1$. From $z = 0$ follows

$$2j_2 + r_1 = 2j_1 + r_3. \tag{2.275}$$

In the case of R-symmetry scalars, i. e., $m_1 = 0$, it follows that $j_2 = j_1 = j$, and then we have

$$\kappa = 8 - 6\delta_{j,0}, \quad d = 2 + 2j. \tag{2.276}$$

In the case of R-symmetry non-scalars, i. e., $m_1 \neq 0$, $i_0 + i'_0 \leq 2$, and thus

$$\kappa = 2 + i_0(1 - \delta_{j_2,0}) + i'_0(1 - \delta_{j_1,0}) \leq 4. \tag{2.277}$$

• **ad** From $z = 0$ follows $r_3 = 2 + 2j_2 + r_1 \Rightarrow r_3 \geq 2 \Rightarrow m_1 \neq 0$, and $i'_0 = 0$, $i_0 \leq 2 \Rightarrow$

$$\begin{aligned} \kappa &= 3 + i_0(1 - \delta_{j_2,0}) \leq 5, \\ d = d^{ad} &= 1 + j_2 + m_1 = 3 + 3j_2 + 2r_1 + r_2, \\ \chi_4 &= \{0;\, r_1, r_2, 2 + 2j_2 + r_1;\, 2j_2\}. \end{aligned} \tag{2.278}$$

• **bc** From $z = 0$ follows $r_1 = 2 + 2j_2 + r_3 \Rightarrow r_1 \geq 2 \Rightarrow m_1 \neq 0$, and $i_0 = 0$, $i'_0 \leq 2 \Rightarrow$

$$\begin{aligned} \kappa &= 3 + i'_0(1 - \delta_{j_1,0}) \leq 5, \\ d = d^{bc} &= 1 + j_2 + m_1 = 3 + 3j_2 + 2r_1 + r_2, \\ \chi_4 &= \{2j_1;\, 2 + 2j_2 + r_3, r_2, r_3;\, 0\}. \end{aligned} \tag{2.279}$$

• **bd** From $z = 0$ follows $r_1 = r_3 = r$, thus, $i_0 = i'_0 = 0, 1$ and then we have

$$\begin{aligned} \kappa &= 4(1 + i_0), \\ d = d^{bd} &= m_1 = 2r + r_2 \neq 0, \quad r, r_2 \in \mathbb{Z}_+, \\ \chi_4 &= \{0;\, r, r_2, r;\, 0\}. \end{aligned} \tag{2.280}$$

From the above BPS states we list now the most interesting ones.

2.4.3.1 $\frac{1}{2}$-BPS states, $\kappa = 8$

These are possible in case **ac**, cf. (2.276), for R-symmetry scalars and non-trivial vector Lorentz spin:

$$d = 2 + 2j \geq 3, \quad j_1 = j_2 = j \geq \frac{1}{2}, \quad m_1 = z = 0, \tag{2.281}$$

or in terms of the signature in (2.17):

$$d = 2 + n, \quad \chi_4 = \{n;\, 0, 0, 0;\, n\}, \quad n = 2j \in \mathbb{N}. \tag{2.282}$$

They are also possible in case **bd**, cf. (2.280), when $i_0 = i'_0 = 1$, i. e., $r_1 = r_3 = 0$, $r_2 \neq 0$:

$$\begin{aligned} &d = r_2 = r \geq 1, \quad r_1 = r_3 = j_1 = j_2 = z = 0, \\ &\text{or:} \quad d = r \in \mathbb{N}, \quad \chi_4 = \{0;\, 0, r, 0;\, 0\}. \end{aligned} \tag{2.283}$$

2.4.3.2 $\frac{1}{4}$-BPS states, $\kappa = 4$

These happen in most cases with appropriate conditions:

Case **a**, cf. (2.266),

$$\begin{aligned} &d = 2 + 2j_2 \geq 3, \quad m_1 = z = 0,\ j_2 \geq \frac{1}{2}, \\ &\text{or:} \quad d = 2 + k, \quad \chi_4 = \{n;\, 0, 0, 0;\, k\}, \quad n = 2j_1 \in \mathbb{Z}_+,\ k = 2j_2 \in \mathbb{N}. \end{aligned} \tag{2.284}$$

Case **c**, cf. (2.271),

$$d = 2 + 2j_1 \geq 3, \quad m_1 = z = 0,\ j_1 \geq \frac{1}{2}, \tag{2.285}$$
$$\text{or:} \quad d = 2 + n + k, \quad \chi_4 = \{n + k;\ 0, 0, 0;\ n\}, \quad n = 2j_2 \in \mathbb{Z}_+,\ k \in \mathbb{N}.$$

In case **ac** we deal with R-symmetry non-scalars with only one non-zero r_i entry, since we have $i_0 + i_0' = 2$, thus, we take $r_{1+i_0} \neq 0$, $0 \leq i_0 \leq 2$, with the condition

$$j_1 - j_2 = m_1 - \frac{1}{2}m = \frac{1}{2}r_{1+i_0}(1 - i_0), \tag{2.286}$$

and then we have

$$d = 2 + j_1 + j_2 + r_{1+i_0} = 2 + 2j + r_{i_0}, \tag{2.287}$$

where case-wise:

$$\begin{aligned} &j = j_2, \quad j_1 = j + \frac{1}{2}r_1, \quad \chi_4 = \{2j + r_1;\ r_1, 0, 0;\ 2j\}, \\ &j = j_1 = j_2, \quad \chi_4 = \{2j;\ 0, r_2, 0;\ 2j\}, \\ &j = j_1, \quad j_2 = j + \frac{1}{2}r_3, \quad \chi_4 = \{2j;\ 0, 0, r_3;\ 2j + r_3\}. \end{aligned} \tag{2.288}$$

Case **ad**, cf. (2.278),

$$d = 3 + 3j_2 + r_2 \geq \frac{9}{2}, \quad r_3 = 2 + 2j_2 \geq 3,\ j_2 \geq \frac{1}{2},\ r_1 = j_1 = z = 0,$$
$$\text{or:} \quad d = 3 + \frac{3}{2}n + r, \quad \chi_4 = \{0;\ 0, r, 2 + n;\ n\}, \quad r, n \in \mathbb{N}. \tag{2.289}$$

Case **bc**, cf. (2.279),

$$d = 3 + 3j_1 + r_2 \geq \frac{9}{2}, \quad r_1 = 2 + 2j_1 \geq 3,\ j_1 \geq \frac{1}{2},\ r_3 = j_2 = z = 0,$$
$$\text{or:} \quad d = 3 + \frac{3}{2}n + r, \quad \chi_4 = \{n;\ 2 + n, r, 0;\ 0\}, \quad r, n \in \mathbb{N}. \tag{2.290}$$

Case **bd**, cf. (2.280), when $i_0 = i_0' = 0$, i. e., $r_1 = r_3 = n \neq 0$:

$$d = r + 2n \geq 2, \quad n \geq 1,\ r = r_2 \geq 0,\ j_1 = j_2 = z = 0, \tag{2.291}$$
$$\text{or:} \quad d = r + 2n, \quad \chi_4 = \{0;\ n, r, n;\ 0\}, \quad n \in \mathbb{N},\ r \in \mathbb{Z}_+.$$

2.4.3.3 $\frac{1}{8}$-BPS states, $\kappa = 2$

Case **a**, cf. (2.266) with $j_2 > 0$, $i_0 = 1$,

$$d = 2 + 2j_2 + r_2 + \frac{1}{2}r_3, \quad 2j_2 > 2j_1 + r_3,\ r_2 > 0,\ r_1 = z = 0,$$

$$\text{or:} \quad d = 2 + k + 2n + r_2 + \frac{3}{2}r_3, \quad \chi_4 = \{n;\, 0, r_2, r_3;\, n + r_3 + k\},$$
$$k, r_2 \in \mathbb{N}, \; n, r_3 \in \mathbb{Z}_+. \tag{2.292}$$

Case **b**, cf. (2.269),

$$d = \frac{1}{2}(3r_1 + 2r_2 + r_3) \equiv \frac{1}{2}m^*, \quad r_1 > 2 + 2j_1 + r_3,\, j_2 = z = 0,$$
$$\text{or:} \quad d = 3 + 2r_3 + r_2 + \frac{3}{2}(n + k), \quad \chi_4 = \{n;\, 2 + n + r_3 + k, r_2, r_3;\, 0\},$$
$$k \in \mathbb{N}, \; n, r_2, r_3 \in \mathbb{Z}_+. \tag{2.293}$$

Case **c**, cf. (2.271) with $j_1 > 0$, $i'_0 = 1$,

$$d = 2 + 2j_1 + r_2 + \frac{1}{2}r_1, \quad 2j_1 > 2j_2 + r_1,\, r_2 > 0,\, r_3 = z = 0,$$
$$\text{or:} \quad d = 2 + k + 2n + r_2 + \frac{3}{2}r_1, \quad \chi_4 = \{n + r_1 + k;\, r_1, r_2, 0;\, n\},$$
$$k, r_2 \in \mathbb{N}, \; n, r_1 \in \mathbb{Z}_+. \tag{2.294}$$

Case **d**, cf. (2.274),

$$d = \frac{1}{2}m = \frac{1}{2}(r_1 + 2r_2 + 3r_3), \quad r_3 > 2 + 2j_2 + r_1,\, j_1 = z = 0,$$
$$\text{or:} \quad d = 3 + 2r_1 + r_2 + \frac{3}{2}(n + k), \quad \chi_4 = \{0;\, r_1, r_2, 2 + n + r_1 + k;\, n\},$$
$$k \in \mathbb{N}, \; n, r_2, r_1 \in \mathbb{Z}_+. \tag{2.295}$$

Case **ac**, cf. (2.276) and (2.277),

$$d = 2 + m_1 \geq 2, \quad j_1 = j_2 = z = 0, \tag{2.296}$$
$$\text{or:} \quad d = 2 + r_1 + r_2 + r_3, \quad \chi_4 = \{0;\, r_1, r_2, r_3;\, 0\}.$$

Some of these BPS cases are extensively studied in the literature, mostly those listed here as cases **ac, bd**, cf. [9, 172, 173, 176, 174, 159, 338, 106, 22, 104, 149].

Finally, we remark that some of the above states would violate the protectedness conditions that we gave in Section 2.2.5. These would be the $\frac{1}{4}$-BPS cases listed as cases **ad, bc**, and in case **bd** for $n > 2$, while for the $\frac{1}{8}$-BPS cases that would be the cases **b, d**, and in case **ac** for $r_1 r_3 \neq 0$.

2.4.4 SU(2, 2/N), $N \leq 3$

We can set $z = 0$ also for $N \neq 4$ though this does not have the same group-theoretical meaning as for $N = 4$. In this section we treat separately the cases $N = 1, 2, 3$, which are more peculiar.

2.4.4.1 SU(2, 2/1)

For $N = 1$ setting $z = 0$ is possible only for three cases **a, c, ac**:

• **a** $d = 2 + 2j_2, j_2 > j_1 \geq 0$,
$\kappa = 1$, $\frac{1}{4}$-BPS;
• **c** $d = 2 + 2j_1, j_1 > j_2 \geq 0$,
$\kappa = 1$, $\frac{1}{4}$-BPS;
• **ac** $d = 2 + 2j, j_1 = j_2 = j$,
$\kappa = 2$, $\frac{1}{2}$-BPS.

Note that according to the result of Section 2.2.5 the first two cases would not be protected.

2.4.4.2 SU(2, 2/2)

For $N = 2$ we have $i_0 = i'_0 = 0, 1$. Setting $z = 0$ is possible for the four cases **a, c, ac, bd** when we have

• **a** $d = 2 + 2j_2 + r_1, j_2 > j_1 \geq 0$,
$\kappa = 1 + i_0 \leq 2$;
• **c** $d = 2 + 2j_1 + r_1, j_1 > j_2 \geq 0$,
$\kappa = 1 + i'_0 \leq 2$;
• **ac** $d = 2 + 2j + r_1, j_1 = j_2 = j$,
$\kappa = 2 + 2\delta_{i_0 j,0} \leq 4$;
• **bd** $d = r_1 \neq 0, j_1 = j_2 = 0$ (here $z = 0$ holds in all cases),
$\kappa = 4$, $\frac{1}{2}$-BPS.

Note that according to the result of Section 2.2.5 the first three cases would not be protected when $r_1 \neq 0$, i. e., when $i_0 = i'_0 = 0$. In contradistinction, when $r_1 = 0$, i. e., $i_0 = i'_0 = 1$, the first two are $\frac{1}{4}$-BPS, and the third, when $j > 0$, a $\frac{1}{2}$-BPS. The fourth case would not be protected if $r_1 > 4$.

2.4.4.3 SU(2, 2/3)

In fact, the case $N = 3$ is similar in these considerations to $N = 4$ (though some results differ), so we present it telegraphically.

$$\bullet\, \mathbf{a} \quad d = d^1_{31} = 2 + 2j_2 + 2m_1 - 2m/3 > d^3_{33} \quad \Rightarrow \quad j_2 + \frac{1}{3}r_1 > j_1 + \frac{1}{3}r_2. \tag{2.297}$$

For $m_1 = 0 \Rightarrow j_2 > j_1 \Rightarrow j_2 > 0 \Rightarrow$

$$\Rightarrow \quad \kappa = 3, \quad m_1 = 0, j_2 > 0. \tag{2.298}$$

For $m_1 \neq 0, \Rightarrow i_0 + i'_0 \leq 1 \Rightarrow$

$$\Rightarrow \kappa = 1 + i_0(1 - \delta_{j_2,0}) \leq 2. \tag{2.299}$$

$$\bullet\, \mathbf{b} \quad d = d^2_{31} = 2m_1 - 2m/3 > d^3_{33}, \quad j_2 = 0 \quad \Rightarrow \tag{2.300}$$

$$\Rightarrow \quad r_1 > 3 + 3j_1 + r_2 \quad \Rightarrow \quad r_1 > 3 \quad \Rightarrow \quad m_1 \neq 0 \text{ and } i_0 = 0 \quad \Rightarrow \quad \kappa = 2.$$

• **c** $d = d^3_{33} = 2 + 2j_1 + 2m/3 > d^1_{31} \quad \Rightarrow \quad j_1 + \frac{1}{3}r_2 > j_2 + \frac{1}{3}r_1,$

$$m_1 = 0 \quad \Rightarrow \quad j_1 > j_2 \quad \Rightarrow \quad j_1 > 0 \quad \Rightarrow \quad \kappa = 3, \quad m_1 = 0,\ j_1 > 0,$$
$$m_1 \neq 0 \quad \Rightarrow \quad i_0 + i'_0 \leq 1 \quad \Rightarrow \quad \kappa = 1 + i'_0(1 - \delta_{j_1,0}) \leq 2. \tag{2.301}$$

• **d** $d = d^4_{33} = 2m/3 > d^1_{31}, \quad j_1 = 0, \quad \Rightarrow \quad r_2 > 3 + 3j_2 + r_1 \quad \Rightarrow$

$$\Rightarrow \quad r_2 > 3 \quad \Rightarrow \quad m_1 \neq 0, \text{ and } i'_0 = 0 \quad \Rightarrow \quad \kappa = 2. \tag{2.302}$$

• **ac** $d = d^1_{31} = d^3_{33} = 2 + j_1 + j_2 + m_1$. From $z = 0$ follows

$$j_2 + \frac{1}{3}r_1 = j_1 + \frac{1}{3}r_2. \tag{2.303}$$

In the case of R-symmetry scalars, i. e., $m_1 = 0$, it follows that $j_2 = j_1 = j$, and then we have

$$\kappa = 6 - 4\delta_{j,0}. \tag{2.304}$$

Thus, for $j \neq 0$ we have $\frac{1}{2}$-BPS state:

$$\kappa = 6, \quad d = 2 + 2j \geq 3, \quad \chi_3 = \{2j;\ 0, 0;\ 2j\}. \tag{2.305}$$

In the case of R-symmetry non-scalars, i. e., $m_1 \neq 0$, $i_0 + i'_0 \leq 1$, and thus

$$\kappa = 2 + i_0(1 - \delta_{j_2,0}) + i'_0(1 - \delta_{j_1,0}) \leq 3. \tag{2.306}$$

Thus, when $i_0 j_2 \neq 0$ or $i'_0 j_1 \neq 0$ we have $\frac{1}{4}$-BPS state since $\kappa = 3$.

• **ad** From $z = 0$ follows $r_2 = 3 + 3j_2 + r_1 \Rightarrow r_2 \geq 3 \Rightarrow m_1 \neq 0$, and $i'_0 = 0$, $i_0 \leq 1 \Rightarrow$

$$\begin{aligned} &\kappa = 3 + i_0(1 - \delta_{j_2,0}) \leq 4, \\ &d = d^{ad} = 1 + j_2 + m_1 = 4 + 4j_2 + 2r_1, \\ &\chi_3 = \{0;\ r_1, 3 + 3j_2 + r_1;\ 2j_2\}. \end{aligned} \tag{2.307}$$

Thus, when $i_0 j_2 = 0$ we have $\frac{1}{4}$-BPS state since $\kappa = 3$.

• **bc** From $z = 0$ follows $r_1 = 3 + 3j_1 + r_2 \Rightarrow r_1 \geq 3 \Rightarrow m_1 \neq 0$, and $i_0 = 0$, $i'_0 \leq 1 \Rightarrow$

$$\begin{aligned} &\kappa = 3 + i'_0(1 - \delta_{j_1,0}) \leq 4, \\ &d = d^{bc} = 1 + j_1 + m_1 = 4 + 4j_1 + 2r_2, \\ &\chi_3 = \{2j_1;\ 3 + 3j_1 + r_2, r_2;\ 0\}. \end{aligned} \tag{2.308}$$

Thus, when $i'_0 j_1 = 0$ we have $\frac{1}{4}$-BPS state since $\kappa = 3$.

• **bd** From $z = 0$ follows $r_1 = r_2 = \frac{1}{2}m_1 = r \in \mathbb{N}$, thus, $i_0 = i'_0 = 0$ and then we have

$$\kappa = 4, \quad d = d^{bd} = 2r \neq 0, \quad \chi_3 = \{0;\ r, r;\ 0\}. \tag{2.309}$$

Note that according to the result of Section 2.2.5 the following cases would not be protected: cases **a, ad** when $r_1 \neq 0$; cases **c, bc** when $r_2 \neq 0$; cases **b, d**; case **ac** when $r_1 r_2 \neq 0$ (i. e., $i_0 = i'_0 = 0$); case **bd** when $r > 2$.

2.4.5 SU(2, 2/N), $N \geq 5$

The cases $N \geq 5$ are described adequately by the general exposition in the beginning of this section, though some cases are excluded for $z = 0$. Thus, we shall give only the special cases.

2.4.5.1 $\frac{1}{2}$-BPS states, $\kappa = 2N$

These are possible only in cases **ac, bd**, and they appear as for $N = 4$.

In case **ac** we deal with R-symmetry scalars and $j_1 = j_2 = j \geq \frac{1}{2}$:

$$d = 2 + n, \quad n = 2j \in \mathbb{N}, \quad \chi_N = \{n;\, 0, \dots, 0;\, n\}. \tag{2.310}$$

In case **bd** this is possible when N is *even*, and there is only one non-zero r_i, namely, the middle one, i. e.,

$$d = m_1 = r_{\frac{1}{2}N} \neq 0, \quad \chi_N = \{0;\, 0, \dots, 0, r_{\frac{1}{2}N}, 0, \dots, 0;\, 0\}. \tag{2.311}$$

Note that according to the result of Section 2.2.5 these $\frac{1}{2}$-BPS cases would be protected.

2.4.5.2 $\frac{1}{4}$-BPS states, $\kappa = N$

In case **a** we deal with R-symmetry scalars and $j_2 > j_1$,

$$d = 2 + k, \quad n = 2j_1 \in \mathbb{Z}_+, \quad k = 2j_2 \in \mathbb{N}, \quad \chi_N = \{n;\, 0, \dots, 0;\, k\}. \tag{2.312}$$

In the conjugate case **c**, $j_1 > j_2$,

$$d = 2 + k, \quad n = 2j_2 \in \mathbb{Z}_+, \quad k = 2j_1 \in \mathbb{N}, \quad \chi_N = \{k;\, 0, \dots, 0;\, n\}. \tag{2.313}$$

In case **b** we would deal with R-symmetry non-scalars for N-*even*, and we must have $i_0 = \frac{N}{2} - 1$; this means that $r_i = 0$ for $i = 1, \dots, \frac{N}{2} - 1$. On the other hand we must satisfy the condition

$$1 + j_1 < m_1 - 2m/N = \sum_{k=1}^{\frac{N}{2}-1} (r_k - r_{N-k})(1 - 2k/N) = - \sum_{k=1}^{\frac{N}{2}-1} r_{N-k}(1 - 2k/N) \leq 0,$$

which is not possible.

For the same reason the conjugate case **d** is not possible.

In case **ac** we deal with R-symmetry non-scalars with only one non-zero r_i entry, since we have $i_0 + i_0' = N - 2$, thus, we take $r_{1+i_0} \neq 0$, $0 \leq i_0 \leq N - 2$, with the condition

$$j_1 - j_2 = m_1 - 2m/N = r_{1+i_0}\left(1 - \frac{2}{N}(1 + i_0)\right), \tag{2.314}$$

and then we have

$$d = 2 + j_1 + j_2 + r_{1+i_0}, \quad \chi_N = \{2j_1; 0, \ldots, 0, r_{i_0}, 0, \ldots, 0; 2j_2\}. \tag{2.315}$$

Note that

$$\begin{aligned} i_0 < \frac{N}{2} - 1 \quad &\Rightarrow \quad j_1 - j_2 > 0, \\ i_0 > \frac{N}{2} - 1 \quad &\Rightarrow \quad j_1 - j_2 < 0, \\ i_0 = \frac{N}{2} - 1 \quad &\Rightarrow \quad j_1 - j_2 = 0, \quad \text{only for } N\text{-even}. \end{aligned}$$

In case **ad** we deal with R-symmetry non-scalars with two subcases depending whether j_2 is zero or not.

- When $j_2 = 0$ we have $\kappa = N = 3 + 2i'_0$, i. e., N is odd, and $i'_0 = \frac{1}{2}(N-3)$ ($i_0 \leq \frac{1}{2}(N-1)$). On the other hand because of $z = 0$ we must satisfy the condition (rescaling by N)

$$N = 2m - Nm_1 = \sum_{k=1}^{\frac{1}{2}(N-1)} (r_{N-k} - r_k)(N-2k) = r_{\frac{1}{2}(N+1)} - \sum_{k=1}^{\frac{1}{2}(N-1)} r_k(N-2k).$$

Thus, we have

$$d = 1 + m_1 = 1 + \sum_{k=1}^{\frac{1}{2}(N+1)} r_k, \quad \chi_N = \{0; r_1, \ldots, r_{\frac{1}{2}(N+1)}, 0, \ldots, 0; 0\} \tag{2.316}$$

- When $j_2 \neq 0$ we have $\kappa = N = 3 + 2i'_0 + i_0$, from which it follows that $i_0 + i'_0 = N-2$ is not possible, thus $i_0 + i'_0 \leq N-3$, and $i'_0 \leq \frac{1}{2}(N-3)$. Also the following condition must hold:

$$1 + j_2 = \frac{2}{N}m - m_1.$$

Thus, we have

$$d = 1 + j_2 + m_1 = \frac{2}{N}m, \quad j_1 = 0. \tag{2.317}$$

In the conjugate case **bc** we write briefly

- $j_1 = 0 \Rightarrow \kappa = N = 3 + 2i_0 \Rightarrow N$ is odd, $\Rightarrow i_0 = \frac{1}{2}(N-3)$ ($i'_0 \leq \frac{1}{2}(N-1)$). On the other hand we must have

$$N = Nm_1 - 2m = \sum_{k=1}^{\frac{1}{2}(N-1)} (r_k - r_{N-k})(N-2k) = r_{\frac{1}{2}(N-1)} - \sum_{k=1}^{\frac{1}{2}(N-1)} r_{N-k}(N-2k)$$

Thus, we have

$$
\begin{aligned}
d = 1 + m_1 = 1 + \sum_{k=\frac{1}{2}(N-1)}^{N-1} r_k = 1 + N + \sum_{k=\frac{1}{2}(N+1)}^{N-1} r_k(1 + 2k - N),\\
\chi_N = \{0; 0, \ldots, 0, r_{\frac{1}{2}(N-1)}, r_{\frac{1}{2}(N+1)}, \ldots, r_{N-1}; 0\}.
\end{aligned}
\tag{2.318}
$$

- $j_1 \neq 0 \Rightarrow \kappa = N = 3 + 2i_0 + i'_0 \Rightarrow i_0 + i'_0 = N - 2$ is not possible, thus $i_0 + i'_0 \leq N - 3$, also $i_0 \leq \frac{1}{2}(N - 3)$. Also the following condition must hold:

$$
1 + j_1 = m_1 - \frac{2}{N} m.
$$

Thus, we have

$$
d = 1 + j_1 + m_1 = 2m_1 - \frac{2}{N} m = \frac{2}{N} m^*, \quad j_2 = 0. \tag{2.319}
$$

Case **bd** is possible only for N-even with R-symmetry non-scalars, and from $\kappa = N$ and $z = 0$ it follows that

$$
i_0 + i'_0 = \frac{N}{2} - 2, \quad m_1 = \frac{N}{2} m \neq 0.
$$

Thus, we have

$$
d = m_1 = \sum_{k=1+i_0}^{1+i_0+\frac{N}{2}} r_k, \quad j_1 = j_2 = 0. \tag{2.320}
$$

Note that according to the result of Section 2.2.5 the following $\frac{1}{4}$-BPS cases would not be protected: case **ad** when $r_1 \neq 0$; case **bc** when $r_{N-1} \neq 0$; case **bd** when $r_1, r_{N-1} > 2$.

2.4.5.3 $\frac{1}{8}$-BPS states, $\kappa = N/2$, N-even

In all possible cases we deal with R-symmetry non-scalars.

In case **a** to achieve $\kappa = \frac{N}{2}$, we need $j_2 \neq 0$, and $i_0 = \frac{N}{2} - 1$. We also have the defining restriction (with $z = 0$): $j_2 > j_1 + (m - m^*)/N$. Combining all this, we have

$$
\begin{aligned}
d = 2 + 2j_2 + \frac{2}{N} m^*,\\
\chi_N = \{2j_1; 0, \ldots, 0, r_{\frac{N}{2}}, \ldots, r_{N-1}; 2j_2\},\\
j_2 > j_1 + \sum_{k=\frac{N}{2}+1}^{N-1} \left(\frac{2}{N} k - 1\right) r_k.
\end{aligned}
\tag{2.321}
$$

In case **b** to achieve $\kappa = \frac{N}{2}$, we need $i_0 = \frac{N}{4} - 1$. Thus, this case is possible only if N is divisible by 4. We also have the defining restriction (with $z = 0$): $(m^* - m)/N > j_1 + 1$. Combining all this, we have

$$d = \frac{2}{N}m^*, \quad \chi_N = \{2j_1;\, 0, \dots, 0, r_{\frac{N}{4}}, \dots, r_{N-1};\, 0\},$$
$$\sum_{k=\frac{N}{4}}^{\frac{N}{2}-1} \left(1 - \frac{2}{N}k\right) r_k > 1 + j_1 + \sum_{k=\frac{N}{2}+1}^{N-1} \left(\frac{2}{N}k - 1\right) r_k. \tag{2.322}$$

The conjugated cases **c, d** are presented briefly:
In case **c:** $j_1 \neq 0$, $i'_0 = \frac{N}{2} - 1$, $j_1 > j_2 + (m^* - m)/N \Rightarrow$

$$d = 2 + 2j_1 + \frac{2}{N}m,$$
$$\chi_N = \{2j_1;\, r_1, \dots, r_{\frac{N}{2}}, 0, \dots, 0;\, 2j_2\},$$
$$j_1 > j_2 + \sum_{k=1}^{\frac{N}{2}-1} \left(1 - \frac{2}{N}k\right) r_k. \tag{2.323}$$

In case **d:** $i'_0 = \frac{N}{4} - 1$, N is divisible by 4, $(m - m^*)/N > j_2 + 1$. Combining all this, we have

$$d = \frac{2}{N}m, \quad \chi_N = \{0;\, r_1, \dots, 0, r_{\frac{3N}{4}}, 0, \dots, 0;\, 2j_2\},$$
$$\sum_{k=\frac{N}{2}+1}^{\frac{3N}{4}} \left(\frac{2}{N}k - 1\right) r_k > 1 + j_2 + \sum_{k=1}^{\frac{2}{N}-1} \left(1 - \frac{2}{N}k\right) r_k. \tag{2.324}$$

The case **ac** has several subcases depending on j_1, j_2 being zero or not:

- The subcase $j_1 = j_2 = 0$ is possible only for $N = 4$ considered above.
- In the subcase $j_1 = 0, j_2 \neq 0$ should hold $i_0 = \frac{N}{2} - 2$, $i'_0 \leq \frac{N}{2} - 2$. Altogether we

have

$$d = 2 + j_2 + m_1,$$
$$\chi_N = \{0;\, 0, \dots, 0, r_{\frac{N}{2}-1}, \dots, r_{N-1};\, 2j_2\},$$
$$j_2 + \frac{2}{N} r_{\frac{N}{2}-1} = \sum_{k=\frac{N}{2}+1}^{N-1} r_k \left(\frac{2}{N}k - 1\right). \tag{2.325}$$

- In the conjugate subcase $j_1 \neq 0, j_2 = 0$ should hold $i_0 \leq \frac{N}{2} - 2$, $i'_0 = \frac{N}{2} - 2 \Rightarrow$

$$d = 2 + j_1 + m_1,$$
$$\chi_N = \{2j_1;\, r_1, \dots, r_{\frac{N}{2}+1}, 0, \dots, 0;\, 0\}, \tag{2.326}$$

$$j_1 + \frac{2}{N} r_{\frac{N}{2}+1} = \sum_{k=1}^{\frac{N}{2}-1} r_k \left(1 - \frac{2}{N}k\right).$$

- In the subcase $j_1 j_2 \neq 0$ we should have $i_0 + i_0' = \frac{N}{2} - 2$, χ_N is in a general position, and we have

$$d = 2 + j_1 + j_2 + m_1, \qquad (2.327)$$
$$j_2 - j_1 = (m - m^*)/N.$$

In case **ad** we need $\kappa = \frac{N}{2} = 3 + i_0(1 - \delta_{j_2,0}) + 2i_0'$, while from the condition $z = 0$ follows

$$1 + j_2 = (m - m^*)/N = \sum_{k=1+i_0}^{N-1-i_0'} r_k \left(\frac{2}{N}k - 1\right), \qquad (2.328)$$

and then we have

$$d = \frac{2}{N} m = \frac{2}{N} \sum_{k=1+i_0}^{N-1-i_0'} k r_k. \qquad (2.329)$$

The subcase $j_2 = 0$ leads to the restriction that $N = 6, 10, \ldots$, and $i_0' = \frac{1}{2}(\frac{N}{2} - 3)$, and then

$$j_2 = 0 \quad \Rightarrow \quad 1 = \sum_{k=1+i_0}^{\frac{1}{4}(3N+2)} r_k \left(\frac{2}{N}k - 1\right), \quad d = \frac{2}{N} \sum_{k=1+i_0}^{\frac{1}{4}(3N+2)} k r_k. \qquad (2.330)$$

In case **bc** we need $\kappa = \frac{N}{2} = 3 + i_0'(1 - \delta_{j_1,0}) + 2i_0$, while from the condition $z = 0$ follows

$$1 + j_1 = (m^* - m)/N = \sum_{k=1+i_0}^{N-1-i_0'} r_k \left(1 - \frac{2}{N}k\right), \qquad (2.331)$$

and then we have

$$d = \frac{2}{N} m^* = \frac{2}{N} \sum_{k=1+i_0}^{N-1-i_0'} (N - k) r_k. \qquad (2.332)$$

The subcase $j_1 = 0$ leads to the restriction that $N = 6, 10, \ldots$, and $i_0 = \frac{1}{2}(\frac{N}{2} - 3)$, and then

$$j_1 = 0 \quad \Rightarrow \quad 1 = \sum_{k=\frac{1}{4}(3N+2)}^{N-1-i_0'} r_k \left(1 - \frac{2}{N}k\right), \quad d = \frac{2}{N} \sum_{k=\frac{1}{4}(3N+2)}^{N-1-i_0'} (N - k) r_k. \qquad (2.333)$$

In case **bd** we need $\kappa = \frac{N}{2} = 4 + 2i_0 + 2i'_0$, thus $i_0 + i'_0 = \frac{N}{4} - 2$, thus $N = 8, 12, \ldots$. From $z = 0$ it follows that $m = m^* = \frac{N}{2}m_1$, and then

$$d = m_1 = \sum_{k=1+i_0}^{1+i_0+\frac{3N}{4}} r_k, \quad j_1 = j_2 = 0. \tag{2.334}$$

Note that according to the result of Section 2.2.5 the following $\frac{1}{8}$-BPS cases would not be protected: case **ad** when $r_1 \neq 0$; case **bc** when $r_{N-1} \neq 0$; case **bd** when $r_1, r_{N-1} > 2$.

In the previous sections we gave the complete classification of the BPS states, but not all interesting cases were given in detail. In the next subsection motivated by the paper [63] we spell out the interesting case of $\frac{1}{N}$-BPS states, i. e., the cases when $\kappa = 4$.

2.4.6 $\frac{1}{N}$-BPS states

It is convenient to consider the case of general N while treating separately R-symmetry scalars and R-symmetry non-scalars.

2.4.6.1 R-symmetry scalars

We start with the simpler cases of R-symmetry scalars when $r_i = 0$ for all i, which means also that $m_1 = m = m^* = 0$.

These cases are valid also for $N = 1$; however, for $N = 1$ in all cases we have $\kappa < 4$; see above.

In fact only three cases are relevant for $\kappa = 4$.

• **a** $d = (d^1_{N1})_{|m=0=z} = 2 + 2j_2 > 2 + 2j_1 = (d^3_{NN})_{|m=0=z}$. The last inequality leads to the restriction $j_2 > j_1$, i. e., $j_2 > 0$, and then we have

$$\kappa = N, \quad m_1 = m = 0, \quad j_2 > 0. \tag{2.335}$$

These semi-short UIRs may be called semi-chiral since they lack half of the anti-chiral generators: $X^+_{3,4+k}$, $k = 1, \ldots, N$.

• **c** $d = (d^3_{NN})_{|m=0=z} = 2 + 2j_1 > (d^1_{N1})_{|m=0=z} \Rightarrow$

$$\kappa = N, \quad m_1 = m = 0, \quad j_1 > 0. \tag{2.336}$$

These semi-short UIRs may be called semi-anti-chiral since they lack half of the chiral generators: $X^+_{1,4+k}$, $k = 1, \ldots, N$.

Thus, in both cases above the interesting case $\kappa = 4$ occurs only for $N = 4$, as $\frac{1}{4}$-BPS.

$$\text{•}\mathbf{ac} \quad d = d^{ac}_{|m=0} = 2 + j_1 + j_2, \quad z = j_1 - j_2,$$

$$\kappa = 2N, \quad \text{if } j_1, j_2 > 0,$$
$$\kappa = N+1, \quad \text{if } j_1 > 0,\ j_2 = 0,$$
$$\kappa = N+1, \quad \text{if } j_1 = 0,\ j_2 > 0,$$
$$\kappa = 2, \quad \text{if } j_1 = j_2 = 0.$$

Here, κ is the number of mixed eliminations: chiral generators $X^+_{1,4+k}$, $(k = 1, \ldots, N + (1-N)\delta_{j_1,0})$, and anti-chiral generators $X^+_{3,5+N-k}$ $(k = 1, \ldots, N + (1-N)\delta_{j_2,0})$. Thus, in the cases when $\kappa = 2N$ the semi-short UIRs may be called semi-chiral–anti-chiral since they lack half of the chiral and half of the anti-chiral generators.

The interesting case $\kappa = 4$ occurs only for $N = 2$, as $\frac{1}{2}$-BPS.

2.4.6.2 R-symmetry non-scalars

Below we need some additional notation. Let $N > 1$ and let i_0 be an integer such that $0 \le i_0 \le N-1$, $r_i = 0$ for $i \le i_0$, and if $i_0 < N-1$ then $r_{i_0+1} > 0$. Let now i'_0 be an integer such that $0 \le i'_0 \le N-1$, $r_{N-i} = 0$ for $i \le i'_0$, and if $i'_0 < N-1$ then $r_{N-1-i'_0} > 0$.

The interesting cases of $\frac{1}{N}$-BPS states, i. e., when $\kappa = 4$, are given in the following list:

$$\bullet\mathbf{a} \quad d = d^a = 2 + 2j_2 + 2m^*/N, \quad N \ge 5, \qquad (2.337)$$
$$j_1 \text{ arbitrary},\ j_2 > 0,\ i_0 = 3,\ 0 \le i'_0 \le N-5,$$
$$j_2 > j_1 + \sum_{k=4}^{N-1} (2k/N - 1) r_k.$$

Here are eliminated four anti-chiral generators $X^+_{3,4+k}$, $k \le 4$.

$$\bullet\mathbf{b} \quad d = d^b = 2m^*/N, \quad N \ge 5, \qquad (2.338)$$
$$j_2 = 0,\ j_1 \text{ arbitrary},\ i_0 = 1,\ 0 \le i'_0 \le N-3,$$
$$\sum_{k=2}^{[(N-1)/2]} (1 - 2k/N) r_k > j_1 + \sum_{[(N+1)/2]}^{N-1} (2k/N - 1) r_k.$$

Here are eliminated four anti-chiral generators $X^+_{3,5+N-k}$, $X^+_{4,5+N-k}$, $k \le 2$.

$$\bullet\mathbf{c} \quad d = d^c = 2 + 2j_1 + 2m/N, \quad N \ge 5, \qquad (2.339)$$
$$j_1 > 0,\ j_2 \text{ arbitrary},\ i'_0 = 3,\ 0 \le i_0 \le N-5,$$
$$j_1 > j_2 + \sum_{k=1}^{N-4} (1 - 2k/N) r_k.$$

Here the four chiral generators $X^+_{1,4+k}$, $k \le 4$, are eliminated.

$$\bullet\mathbf{d} \quad d = d^d = 2m/N, \quad N \ge 5, \qquad (2.340)$$

$$j_1 = 0,\ j_2 \text{ arbitrary},\ i'_0 = 1,\ 0 \le i_0 \le N-3,$$
$$\sum_{k=1}^{[(N-1)/2]} (1 - 2k/N)r_k > j_2 + \sum_{[(N+1)/2]}^{N-4} (2k/N - 1)r_k.$$

Here are eliminated four chiral generators $X^+_{1,4+k}$, $X^+_{2,4+k}$, $k \le 2$.

$$\bullet\mathbf{ac}\quad d = d^{ac} = 2 + j_1 + j_2 + m_1, \quad N \ge 4,$$
$$j_1 + m/N = j_2 + m^*/N,$$
$$j_1 j_2 > 0, \quad i_0 + i'_0 = 2, \tag{2.341a}$$
$$j_1 > 0, \quad j_2 = 0, \quad i_0 = 0, i'_0 = 2, \tag{2.341b}$$
$$j_1 = 0, \quad j_2 > 0, \quad i_0 = 2, \quad i'_0 = 0. \tag{2.341c}$$

Here are eliminated four generators: chiral generators $X^+_{1,4+k}$, $k \le 1 + i'_0(1 - \delta_{j_1,0})$, and anti-chiral generators $X^+_{3,5+N-k}$, $k \le 1 + i_0(1 - \delta_{j_2,0})$.

$$\bullet\mathbf{ad}\quad d = d^{ad} = 1 + j_2 + m_1 = 2m/N, \quad N \ge 3, \tag{2.342}$$
$$j_1 = 0, \quad j_2 > 0, \quad i_0 = 1, \quad i'_0 = 0.$$

Here the two chiral generators $X^+_{1,5}$, $X^+_{2,5}$ are eliminated and the two anti-chiral generators $X^+_{3,5+N-k}$, $k = 1, 2$.

$$\bullet\mathbf{bc}\quad d = d^{bc} = 1 + j_1 + m_1 = 2m^*/N, \quad N \ge 3, \tag{2.343}$$
$$j_2 = 0, \quad j_1 > 0, \quad i_0 = 0, \quad i'_0 = 1.$$

Here the two chiral generators $X^+_{1,4+k}$, $k = 1, 2$, are eliminated, and the two anti-chiral generators $X^+_{3,8}$, $X^+_{4,8}$.

$$\bullet\mathbf{bd}\quad d = d^{bd} = m_1, \quad N \ge 2, \tag{2.344}$$
$$j_1 = j_2 = 0, \quad i_0 = i'_0 = 0.$$

Here are eliminated two chiral generators $X^+_{1,5}$, $X^+_{2,5}$, and two anti-chiral generators $X^+_{3,8}$, $X^+_{4,8}$.

Note that according to the results of Section 2.2.5 the following cases would not be protected: **ad** for $r_{N-1} > 2$, **bc** for $r_1 > 2$, **bd** for $r_1, r_{N-1} > 2$ when $N > 2$, and for $r_1 > 4$ when $N = 2$.

3 Examples of conformal supersymmetry for $D > 4$

Summary

In this chapter we study the representation theory of some conformal superalgebras suitable for $D > 4$. First we present the classification of the positive energy UIRs of $D = 6$ conformal supersymmetry following [122]. Then we present the classification of the positive energy UIRs of the superalgebras $\mathrm{osp}(1|2n, \mathbb{R})$ suitable for conformal supersymmetry in $D = 9, 10, 11$ following [148, 135, 143–145].

3.1 Positive energy UIRs of $D = 6$ conformal supersymmetry

The applications of $D = 6$ unitary irreps require firmer theoretical basis. Among the many interesting applications we shall mention the analysis of OPEs and 1/2 BPS operators [174, 157, 172, 173, 176]. In particular, it is important that some general properties of abstract superconformal field theories can be obtained by using the BPS nature of a certain class of superconformal primary operators and the model independent nature of superconformal OPEs. In the classification of UIRs of superconformal algebras an important role is played by the representations with "quantized" conformal dimension, since in the quantum field theory framework they correspond to operators with "protected" scaling dimension and therefore imply "non-renormalization theorems" at the quantum level.

Motivated by the above we decided to reexamine the list of UIRs of the $D = 6$ superconformal algebras that were given in [302] (for $N = 1, 2$) (some of the results being conjectural) in detail. Moreover, we treat the superalgebras $\mathrm{osp}(8^*/2N)$ for arbitrary N. Thus, we give the final list of UIRs for $D = 6$. Our main tool is the explicit construction of the norms. This, on the one hand, enables us to prove the unitarity list, and, on the other hand, enables us to give explicitly the states of the irreps.

Accordingly, this section is organized as follows. First we discuss in detail the lowest weight representations of the superalgebras $\mathrm{osp}(8^*/2N)$. In particular, we define explicitly the norm squared of the states that has to be checked for positivity. Then we state the main result (as a theorem) on the lowest weight (positive energy) UIRs and show explicitly the proof of necessity. We also give the general form of the norms, which is enough for the proof of sufficiency. For part of the states (the fully factorizable ones) we give the norms explicitly, for the rest (the unfactorizable ones) the formulas are very involved and in general only recursive. These results are in the generic situation. Then we show the unitarity at the four exceptional points. We give explicitly the states of zero norm (though not all for $N > 1$), which have to be decoupled for the unitary irrep.

https://doi.org/10.1515/9783110527490-003

3.1.1 Representations of $D = 6$ conformal supersymmetry

3.1.1.1 The setting

Our basic reference for Lie superalgebras is [241]. The superconformal algebras in $D = 6$ are $\mathcal{G} = \mathrm{osp}(8^*/2N)$. We label their physically relevant representations by the signature:

$$\chi = [d;\ n_1, n_2, n_3;\ a_1, \ldots, a_N] \tag{3.1}$$

where d is the conformal weight, n_1, n_2, n_3 are non-negative integers which are Dynkin labels of the finite-dimensional irreps of the $D = 6$ Lorentz algebra $\mathrm{so}(5,1)$, and $a_1, \ldots, a_N$ are non-negative integers which are Dynkin labels of the finite-dimensional irreps of the internal (or R) symmetry algebra $\mathrm{usp}(2N)$. The even subalgebra of $\mathrm{osp}(8^*/2N)$ is the algebra $\mathrm{so}^*(8) \oplus \mathrm{usp}(2N)$, and $\mathrm{so}^*(8) \cong \mathrm{so}(6,2)$ is the $D = 6$ conformal algebra.

Our aim is to give a constructive proof for the UIRs of $\mathrm{osp}(8^*/2N)$ following the methods used for the $D = 4$ superconformal algebras $\mathrm{su}(2,2/N)$, cf. [140–142, 139]. The main tool is an adaptation of the Shapovalov form on the Verma modules V^χ over the complexification $\mathcal{G}^{\mathbb{C}} = \mathrm{osp}(8/2N)$ of $\mathcal{G}$.

3.1.1.2 Verma modules

To introduce Verma modules we use the standard triangular decomposition:

$$\mathcal{G}^{\mathbb{C}} = \mathcal{G}^+ \oplus \mathcal{H} \oplus \mathcal{G}^- \tag{3.2}$$

where $\mathcal{G}^+$, $\mathcal{G}^-$, respectively, are the subalgebras corresponding to the positive, negative, roots, respectively, and $\mathcal{H}$ denotes the Cartan subalgebra.

We consider lowest weight Verma modules, so that $V^\Lambda \cong U(\mathcal{G}^+) \otimes v_0$, where $U(\mathcal{G}^+)$ is the universal enveloping algebra of $\mathcal{G}^+$, and v_0 is a lowest weight vector v_0 such that

$$\begin{aligned} Z v_0 &= 0, \quad Z \in \mathcal{G}^- \\ H v_0 &= \Lambda(H) v_0, \quad H \in \mathcal{H}. \end{aligned} \tag{3.3}$$

Furthermore, for simplicity we omit the sign $\otimes$, i. e., we write $p v_0 \in V^\Lambda$ with $p \in U(\mathcal{G}^+)$.

The lowest weight Λ is characterized by its values on the Cartan subalgebra $\mathcal{H}$. In order to have Λ corresponding to χ, one can choose a basis in $\mathcal{H}$ so that to obtain the entries in the signature χ by evaluating Λ on the basis elements of $\mathcal{H}$.

3.1.1.3 Root systems

In order to explain how the above is done we recall some facts about $\mathrm{osp}(8/2N)$ (denoted $D(4,N)$ in [241]).[1] Their root systems are given in terms of $\epsilon_1 \ldots, \epsilon_4$, $\delta_1 \ldots, \delta_N$,

1 These initial facts can be given for $\mathrm{osp}(2M/2N) = D(M,N)$ in a very similar fashion.

$(\epsilon_i, \epsilon_j) = \delta_{ij}$, $(\delta_{\hat{\imath}}, \delta_{\hat{\jmath}}) = -\delta_{\hat{\imath}\hat{\jmath}}$, $(\epsilon_i, \delta_{\hat{\jmath}}) = 0$. The indices $i, j, \ldots$ will take values in the set $\{1,2,3,4\}$, the indices $\hat{\imath}, \hat{\jmath}, \ldots$ will take values in the set $\{1,\ldots,N\}$. The even and odd roots systems are [241]

$$\Delta_{\bar{0}} = \{\pm\epsilon_i \pm \epsilon_j, i < j, \pm\delta_{\hat{\imath}} \pm \delta_{\hat{\jmath}}, \hat{\imath} < \hat{\jmath}, \pm 2\delta_{\hat{\imath}}\}, \quad \Delta_{\bar{1}} = \{\pm\epsilon_i \pm \delta_{\hat{\jmath}}\} \tag{3.4}$$

(we recall that the signs $\pm$ are not correlated).[2] We shall use the following simple root system [241]:

$$\begin{aligned} \Pi &= \{\epsilon_1 - \epsilon_2, \epsilon_2 - \epsilon_3, \epsilon_3 - \epsilon_4, \epsilon_4 - \delta_1, \delta_1 - \delta_2, \ldots, \delta_{N-1} - \delta_N, 2\delta_N\} \\ &= \{\alpha_1, \ldots, \alpha_{4+N}\} \\ \alpha_j &= \epsilon_j - \epsilon_{j+1}, \quad j = 1,2,3, \quad \alpha_4 = \epsilon_4 - \delta_1, \\ \alpha_{4+\hat{\jmath}} &= \delta_{\hat{\jmath}} - \delta_{\hat{\jmath}+1}, \quad \hat{\jmath} = 1,\ldots,N-1, \quad \alpha_{4+N} = 2\delta_N. \end{aligned} \tag{3.5}$$

The root $\alpha_4 = \epsilon_4 - \delta_1$ is odd, the other simple roots are even. For future use we need also the positive root system corresponding to Π:

$$\Delta_{\bar{0}}^{+} = \{\epsilon_i \pm \epsilon_j, i < j, \delta_{\hat{\imath}} \pm \delta_{\hat{\jmath}}, \hat{\imath} < \hat{\jmath}, 2\delta_{\hat{\imath}}\}, \quad \Delta_{\bar{1}}^{+} = \{\epsilon_i \pm \delta_{\hat{\jmath}}\}. \tag{3.6}$$

3.1.1.4 Basis of the Cartan subalgebra

Let us denote by H_A the generators of the Cartan subalgebra, $A = 1,\ldots,4+N$. There is a standard choice for these generators [241]. Namely, to every even simple root α_A we choose a generator H_A so that the following equality is valid for arbitrary $\mu \in \mathcal{H}^*$:

$$\mu(H_A) = (\mu, \alpha_A^\vee), \quad A \neq 4, \tag{3.7}$$

where $\alpha_A^\vee \equiv 2\alpha_A/(\alpha_A, \alpha_A)$. Because these H_A correspond to the simple even roots, which define the Dynkin labeling, we have the following relation with the signature χ:

$$\Lambda(H_A) = \begin{cases} -n_A, & A = 1,2,3 \\ -a_{A-4}, & A = 5,\ldots,N+4. \end{cases} \tag{3.8}$$

The minus signs are related to the fact that we work with lowest weight Verma modules (instead of the highest weight modules used in [241]) and to Verma module reducibility w.r.t. the roots α_A (this is explained in detail in [139]).

We have not fixed only the generator H_4. The standard choice [241] is a generator corresponding to the odd simple root α_4, but we can take any element of the Cartan subalgebra which is not a linear combination of the established already $N + 3$ generators H_A. Our choice is to take the generator H_4 which corresponds to the root

2 The roots $\pm\epsilon_i \pm \epsilon_j$ provide the root system of so$(8;\mathbb{C})$, the roots $\pm\delta_i \pm \delta_j$ and $\pm 2\delta_i$ provide the root system of sp$(2N;\mathbb{C})$.

$\epsilon_3 + \epsilon_4$ and which together with α_1, α_2, α_3 provides the root system of so(8; ℂ).[3] The value $\Lambda(H_4)$ cannot be a non-positive integer like the other $\Lambda(H_A)$ given in (3.8), since then we would obtain finite-dimensional representations of so(8, ℂ) [243] and, thus, non-unitary representations of so(6, 2). In fact, unitarity w.r.t. so(6, 2) would already require that $\Lambda(H_4)$ is a non-negative number related to the physically relevant conformal weight d, which is related to the eigenvalue of the conformal Hamiltonian. That is why the lowest weight UIRs are also called *positive energy* UIRs. We omit here the analysis by which it turns out that $\Lambda(H_4)$ differs from d by the quantity $(n_1 + 2n_2 + n_3)/2$ (which is the value of the conformal Hamiltonian of the algebra so(5, 1) mentioned above). Thus, we set

$$\Lambda(H_4) = d + \frac{1}{2}(n_1 + 2n_2 + n_3) = (\Lambda, (\epsilon_3 + \epsilon_4)^\vee) = (\Lambda, \epsilon_3 + \epsilon_4). \tag{3.9}$$

This choice is consistent with the one in [302], and the usage in [174].

Having in hand the values of Λ on the basis we can recover them for any element of $\mathcal{H}$ and $\mathcal{H}^*$. In particular, for the values on the elementary functionals we have from (3.8) and (3.9)

$$\begin{aligned}
(\Lambda, \epsilon_1) &= \frac{1}{2}d - \frac{1}{4}(3n_1 + 2n_2 + n_3) \\
(\Lambda, \epsilon_2) &= \frac{1}{2}d + \frac{1}{4}(n_1 - 2n_2 - n_3) \\
(\Lambda, \epsilon_3) &= \frac{1}{2}d + \frac{1}{4}(n_1 + 2n_2 - n_3) \\
(\Lambda, \epsilon_4) &= \frac{1}{2}d + \frac{1}{4}(n_1 + 2n_2 + 3n_3) \\
(\Lambda, \delta_{\hat{j}}) &= a_{\hat{j}} + a_{\hat{j}+1} + \cdots + a_N \equiv r_{\hat{j}}.
\end{aligned} \tag{3.10}$$

Using (3.8) and (3.9) one can write easily $\Lambda = \Lambda(\chi)$ as a linear combination of the simple roots or of the elementary functionals ϵ_j, $\delta_{\hat{j}}$, but this is not necessary in what follows.

3.1.1.5 Reducibility of Verma modules

Having established the relation between χ and Λ we turn our attention to the question of unitarity. The conditions of unitarity are intimately related with the conditions for reducibility of the Verma modules w.r.t. to the odd positive roots. A Verma module V^Λ is reducible w.r.t. the odd positive root γ iff the following holds [241]:

$$(\Lambda - \rho, \gamma) = 0, \quad \gamma \in \Delta^+_{\bar{1}}, \tag{3.11}$$

where $\rho \in \mathcal{H}^*$ is the very important in representation theory element given by the difference of the half-sums $\rho_{\bar{0}}, \rho_{\bar{1}}$ of the even, odd, respectively, positive roots (cf. (1.28)):

$$\rho \doteq \rho_{\bar{0}} - \rho_{\bar{1}}$$

3 However, in the osp(8/2N) root system we have $\epsilon_3 + \epsilon_4 = \alpha_3 + 2\alpha_4 + \cdots + 2\alpha_{N+3} + \alpha_{N+4}$.

$$\rho_{\bar{0}} = 3\epsilon_1 + 2\epsilon_2 + \epsilon_3 + N\delta_1 + (N-1)\delta_2 + \cdots + 2\delta_{N-1} + \delta_N$$
$$\rho_{\bar{1}} = N(\epsilon_1 + \epsilon_2 + \epsilon_3 + \epsilon_4). \tag{3.12}$$

To make (3.11) explicit we need the values of Λ and ρ on the positive odd roots $\epsilon_i \pm \delta_j$ (which we obtain from (3.10)):

$$(\Lambda, \epsilon_1 \pm \delta_{\hat{j}}) = \frac{1}{2}d - \frac{1}{4}(3n_1 + 2n_2 + n_3) \pm r_{\hat{j}} \tag{3.13}$$
$$(\Lambda, \epsilon_2 \pm \delta_{\hat{j}}) = \frac{1}{2}d + \frac{1}{4}(n_1 - 2n_2 - n_3) \pm r_{\hat{j}},$$
$$(\Lambda, \epsilon_3 \pm \delta_{\hat{j}}) = \frac{1}{2}d + \frac{1}{4}(n_1 + 2n_2 - n_3) \pm r_{\hat{j}},$$
$$(\Lambda, \epsilon_4 \pm \delta_{\hat{j}}) = \frac{1}{2}d + \frac{1}{4}(n_1 + 2n_2 + 3n_3) \pm r_{\hat{j}},$$
$$(\rho, \epsilon_i \pm \delta_{\hat{j}}) = 4 - i - N \mp (N - \hat{j} + 1). \tag{3.14}$$

Consecutively we find that the Verma module $V^{\Lambda(\chi)}$ is reducible if the conformal weight takes one of the following $8N$ values $d^{\pm}_{i\hat{j}}$ labeled by the respective odd root $\epsilon_i \pm \delta_{\hat{j}}$:

$$d = d^{\pm}_{1\hat{j}} \doteq \frac{1}{2}(3n_1 + 2n_2 + n_3) + 2(3 - N) \mp 2(r_{\hat{j}} + N - \hat{j} + 1),$$
$$d = d^{\pm}_{2\hat{j}} \doteq \frac{1}{2}(n_3 + 2n_2 - n_1) + 2(2 - N) \mp 2(r_{\hat{j}} + N - \hat{j} + 1),$$
$$d = d^{\pm}_{3\hat{j}} \doteq \frac{1}{2}(n_3 - 2n_2 - n_1) + 2(1 - N) \mp 2(r_{\hat{j}} + N - \hat{j} + 1),$$
$$d = d^{\pm}_{4\hat{j}} \doteq -\frac{1}{2}(n_1 + 2n_2 + 3n_3) - 2N \mp 2(r_{\hat{j}} + N - \hat{j} + 1). \tag{3.15}$$

For future use we note the following relations:

$$\frac{1}{2}(d^{-}_{i\hat{j}} - d^{-}_{k\hat{\ell}}) = n_i + \cdots + n_{k-1} + k - i + \hat{\ell} - \hat{j} + a_{\hat{j}} + \cdots + a_{\hat{\ell}-1} > 0,$$
$$i \leq k,\ \hat{j} \leq \hat{\ell},\ i\hat{j} \neq k\hat{\ell},$$
$$\frac{1}{2}(d^{+}_{i\hat{j}} - d^{+}_{k\hat{\ell}}) = n_i + \cdots + n_{k-1} + k - i + \hat{j} - \hat{\ell} + a_{\hat{\ell}} + \cdots + a_{\hat{j}-1} > 0,$$
$$i \leq k,\ \hat{j} \geq \hat{\ell},\ i\hat{j} \neq k\hat{\ell},$$
$$\frac{1}{2}(d^{-}_{i\hat{j}} - d^{+}_{k\hat{\ell}}) = n_i + \cdots + n_{k-1} + k - i + 2N - \hat{j} - \hat{\ell} + r_{\hat{j}} + r_{\hat{\ell}} + 2 > 0,$$
$$i \leq k, \tag{3.16}$$

which introduces some partial ordering between the quantities $d^{\pm}_{i\hat{j}}$ of which the essential would turn out to be the following:

$$d^{-}_{11} > d^{-}_{21} > d^{-}_{31} > d^{-}_{41}. \tag{3.17}$$

The four values in (3.17) play special role in the unitarity formulation. The value d^{-}_{11} is the biggest among all $d^{\pm}_{ij}$; it is called *the first reduction point* in [182].

3.1.1.6 Shapovalov form and unitarity

The Shapovalov form is a bilinear $\mathbb{C}$-valued form on Verma modules [353]. We need also the involutive antiautomorphism ω of $U(\mathcal{G}^+)$ which will provide the real form we are interested in. Thus, an adaptation of the Shapovalov form suitable for our purposes is defined (as in [141, 142]) as follows:

$$\begin{aligned}(u,u') &= (pv_0, p'v_0) \equiv (v_0, \omega(p)p'v_0) \\ &= (\omega(p')pv_0, v_0), \\ u = pv_0, \quad u' &= p'v_0, \quad p,p' \in U(\mathcal{G}^+),\ u,u' \in V^\Lambda,\end{aligned} \tag{3.18}$$

supplemented by the normalization condition $(v_0, v_0) = 1$. The norms squared of the states would be denoted by

$$\|u\|^2 \equiv (u,u). \tag{3.19}$$

We suppose that we consider representations which are unitary when restricted to the even part $\mathcal{G}^+_{\bar{0}}$. This is justified aposteriori since (as in the $D = 4$ case [140–142]) the unitary bounds of the even part are weaker than the supersymmetric ones [167]. Thus, as in [140–142] we shall factorize the even part and we shall consider only the states created by the action of the odd generators, i. e., $\mathcal{F}^\Lambda = (U(\mathcal{G}^+)/U(\mathcal{G}^+_{\bar{0}}))v_0$. We introduce notation $X^+_{i\hat{j}}$ for the odd generator corresponding to the positive root $\epsilon_i - \delta_{\hat{j}}$, and $Y^+_{i\hat{j}}$ shall correspond to $\epsilon_i + \delta_{\hat{j}}$. Since the odd generators are Grassmann there are only 2^{8N} states in $\mathcal{F}$ and choosing an ordering we give these states explicitly as follows:

$$\begin{aligned}\Psi_{\bar{\varepsilon}\bar{\nu}} &= \left(\prod_{i=1}^{4}(Y^+_{i\hat{1}})^{\varepsilon_{i1}}\right)\cdots\left(\prod_{i=1}^{4}(Y^+_{i\hat{N}})^{\varepsilon_{iN}}\right) \\ &\times \left(\prod_{i=1}^{4}(X^+_{i\hat{N}})^{\nu_{iN}}\right)\cdots\left(\prod_{i=1}^{4}(X^+_{i\hat{1}})^{\nu_{i1}}\right)v_0, \quad \varepsilon_{i\hat{j}}, \nu_{i\hat{j}} = 0,1,\end{aligned} \tag{3.20}$$

where $\bar{\varepsilon}$ and $\bar{\nu}$ denote the sets of all $\varepsilon_{i\hat{j}}$ and $\nu_{i\hat{j}}$, respectively For future use we give the notation for the numbers of Y and X:

$$\varepsilon \equiv \sum_{i=1}^{4}\sum_{\hat{j}=1}^{N}\varepsilon_{i\hat{j}}, \quad \nu \equiv \sum_{i=1}^{4}\sum_{\hat{j}=1}^{N}\nu_{i\hat{j}}, \tag{3.21}$$

and through them for the *level* ℓ:

$$\ell(\Psi_{\bar{\varepsilon}\bar{\nu}}) = \varepsilon + \nu. \tag{3.22}$$

3.1.1.7 Explicit realization of the basis of osp(8/2*N*)

To proceed we need the explicit realization of the generators of osp$(8/2N)$. It is obtained from the standard one of [241] by applying a unitary transformation done in order to bring the Cartan subalgebra in diagonal form. The matrices are $(8+2N)\times(8+2N)$

and are in standard supermatrix form, i. e., the even, respectively, the odd ones are of the form

$$\begin{pmatrix} * & 0 \\ 0 & * \end{pmatrix}, \quad \begin{pmatrix} 0 & * \\ * & 0 \end{pmatrix}$$

The description is done very conveniently in terms of the matrices $E_{AB} \in gl(8/2N, \mathbb{C})$, $A, B = 1, \dots, 8 + 2N$. Fix A, B, then the matrix E_{AB} has only non-zero entry, equal to 1, at the intersection of the Ath row and Bth column.

Then the generators H_A are given by

$$\begin{aligned}
H_j &= E_{jj} - E_{j+1,j+1} - E_{j+4,j+4} + E_{j+5,j+5}, \quad j = 1,2,3 \\
H_4 &= E_{33} + E_{44} - E_{77} - E_{88} \\
H_{4+\hat{j}} &= E_{8+\hat{j},8+\hat{j}} - E_{9+\hat{j},9+\hat{j}} - E_{8+N+\hat{j},8+N+\hat{j}} + E_{9+N+\hat{j},9+N+\hat{j}}, \\
&\quad \hat{j} = 1, \dots, N-1 \\
H_{4+N} &= E_{8+N,8+N} - E_{8+2N,8+2N},
\end{aligned} \tag{3.23}$$

the basis of $\mathcal{G}^+$—enumerated by the corresponding roots—is

$$\begin{aligned}
L^+_{ij} &= E_{ij} - E_{4+j,4+i}, \quad \text{roots: } \epsilon_i - \epsilon_j,\ i < j \\
P^+_{ij} &= E_{i,4+j} - E_{j,4+i}, \quad \text{roots: } \epsilon_i + \epsilon_j,\ i < j \\
T^+_{\hat{i}\hat{j}} &= E_{8+\hat{i},8+\hat{j}} - E_{8+N+\hat{j},8+N+\hat{i}}, \quad \text{roots: } \delta_{\hat{i}} - \delta_{\hat{j}},\ \hat{i} < \hat{j} \\
R^+_{\hat{i}\hat{j}} &= E_{8+\hat{i},8+N+\hat{j}} + E_{8+\hat{j},8+N+\hat{i}}, \quad \text{roots: } \delta_{\hat{i}} + \delta_{\hat{j}},\ \hat{i} < \hat{j} \\
R^+_{\hat{i}} &= E_{8+\hat{i},8+N+\hat{i}}, \quad \text{roots: } 2\delta_{\hat{i}} \\
X^+_{i\hat{j}} &= E_{i,8+\hat{j}} + E_{8+N+\hat{j},4+i}, \quad \text{roots: } \epsilon_i - \delta_{\hat{j}} \\
Y^+_{i\hat{j}} &= E_{i,8+N+\hat{j}} - E_{8+\hat{j},4+i}, \quad \text{roots: } \epsilon_i + \delta_{\hat{j}},
\end{aligned} \tag{3.24}$$

while the basis of $\mathcal{G}^-$ is

$$\begin{aligned}
L^-_{ij} &= E_{ij} - E_{4+j,4+i}, \quad \text{roots: } \epsilon_i - \epsilon_j,\ i > j \\
P^-_{ij} &= E_{4+j,i} - E_{4+i,j}, \quad \text{roots: } -(\epsilon_i + \epsilon_j),\ i < j \\
T^-_{\hat{i}\hat{j}} &= E_{8+\hat{i},8+\hat{j}} - E_{8+N+\hat{j},8+N+\hat{i}}, \quad \text{roots: } \delta_{\hat{i}} - \delta_{\hat{j}},\ \hat{i} > \hat{j} \\
R^-_{\hat{i}\hat{j}} &= E_{8+N+\hat{i},8+\hat{j}} + E_{8+N+\hat{j},8+\hat{i}}, \quad \text{roots: } -(\delta_{\hat{i}} + \delta_{\hat{j}}),\ \hat{i} < \hat{j} \\
R^-_{\hat{i}} &= E_{8+N+\hat{i},8+\hat{i}}, \quad \text{roots: } -2\delta_{\hat{i}} \\
X^-_{i\hat{j}} &= E_{4+\hat{i},8+N+\hat{j}} - E_{8+\hat{j},i}, \quad \text{roots: } -\epsilon_i + \delta_{\hat{j}} \\
Y^-_{i\hat{j}} &= E_{4+i,8+\hat{j}} + E_{8+N+\hat{j},i}, \quad \text{roots: } -(\epsilon_i + \delta_{\hat{j}}).
\end{aligned} \tag{3.25}$$

From the explicit matrix realization above one easily obtains all commutation relations. We shall write down only some more important ones:

$$[X^+_{i\hat{j}}, X^-_{i\hat{j}}]_+ = -E_{ii} + E_{4+i,4+i} - E_{8+\hat{j},8+\hat{j}} + E_{8+N+\hat{j},8+N+\hat{j}}$$

$$\begin{aligned}
&= -\hat{H}_i - \tilde{H}_{\hat{j}},\\
[Y^+_{i\hat{j}}, Y^-_{i\hat{j}}]_+ &= E_{ii} - E_{4+i,4+i} - E_{8+\hat{j},8+\hat{j}} + E_{8+N+\hat{j},8+N+\hat{j}}\\
&= \hat{H}_i - \tilde{H}_{\hat{j}},\\
\hat{H}_i &\equiv E_{ii} - E_{4+i,4+i}, \quad i = 1,\dots,4,\\
\tilde{H}_{\hat{j}} &\equiv E_{8+\hat{j},8+\hat{j}} - E_{8+N+\hat{j},8+N+\hat{j}}, \quad \hat{j} = 1,\dots,N,
\end{aligned} \tag{3.26}$$

where we have introduced notation for an alternative basis of $\mathcal{H}$ which actually is used in the calculation of the scalar products. In particular, we shall use continuously

$$[\hat{H}_k, X^+_{i\hat{j}}] = \delta_{ki} X^+_{i\hat{j}} \tag{3.27a}$$

$$[\hat{H}_k, Y^+_{i\hat{j}}] = \delta_{ki} Y^+_{i\hat{j}} \tag{3.27b}$$

$$[\tilde{H}_{\hat{\ell}}, X^+_{i\hat{j}}] = -\delta_{\hat{\ell}\hat{j}} X^+_{i\hat{j}} \tag{3.27c}$$

$$[\tilde{H}_{\hat{\ell}}, Y^+_{i\hat{j}}] = \delta_{\hat{\ell}\hat{j}} Y^+_{i\hat{j}}. \tag{3.27d}$$

We also give the generators $\hat{H}_A$ in terms of H_A

$$\begin{aligned}
&\hat{H}_1 = H_1 + H_2 + \frac{1}{2}(H_4 + H_3), \quad \hat{H}_2 = H_2 + \frac{1}{2}(H_4 + H_3),\\
&\hat{H}_3 = \frac{1}{2}(H_4 + H_3), \quad \hat{H}_4 = \frac{1}{2}(H_4 - H_3),\\
&\tilde{H}_{\hat{j}} = H_{4+\hat{j}} + \cdots + H_{4+N}, \quad \hat{j} = 1,\dots,N.
\end{aligned} \tag{3.28}$$

3.1.2 Unitarity

In this section we state our main result (in the theorem) on the lowest weight (positive energy) UIRs and give the proof of necessity in general and the proof of sufficiency at generic points (the reduction points are dealt with in the next section).

3.1.2.1 Calculation of some norms

In this subsection we show how to use the form (3.18) to calculate the norms of the states from $\mathcal{F}$.

We first need explicitly the conjugation ω on the odd generators:

$$\omega(X^+_{i\hat{j}}) = -X^-_{i\hat{j}}, \quad \omega(Y^+_{i\hat{j}}) = Y^-_{i\hat{j}}. \tag{3.29}$$

(In matrix notation this would follow from $\omega(E_{i,8+\hat{j}}) = E_{8+\hat{j},i}$, $\omega(E_{i+4,8+\hat{j}}) = -E_{8+\hat{j},i+4}$.)

We give now explicitly the norms of the one-particle states from $\mathcal{F}$ introducing also notation for future use:

$$x_{i\hat{j}} \equiv \|X^+_{i\hat{j}} v_0\|^2 = (X^+_{i\hat{j}} v_0, X^+_{i\hat{j}} v_0)$$

$$= -(v_0, X^-_{i\hat{j}} X^+_{i\hat{j}} v_0) = (v_0, (\hat{H}_i + \tilde{H}_{\hat{j}}) v_0)$$
$$= \Lambda(\hat{H}_i + \tilde{H}_{\hat{j}}) \qquad (3.30a)$$
$$y_{i\hat{j}} \equiv \|Y^+_{i\hat{j}} v_0\|^2 = (Y^+_{i\hat{j}} v_0, Y^+_{i\hat{j}} v_0)$$
$$= (v_0, Y^-_{i\hat{j}} Y^+_{i\hat{j}} v_0) = (v_0, (\hat{H}_i - \tilde{H}_{\hat{j}}) v_0)$$
$$= \Lambda(\hat{H}_i - \tilde{H}_{\hat{j}}). \qquad (3.30b)$$

Using (3.28), (3.8) and (3.9) we get

$$x_{i\hat{j}} = (\Lambda, \epsilon_i - \delta_{\hat{j}}) = \frac{1}{2}(d - d^-_{i\hat{j}}) + 5 - i - \hat{j} \qquad (3.31a)$$
$$y_{i\hat{j}} = (\Lambda, \epsilon_i + \delta_{\hat{j}}) = \frac{1}{2}(d - d^+_{i\hat{j}}) + 3 - i + \hat{j} - 2N. \qquad (3.31b)$$

Also we note that

$$x_{i+1,\hat{j}} - x_{i\hat{j}} = \frac{1}{2}(d^-_{i\hat{j}} - d^-_{i+1,\hat{j}}) - 1 = n_i \geq 0, \qquad (3.32a)$$
$$x_{i,\hat{j}+1} - x_{i\hat{j}} = \frac{1}{2}(d^-_{i\hat{j}} - d^-_{i,\hat{j}+1}) - 1 = a_{\hat{j}} \geq 0, \qquad (3.32b)$$
$$y_{i+1,\hat{j}} - y_{i\hat{j}} = \frac{1}{2}(d^+_{i\hat{j}} - d^+_{i+1,\hat{j}}) - 1 = n_i \geq 0, \qquad (3.32c)$$
$$y_{i\hat{j}} - y_{i,\hat{j}+1} = \frac{1}{2}(d^+_{i,\hat{j}+1} - d^+_{i\hat{j}}) - 1 = a_{\hat{j}} \geq 0, \qquad (3.32d)$$
$$y_{i,\hat{\ell}} - x_{i\hat{j}} = \frac{1}{2}(d^-_{i\hat{j}} - d^+_{i,\hat{\ell}}) + \hat{j} + \hat{\ell} - 2N - 2 = r_{\hat{\ell}} + r_{\hat{j}} \geq 0. \qquad (3.32e)$$

Thus, x_{11} is the smallest among all $x_{i\hat{j}}$ and $y_{i\hat{j}}$.

3.1.2.2 Statement of main result and proof of necessity

In this subsection we state our main result in the theorem and give the proof of necessity via two propositions (Propositions 1 & 2).

First we give the norms which actually determine all of the unitarity conditions. In order to simplify the exposition we shall also use the notation

$$X^+_j \equiv X^+_{j1}$$
$$x_j \equiv x_{j1}. \qquad (3.33)$$

We note in these terms a subset of (3.32a),

$$x_{i+1} - x_i = n_i \geq 0. \qquad (3.32a')$$

Next we calculate

$$\|X^+_j X^+_k v_0\|^2 = (x_j - 1)x_k, \quad j < k, \qquad (3.34a)$$

$$\|X_j^+ X_k^+ X_\ell^+ v_0\|^2 = (x_j - 2)(x_k - 1)x_\ell, \quad j < k < \ell, \tag{3.34b}$$

$$\|X_1^+ X_2^+ X_3^+ X_4^+ v_0\|^2 = (x_1 - 3)(x_2 - 2)(x_3 - 1)x_4. \tag{3.34c}$$

The norms (3.30a) and (3.34) are all strictly positive iff $x_j > 4 - j$, $j = 1, 2, 3, 4$, which are all fulfilled if $x_1 > 3$, since x_1 is the smallest among the x_j. Thus, these norms are strictly positive iff

$$x_1 > 3 \quad \Longleftrightarrow \quad d > d_{11}^-. \tag{3.35}$$

It turns out that this restriction is sufficient to guarantee unitarity of the whole representation. This is not unexpected: in all cases studied so far it was always so that if d is bigger than the first odd reduction point then the module is unitary.

Of course, the condition (3.35) is not necessary for unitarity. Based on the experience so far it is expected that when d is equal to some of the reducibility values then unitarity is also possible, though in these cases there would be some conditions on the representation parameters, and one has to factor out the resulting zero norm states. Now we can formulate the main result:

Theorem. *All positive energy unitary irreducible representations of the conformal superalgebra* osp$(8^*/2N)$ *characterized by the signature* χ *in* (3.1) *are obtained for real d and are given in the following list:*

$$d \geq d_{11}^- = \frac{1}{2}(3n_1 + 2n_2 + n_3) + 2r_1 + 6, \quad \text{any } n_j \in \mathbb{Z}_+ \tag{3.36a}$$

$$d = d_{21}^- = \frac{1}{2}(n_3 + 2n_2) + 2r_1 + 4, \quad n_1 = 0 \tag{3.36b}$$

$$d = d_{31}^- = \frac{1}{2}n_3 + 2r_1 + 2, \quad n_1 = n_2 = 0 \tag{3.36c}$$

$$d = d_{41}^- = 2r_1, \quad n_1 = n_2 = n_3 = 0. \tag{3.36d}$$

◇

Remark. For $N = 1, 2$ the theorem was conjectured by Minwalla [302], except that he conjectured unitarity also for the open interval (d_{31}^-, d_{21}^-) with conditions on n_j as in (3.36c). We should note that this conjecture could be reproduced neither by the methods of conformal field theory [174], nor by the oscillator method [216] (cf. [302]), and thus it was in doubt. To compare with the notations of [302] one should use the following substitutions: $n_1 = h_2 - h_3$, $n_2 = h_1 - h_2$, $n_3 = h_2 + h_3$, $r_1 = k$, and h_j are all integer or all half-integer. The fact that $n_j \geq 0$ for $j = 1, 2, 3$ translates into $h_1 \geq h_2 \geq |h_3|$, i. e., the parameters h_j are of the type often used for representations of so$(2N)$ (though usually for $N \geq 4$). Note also that the statement of the theorem is arranged in [302] according to the possible values of n_i first and then the possible values of d. To compare with the notation of [174] we use the substitution $(n_1, n_2, n_3) \to (J_3, J_2, J_1)$. Some UIRs at the four exceptional points d_{i1}^- were constructed in [216] by the oscillator method (some of these were identified with Cartan-type signatures like (3.1) in, e. g., [302], [157]). ◇

The proof of the theorem requires one to show that there is unitarity as claimed, i. e., that the conditions are *sufficient* and that there is no unitarity otherwise, i. e., that the conditions are *necessary*. For sufficiency we need all norms, but for the necessity part we need only the knowledge of a few norms. We give the necessity part in two propositions.

Proposition 1. *There is no unitarity in any of the open intervals:* $(d^-_{j+1,1}, d^-_{j1})$, $j = 1, 2, 3$, *and if* $d < d^-_{41}$. ◇

Proof. • Consider d in the open interval (d^-_{21}, d^-_{11}), which means that $3 > x_1 > 2 - n_1$. Consider the norm (3.34c) and using (3.32a) express all x_i in terms of x_1. We have

$$(x_1 - 3)(x_2 - 2)(x_3 - 1)x_4 = (x_1 - 3)(x_1 + n_1 - 2)(x_1 + n_1 + n_2 - 1)(x_1 + n_1 + n_2 + n_3). \quad (3.37)$$

The first term is strictly negative while the other three terms are strictly positive, independently of the values of n_i. Thus, the norm (3.34c) is negative in the open interval (d^-_{21}, d^-_{11}).

• Consider d in the open interval (d^-_{31}, d^-_{21}), which means that $2 > x_1 + n_1 > 1 - n_2$. Consider the norm (3.34b) for $(j, k, \ell) = (1, 3, 4)$ and using (3.32a) express all x_i in terms of x_1. We have

$$(x_1 - 2)(x_3 - 1)x_4 = (x_1 - 2)(x_1 + n_1 + n_2 - 1)(x_1 + n_1 + n_2 + n_3). \quad (3.38)$$

The first term is strictly negative while the other two terms are strictly positive, independently of the values of n_i. Thus, the norm of the state $X_1^+ X_3^+ X_4^+ v_0$ is negative in the open interval (d^-_{32}, d^-_{21}).

• Consider d in the open interval (d^-_{41}, d^-_{31}), which means that $1 > x_1 + n_1 + n_2 > -n_3$. Consider the norm (3.34a) for $(j, k) = (1, 4)$ and using (3.32a) express all x_i in terms of x_1. We have

$$(x_1 - 1)x_4 = (x_1 - 1)(x_1 + n_1 + n_2 + n_3). \quad (3.39)$$

The first term is strictly negative while the second is strictly positive, independently of the values of n_i. Thus, the norm of the state $X_1^+ X_4^+ v_0$ is negative in the open interval (d^-_{31}, d^-_{21}).

• Consider d in the infinite open interval $d < d^-_{41}$. Then the norm of $X_{41}^+ v_0$ is negative using (3.31a):

$$x_4 = x_{41} = \frac{1}{2}(d - d^-_{41}) < 0.$$

Thus, the proposition is proved. □

Thus, we have shown the exclusion of the open intervals in the statement of the theorem. It remains to show the necessity of the restrictions on n_i in cases b, c, d of the theorem.

Proposition 2. *There is no unitarity in the following cases:*

$$d = d_{21}^-, \quad n_1 > 0 \tag{3.36b'}$$

$$d = d_{31}^-, \quad n_1 + n_2 > 0 \tag{3.36c'}$$

$$d = d_{41}^-, \quad n_1 + n_2 + n_3 > 0 \tag{3.36d'}$$

◇

Proof. • Let $d = d_{21}^-$, which means $x_2 = 2$ and $x_1 = 2 - n_1$. Consider again the norm of $X_1^+ X_3^+ X_4^+ v_0$ and substitute the value of x_1 in 3.38 to get

$$(x_1 - 2)(x_3 - 1)x_4 = (-n_1)(1 + n_2)(2 + n_2 + n_3). \tag{3.40}$$

This norm is negative if $n_1 > 0$.

• Let $d = d_{31}^-$, which means $x_3 = 1$ and $x_1 = 1 - n_1 - n_2$. Consider again the norm of $X_1^+ X_4^+ v_0$ and substitute the value of x_1 in 3.39 to get

$$(x_1 - 1)x_4 = (-n_1 - n_2)(1 + n_3). \tag{3.41}$$

This norm is negative if $n_1 + n_2 > 0$.

• Let $d = d_{41}^-$, which means $x_4 = 0$ and $x_1 = -n_1 - n_2 - n_3$. But the latter is the norm of $X_1^+ v_0$ and it is negative if $n_1 + n_2 + n_3 > 0$.

Thus, the proposition is proved. □

With this we have shown that the conditions of the theorem are *necessary*. □

The proof of sufficiency is postponed to the next subsection.

Remark. The reader may wonder why the other reducibility points are not playing such an important role as the quartet appearing in the theorem.

First we note that the analogous calculations involving other quartets of operators, $X_{i\hat{j}}^+$ ($\hat{j} \neq 1$ fixed, $i = 1,2,3,4$) or $Y_{i\hat{j}}^+$ ($\hat{j}$ fixed, $i = 1,2,3,4$) give the same results as (3.34) with just replacing $x_i \to x_{i\hat{j}}$ or $x_i \to y_{i\hat{j}}$. This brings about the conditions $x_{i\hat{j}} > 4-i$ or $y_{i\hat{j}} > 4 - i$ which all follow from $x_1 > 3$ because of (3.32). This is related to the fact that d_{11}^- is the largest reduction point.

Furthermore, we may look for the analog of Proposition 1 and we can prove the same results involving $d_{i\hat{j}}^-$ ($\hat{j} \neq 1$ fixed), or $d_{i\hat{j}}^+$ ($\hat{j}$ fixed). However, these results may be relevant only if the exceptional points d_{21}^-, d_{31}^-, d_{41}^-, together with the respective conditions, would happen to be in some of the open intervals defined by some other quartet, which would prove their non-unitarity. The reason that this does not happen is the following.

Let $n_1 = 0$. Then one can easily see that d_{11}^- and d_{21}^- are the two largest reduction points (for $N = 1,2$ cf. [302]), i. e., all other reduction points are smaller, and thus, d_{21}^- cannot be in any open interval defined by some other quartet.

Analogously, for $n_1 = n_2 = 0$ one can easily see that d_{11}^-, d_{21}^-, and d_{31}^- are the three largest reduction points (for $N = 1,2$ cf. [302]), and so d_{31}^- cannot be in any open interval defined by some other quartet. Finally, for $n_1 = n_2 = n_3 = 0$ the points d_{11}^-, d_{21}^-, d_{31}^-, and d_{41}^- are the four largest reduction points (for $N = 1, 2$ cf. [302]), and d_{41}^- cannot be in any open interval defined by some other quartet. ◇

3.1.2.3 General form of the norms and unitarity in the generic case

In this subsection we give the proof of sufficiency of the Theorem in the generic case. This requires the general form of the norms. The states are divided into classes and the norms are given for the different cases in Propositions 3–7. At the end we finish the proof of sufficiency utilizing these propositions.

To present the general formulas for the norms we first we divide the states into factorizable and unfactorizable as follows. Let the first generator in $\Psi_{\bar{\varepsilon}\bar{\nu}}$ be $Y_{i\hat{j}}^+$, i. e., $\Psi_{\bar{\varepsilon}\bar{\nu}} = Y_{i\hat{j}}^+ \dots v_0$. Then $\Psi_{\bar{\varepsilon}\bar{\nu}}$ is called *factorizable states* if the following three statements hold:

$$\epsilon_{k\hat{j}}\epsilon_{i\hat{\ell}} = 0 \quad \text{or} \quad \epsilon_{k\hat{\ell}} = 1, \quad \forall (k, \hat{\ell}) : k > i, \hat{\ell} > \hat{j} \tag{3.42a}$$

$$\epsilon_{k\hat{j}}\nu_{i\hat{\ell}} = 0 \quad \text{or} \quad \nu_{k\hat{\ell}} = 1, \quad \forall k > i, \text{ and all } \hat{\ell} \tag{3.42b}$$

$$\epsilon_{i\hat{\ell}}\nu_{j\hat{\ell}} = 0 \quad \text{or} \quad \nu_{j\hat{j}} = 1, \quad \forall \hat{\ell} > \hat{j}, \text{ and all } j. \tag{3.42c}$$

If the first generator in $\Psi_{\bar{\varepsilon}\bar{\nu}}$ is $X_{i\hat{j}}^+$, so that $\bar{\varepsilon} = 0$, then $\Psi_{0,\bar{\nu}}$ is called *factorizable* if the following statement holds:

$$\nu_{j\hat{j}}\nu_{i\hat{\ell}} = 0 \quad \text{or} \quad \nu_{j\hat{\ell}} = 1, \quad \text{for all pairs } (j, \hat{\ell}) \text{ such that } j > i, \ \hat{\ell} < \hat{j}. \tag{3.43}$$

Our first result on the norms is

Proposition 3. *For factorizable states starting with $X_{i\hat{j}}^+$ the following relation holds:*

$$\|\Psi_{0,\bar{\nu}}\|^2 = (x_{i\hat{j}} + \tilde{\nu}_{i\hat{j}})\|\Psi_{0,\bar{\nu}'}\|^2, \tag{3.44}$$
$$\tilde{\nu}_{i\hat{j}} = \nu_{i,\hat{j}-1} + \dots + \nu_{i,1} - \nu_{i+1,\hat{j}} - \dots - \nu_{4,\hat{j}}, \quad \nu'_{j\hat{\ell}} = \nu_{j\hat{\ell}} - \delta_{ji}\delta_{\hat{\ell}\hat{j}}.$$

For factorizable states starting with $Y_{i\hat{j}}^+$ the following relation holds:

$$\|\Psi_{\bar{\varepsilon},\bar{\nu}}\|^2 = (y_{i\hat{j}} + \tilde{\varepsilon}_{i\hat{j}} + \nu_i + \hat{\nu}_{\hat{j}})\|\Psi_{\bar{\varepsilon}'\bar{\nu}}\|^2 \tag{3.45}$$
$$\tilde{\varepsilon}_{i\hat{j}} = \varepsilon_{i,\hat{j}+1} + \dots + \varepsilon_{i,N} - \varepsilon_{i+1,\hat{j}} - \dots - \varepsilon_{4,\hat{j}}$$
$$\nu_i = \nu_{i,1} + \dots + \nu_{i,N}, \quad \hat{\nu}_{\hat{j}} = \nu_{1,\hat{j}} + \dots + \nu_{4,\hat{j}},$$
$$\varepsilon'_{j\hat{\ell}} = \varepsilon_{j\hat{\ell}} - \delta_{ji}\delta_{\hat{\ell}\hat{j}}.$$
◇

Proof. We start with (3.44). Clearly, $\Psi_{0,\bar{\nu}} = X_{i\hat{j}}^+\Psi_{0,\bar{\nu}'}$. Then the norm squared is

$$\|\Psi_{0,\bar{\nu}}\|^2 = (X_{i\hat{j}}^+\Psi_{0,\bar{\nu}'}, X_{i\hat{j}}^+\Psi_{0,\bar{\nu}'}) = -(\Psi_{0,\bar{\nu}'}, X_{i\hat{j}}^- X_{i\hat{j}}^+\Psi_{0,\bar{\nu}'})$$
$$= -(\Psi_{0,\bar{\nu}'}, (-X_{i\hat{j}}^+ X_{i\hat{j}}^- - \hat{H}_i - \tilde{H}_{\hat{j}})\Psi_{0,\bar{\nu}'}) \tag{3.46a}$$

$$= (\Psi_{0,\bar{\nu}'}, (\hat{H}_i + \tilde{H}_{\hat{j}})\Psi_{0,\bar{\nu}'}) \tag{3.46b}$$
$$= (\Lambda(\hat{H}_i + \tilde{H}_{\hat{j}}) + \tilde{\nu}_{i\hat{j}})(\Psi_{0,\bar{\nu}'}, \Psi_{0,\bar{\nu}'}) \tag{3.46c}$$
$$= (x_{i\hat{j}} + \tilde{\nu}_{i\hat{j}})\|\Psi_{0,\bar{\nu}'}\|^2. \tag{3.46d}$$

Note that the term $X^+_{i\hat{j}}X^-_{i\hat{j}}$ in (3.46a) gives no contribution: due to the conditions 3.43 the operator $X^-_{i\hat{j}}$ anticommutes with the operators in $\Psi_{0,\bar{\nu}'}$ or produces terms like $(X^+_{\hat{j}\hat{\ell}})^2 = 0$, thus it reaches v_0 without additional terms. Moving the operator $\hat{H}_i + \tilde{H}_{\hat{j}}$ through $\Psi_{0,\bar{\nu}'}$ produces the addition $\tilde{\nu}_{i\hat{j}}$—the terms $\nu_{i,\hat{j}-1} + \cdots + \nu_{i,1}$ are due to (3.27a), and the terms $-\nu_{i+1,\hat{j}} - \cdots - \nu_{4,\hat{j}}$ are due to (3.27c). Analogously, we consider (3.45). Clearly, $\Psi_{\bar{\varepsilon},\bar{\nu}} = Y^+_{i\hat{j}}\Psi_{\bar{\varepsilon}',\bar{\nu}}$. The norm squared is

$$\begin{aligned}
\|\Psi_{\bar{\varepsilon}\bar{\nu}}\|^2 &= (Y^+_{i\hat{j}}\Psi_{\bar{\varepsilon}',\bar{\nu}}, Y^+_{i\hat{j}}\Psi_{\bar{\varepsilon}',\bar{\nu}}) = (\Psi_{\bar{\varepsilon}',\bar{\nu}}, Y^-_{i\hat{j}}Y^+_{i\hat{j}}\Psi_{\bar{\varepsilon}',\bar{\nu}}) \\
&= (\Psi_{\bar{\varepsilon}',\bar{\nu}}, (-Y^+_{i\hat{j}}Y^-_{i\hat{j}} + \hat{H}_i - \tilde{H}_{\hat{j}})\Psi_{\bar{\varepsilon}',\bar{\nu}}) \\
&= (\Psi_{\bar{\varepsilon}',\bar{\nu}}, (\hat{H}_i - \tilde{H}_{\hat{j}})\Psi_{\bar{\varepsilon}',\bar{\nu}}) \\
&= (\Lambda(\hat{H}_i - \tilde{H}_{\hat{j}}) + \tilde{\varepsilon}_{i\hat{j}} + \nu_i + \hat{\nu}_{\hat{j}})(\Psi_{\bar{\varepsilon}',\bar{\nu}}, \Psi_{\bar{\varepsilon}',\bar{\nu}}) \\
&= (y_{i\hat{j}} + \tilde{\varepsilon}_{i\hat{j}} + \nu_i + \hat{\nu}_{\hat{j}})\|\Psi_{\bar{\varepsilon}',\bar{\nu}}\|^2.
\end{aligned} \tag{3.47}$$

Note that to produce the additional terms $\tilde{\varepsilon}_{i\hat{j}} + \nu_i + \hat{\nu}_{\hat{j}}$ we need all of (3.27). □

The states $\Psi_{\bar{\varepsilon}'\bar{\nu}}$ and $\Psi_{0,\bar{\nu}'}$ may still be factorizable and so on. The state $\Psi_{0,\bar{\nu}}$ is called a *fully factorizable state* if the process of factorization can be repeated ν times. The state $\Psi_{\bar{\varepsilon},\bar{\nu}}$ is called *fully factorizable* if the process of factorization can be repeated ε times and the resulting state $\Psi_{0,\bar{\nu}}$ is fully factorizable.

Our first main result on the norms now follows.

Proposition 4. *The norm of a fully factorizable state $\Psi_{\bar{\varepsilon}\bar{\nu}}$ is given by the following formula:*

$$\|\Psi_{\bar{\varepsilon}\bar{\nu}}\|^2 = \mathcal{N}_{\bar{\varepsilon}\bar{\nu}}, \tag{3.48}$$

where

$$\mathcal{N}_{\bar{\varepsilon}\bar{\nu}} = \prod_{i=1}^{4}\prod_{\hat{j}=1}^{N}(y_{i\hat{j}} + \tilde{\varepsilon}_{i\hat{j}} + \nu_i + \tilde{\nu}_{\hat{j}})^{\varepsilon_{i\hat{j}}}(x_{i\hat{j}} + \tilde{\nu}_{i\hat{j}})^{\nu_{i\hat{j}}}. \tag{3.49}$$

◇

Proof. By direct iteration of (3.45) and (3.44). □

Naturally, the norms in (3.34) are special cases of (3.48).

Note that the norm squared of a state is a polynomial in d of degree the level ℓ of the state.

We shall now discuss states which are not fully factorizable. It is enough to consider unfactorizable states, since if a state is factorizable then we apply (3.45) or (3.44) one or more times until we are left with the norm squared of an unfactorizable state. We shall have two propositions, the first of which is

Proposition 5. *Let $\Psi_{0,\bar{v}}$ be an unfactorizable state starting with the generator $X^+_{i\hat{j}}$. This means that there is at least one pair of integers $(k, \hat{\ell})$ so that (3.43) is violated. Let us enumerate the pairs violating (3.43) as*

$$(k_m, \hat{\ell}_{m,n}), \quad i < k_1 < \cdots < k_p, \quad \hat{j} > \hat{\ell}_{m,1} > \cdots > \hat{\ell}_{m,q(m)}, \tag{3.50}$$

so that the following holds:

$$v_{k_m,\hat{j}} = v_{i,\hat{\ell}_{m,n}} = 1 \quad \text{and} \quad v_{k_m,\hat{\ell}_{m,n}} = 0. \tag{3.51}$$

Then the norm of $\Psi_{0,\bar{v}}$ is given by the following formula:

$$\|\Psi_{0,\bar{v}}\|^2 = (x_{i,\hat{j}} + \tilde{v}_{i,\hat{j}})\|\Psi_{0,\bar{v}'}\|^2 - \sum_{m=1}^{p}\sum_{n=1}^{q(m)} c^{m,n}_{0,\bar{v}}, \tag{3.52a}$$

$$c^{1,n}_{0,\bar{v}} = \left(\prod_{s=1}^{n-1}(x_{i,\hat{\ell}_{1,s}} + v_i - \hat{v}_{\hat{\ell}_{1,s}} - s + 1)\right)\|\Psi_{0,\bar{v}^{1,n}}\|^2, \tag{3.52b}$$

$$v^{1,n}_{i\hat{j}} = v^{1,n}_{k_1,\hat{j}} = v^{1,n}_{i,\hat{\ell}_{1,1}} = \cdots = v^{1,n}_{i,\hat{\ell}_{1,n}} = 0, \quad v^{1,n}_{k_1,\hat{\ell}_{1,n}} = 1,$$

$$\text{the rest of } v^{1,n}_{k,\hat{\ell}} \text{ are as } v_{k,\hat{\ell}}$$

$$c^{2,n}_{0,\bar{v}} = (x_{k_1,\hat{j}} + v_{k_1} - \hat{v}_{\hat{j}})\left(\prod_{s=1}^{n-1}(x_{i,\hat{\ell}_{2,s}} + v_i - \hat{v}_{\hat{\ell}_{2,s}} - s + 1)\right)\|\Psi_{0,\bar{v}^{2,n}}\|^2, \tag{3.52c}$$

$$v^{2,n}_{i\hat{j}} = v^{2,n}_{k_1,\hat{j}} = v^{2,n}_{k_2,\hat{j}} = v^{2,n}_{i,\hat{\ell}_{2,1}} = \cdots = v^{2,n}_{i,\hat{\ell}_{2,n}} = 0, \quad v^{2,n}_{k_2,\hat{\ell}_{2,n}} = 1,$$

$$\text{the rest of } v^{2,n}_{k,\hat{\ell}} \text{ are as } v_{k,\hat{\ell}}$$

$$\begin{aligned} c^{3,n}_{0,\bar{v}} &= (x_{k_1,\hat{j}} + v_{k_1} - \hat{v}_{\hat{j}})(x_{k_2,\hat{j}} + v_{k_2} - \hat{v}_{\hat{j}} + 1) \\ &\quad \times \left(\prod_{s=1}^{n-1}(x_{i,\hat{\ell}_{3,s}} + v_i - \hat{v}_{\hat{\ell}_{3,s}} - s + 1)\right)\|\Psi_{0,\bar{v}^{3,n}}\|^2, \end{aligned} \tag{3.52d}$$

$$v^{3,n}_{i\hat{j}} = v^{3,n}_{k_1,\hat{j}} = v^{3,n}_{k_2,\hat{j}} = v^{3,n}_{k_3,\hat{j}} = v^{3,n}_{i,\hat{\ell}_{3,1}} = \cdots = v^{3,n}_{i,\hat{\ell}_{3,n}} = 0,$$

$$v^{3,n}_{k_3,\hat{\ell}_{3,n}} = 1, \quad \text{the rest of } v^{3,n}_{k,\hat{\ell}} \text{ are as } v_{k,\hat{\ell}}.$$ ◇

Proof. The reason for the counterterms is in the transmutation of generators which happens for every pair from (3.50), (3.51) by the following mechanism. Let us take one such pair for fixed (m, n). This means that $\Psi_{0,\bar{v}}$ contains the operators

$$\Psi_{0,\bar{v}} = X^+_{i\hat{j}} \ldots X^+_{k_m,\hat{j}} \ldots X^+_{i,\hat{\ell}_{m,n}} \ldots v_0 \tag{3.53a}$$

and its norm squared is

$$\|\Psi_{0,\bar{v}}\|^2 = (X^+_{i\hat{j}} \ldots X^+_{k_m,\hat{j}} \ldots X^+_{i,\hat{\ell}_{m,n}} \ldots v_0, X^+_{i\hat{j}} \ldots X^+_{k_m,\hat{j}} \ldots X^+_{i,\hat{\ell}_{m,n}} \ldots v_0)$$

$$= (-1)^{\nu}(v_0, \ldots X^-_{i,\hat{\ell}_{m,n}} \ldots X^-_{k_m,\hat{\jmath}} \ldots X^-_{i\hat{\jmath}} X^+_{i\hat{\jmath}} \ldots X^+_{k_m,\hat{\jmath}} \ldots X^+_{i,\hat{\ell}_{m,n}} \ldots v_0) \tag{3.53b}$$

Furthermore, we shall give only the term of $\|\Psi_{0,\bar{\nu}}\|^2$, which will turn into the discussed counterterm:

$$\begin{aligned}
\|\Psi_{0,\bar{\nu}}\|^2 &\approx (-1)^{\nu+1}(v_0, \ldots X^-_{i,\hat{\ell}_{m,n}} \ldots X^-_{k_m,\hat{\jmath}} \ldots X^+_{i\hat{\jmath}} X^-_{i\hat{\jmath}} \ldots X^+_{k_m,\hat{\jmath}} \ldots X^+_{i,\hat{\ell}_{m,n}} \ldots v_0) \\
&\approx (-1)^{\nu+1}(v_0, \ldots X^-_{i,\hat{\ell}_{m,n}} \ldots X^-_{k_m,\hat{\jmath}} X^+_{i\hat{\jmath}} \ldots X^-_{i\hat{\jmath}} X^+_{k_m,\hat{\jmath}} \ldots X^+_{i,\hat{\ell}_{m,n}} \ldots v_0) \\
&= (-1)^{\nu+1}(v_0, \ldots X^-_{i,\hat{\ell}_{m,n}} \ldots (-X^+_{i\hat{\jmath}} X^-_{k_m,\hat{\jmath}} - L^+_{i,k_m}) \ldots \\
&\quad \times \ldots (-X^+_{k_m,\hat{\jmath}} X^-_{i\hat{\jmath}} - L^-_{k_m,i}) \ldots X^+_{i,\hat{\ell}_{m,n}} \ldots v_0) && (3.53c) \\
&\approx (-1)^{\nu+1}(v_0, \ldots X^-_{i,\hat{\ell}_{m,n}} L^+_{i,k_m} \ldots L^-_{k_m,i} X^+_{i,\hat{\ell}_{m,n}} \ldots v_0) && (3.53d) \\
&= (-1)^{\nu+1}(v_0, \ldots (L^+_{i,k_m} X^-_{i,\hat{\ell}_{m,n}} + X^-_{k_m,\hat{\ell}_{m,n}}) \ldots \\
&\quad \times \ldots (X^+_{i,\hat{\ell}_{m,n}} L^-_{k_m,i} + X^+_{k_m,\hat{\ell}_{m,n}}) \ldots v_0) && (3.53e) \\
&\approx (-1)^{\nu+1}(v_0, \ldots X^-_{k_m,\hat{\ell}_{m,n}} \ldots X^+_{k_m,\hat{\ell}_{m,n}} \ldots v_0) && (3.53f) \\
&= -\| \ldots X^+_{k_m,\hat{\ell}_{m,n}} \ldots v_0 \|^2 && (3.53g)
\end{aligned}$$

Thus, we have shown that the norm squared of $\Psi_{0,\bar{\nu}}$ contains a term which is the norm squared (with sign 'minus' – hence the word 'counterterm') of a state obtained from $\Psi_{0,\bar{\nu}}$ by replacing the operators $X^+_{i\hat{\jmath}}$, $X^+_{k_m,\hat{\jmath}}$ and $X^+_{i,\hat{\ell}_{m,n}}$ by the operator $X^+_{k_m,\hat{\ell}_{m,n}}$. Note that the latter was not present in $\Psi_{0,\bar{\nu}}$ due to the condition $\nu_{k_m,\hat{\ell}_{m,n}} = 0$ in (3.51). Note also that the counterterm state is of level $\nu - 2$, which brings about the factor $(-1)^{\nu-2}$ in the passage from (3.53f) to (3.53g) which together with the factor $(-1)^{\nu+1}$ results in the overall minus sign in (3.53g). The described transmutation explains totally only the first counterterm in (3.52b) obtained for $(m, n) = (1, 1)$. The other counterterms get additional contributions, in particular, from terms which we neglected in (3.53). For the rest of the counterterms with $(m = 1, n > 1)$ this affects the contributions of the operators $X^+_{i,\hat{\ell}_{1,s}}$, $s < n$. Analogously, for $m > 1$ this affects in addition the operators $X^+_{k_s,\hat{\jmath}}$, $s < m$. In all cases, every counterterm is a polynomial in d of degree $\nu-2$. It remains only to explain the overall restrictions on the number of counterterms: since $i < 4$, $\hat{\jmath} > 1$, it follows that $p \le 4 - i \le 3$, $q(m) \le \hat{\jmath} - 1$. □

Our next main result on the norms is

Proposition 6. *Let $\Psi_{\bar{\varepsilon}\bar{\nu}}$ be a unfactorizable state starting with the generator $Y^+_{i\hat{\jmath}}$. This means that there are one or more pairs of integers $(k, \hat{\ell})$ so that (3.42) is violated. Let us enumerate the pairs violating (3.42a) as*

$$(j_m, \hat{\jmath}_{m,n}), \quad i < j_1 < \cdots < j_p, \quad \hat{\jmath} < \hat{\jmath}_{m,1} < \cdots < \hat{\jmath}_{m,q(m)} \tag{3.54}$$

(note that $i < 4$, $\hat{\jmath} < N$, $p \le 4 - i \le 3$, $q(m) \le N - \hat{\jmath}$) so that the following holds:

$$\varepsilon_{j_m,\hat{\jmath}} = \varepsilon_{i,\hat{\jmath}_{m,n}} = 1 \quad \text{and} \quad \varepsilon_{j_m,\hat{\jmath}_{m,n}} = 0. \tag{3.55}$$

Let us enumerate the pairs violating (3.42b) *as*

$$(k_m, \hat{k}_{m,n}), \quad i < k_1 < \cdots < k_{p'}, \quad \hat{k}_{m,1} > \cdots > \hat{k}_{m,q'(m)} \tag{3.56}$$

(note that $i < 4$, $p' \leq 4 - i \leq 3$, $q'(m) \leq N$*) so that the following holds:*

$$\varepsilon_{k_m,\hat{j}} = \nu_{i,\hat{k}_{m,n}} = 1 \quad \text{and} \quad \nu_{k_m,\hat{k}_{m,n}} = 0. \tag{3.57}$$

Let us enumerate the pairs violating (3.42c) *as*

$$(\ell_m, \hat{\ell}_{m,n}), \quad \ell_1 < \cdots < \ell_{p''}, \quad \hat{j} < \hat{\ell}_{m,1} < \cdots < \hat{\ell}_{m,q''(m)} \tag{3.58}$$

(note that $\hat{j} < N$, $p'' \leq 4$, $q''(m) \leq N - \hat{j}$*) so that the following holds:*

$$\varepsilon_{i,\hat{\ell}_{m,n}} = \nu_{\ell_m,\hat{\ell}_{m,n}} = 1 \quad \text{and} \quad \nu_{\ell_m,\hat{j}} = 0. \tag{3.59}$$

Then the norm is given by the following formula:

$$\begin{aligned} \|\Psi_{\bar{\varepsilon}\bar{\nu}}\|^2 &= (y_{i,\hat{j}} + \tilde{\varepsilon}_{i,\hat{j}} + \nu_i + \tilde{\nu}_{\hat{j}})\|\Psi_{\bar{\varepsilon}'\bar{\nu}}\|^2 - \sum_{m=1}^{p}\sum_{n=1}^{q(m)} C^{m,n}_{\bar{\varepsilon},\bar{\nu}} \\ &\quad - \sum_{m=1}^{p'}\sum_{n=1}^{q'(m)} C'^{m,n}_{\bar{\varepsilon},\bar{\nu}} - \sum_{m=1}^{p''}\sum_{n=1}^{q''(m)} C''^{m,n}_{\bar{\varepsilon},\bar{\nu}}, \end{aligned} \tag{3.60a}$$

$$\begin{aligned} C^{1,n}_{\bar{\varepsilon},\bar{\nu}} &= \left(\prod_{s=1}^{n-1}(y_{i,\hat{j}_{1,s}} + \varepsilon_i - \hat{\varepsilon}_{\hat{j}_{1,s}} - s + 1 + \nu_i + \hat{\nu}_{\hat{j}_{1,s}})\right)\|\Psi_{\bar{\varepsilon}^{1,n},\bar{\nu}}\|^2, \\ \varepsilon_i &= \varepsilon_{i,1} + \cdots + \varepsilon_{i,N}, \quad \hat{\varepsilon}_{\hat{\ell}} = \varepsilon_{1,\hat{\ell}} + \cdots + \varepsilon_{4,\hat{\ell}}, \\ \varepsilon^{1,n}_{i,\hat{j}} &= \varepsilon^{1,n}_{j_1,\hat{j}} = \varepsilon^{1,n}_{i,\hat{j}_{1,1}} = \cdots = \varepsilon^{1,n}_{i,\hat{j}_{1,n}} = 0, \quad \varepsilon^{1,n}_{j_1,\hat{j}_{1,n}} = 1, \\ &\quad \textit{the rest of } \varepsilon^{1,n}_{k,\hat{\ell}} \textit{ are as } \varepsilon_{k,\hat{\ell}} \end{aligned} \tag{3.60b}$$

$$\begin{aligned} C^{2,n}_{\bar{\varepsilon},\bar{\nu}} &= (y_{j_1,\hat{j}} + \varepsilon_{j_1} - \hat{\varepsilon}_{\hat{j}} + \nu_{j_1} + \hat{\nu}_{\hat{j}}) \\ &\quad \times \left(\prod_{s=1}^{n-1}(y_{i,\hat{j}_{2,s}} + \varepsilon_i - \hat{\varepsilon}_{\hat{j}_{2,s}} - s + 1 + \nu_i + \hat{\nu}_{\hat{j}_{2,s}})\right)\|\Psi_{\bar{\varepsilon}^{2,n},\bar{\nu}}\|^2, \\ \varepsilon^{2,n}_{i,\hat{j}} &= \varepsilon^{2,n}_{j_1,\hat{j}} = \varepsilon^{2,n}_{j_2,\hat{j}} = \varepsilon^{2,n}_{i,\hat{j}_{2,1}} = \cdots = \varepsilon^{2,n}_{i,\hat{j}_{2,n}} = 0, \\ \varepsilon^{2,n}_{j_2,\hat{j}_{2,n}} &= 1, \quad \textit{the rest of } \varepsilon^{2,n}_{k,\hat{\ell}} \textit{ are as } \varepsilon_{k,\hat{\ell}}, \end{aligned} \tag{3.60c}$$

$$\begin{aligned} C^{3,n}_{\bar{\varepsilon},\bar{\nu}} &= (y_{j_1,\hat{j}} + \varepsilon_{j_1} - \hat{\varepsilon}_{\hat{j}} + \nu_{j_1} + \hat{\nu}_{\hat{j}})(y_{j_2,\hat{j}} + \varepsilon_{j_2} - \hat{\varepsilon}_{\hat{j}} + 1 + \nu_{j_2} + \hat{\nu}_{\hat{j}}) \\ &\quad \times \left(\prod_{s=1}^{n-1}(y_{i,\hat{j}_{3,s}} + \varepsilon_i - \hat{\varepsilon}_{\hat{j}_{3,s}} - s + 1 + \nu_i + \hat{\nu}_{\hat{j}_{3,s}})\right)\|\Psi_{\bar{\varepsilon}^{3,n},\bar{\nu}}\|^2, \\ \varepsilon^{3,n}_{i,\hat{j}} &= \varepsilon^{3,n}_{j_1,\hat{j}} = \varepsilon^{3,n}_{j_2,\hat{j}} = \varepsilon^{3,n}_{j_3,\hat{j}} = \varepsilon^{3,n}_{i,\hat{j}_{3,1}} = \cdots = \varepsilon^{3,n}_{i,\hat{j}_{3,n}} = 0, \\ \varepsilon^{3,n}_{j_3,\hat{j}_{3,n}} &= 1, \quad \textit{the rest of } \varepsilon^{3,n}_{k,\hat{\ell}} \textit{ are as } \varepsilon_{k,\hat{\ell}}, \end{aligned} \tag{3.60d}$$

$$C'^{m,n}_{\bar{\varepsilon},\bar{\nu}} = \Bigg(\prod_{1\le j\le 4}\ \prod_{\substack{\hat{\jmath}\le\hat{m}\le N\\ (j,\hat{m})\neq(i,\hat{\jmath}),(k_m,\hat{\jmath})}} (y_{j\hat{m}} + \varepsilon'^{m}_{j} - \hat{\varepsilon}'_{\hat{m}} + \nu'_j + \tilde{\nu}'^{m,n}_{\hat{m}})^{\varepsilon_{j\hat{m}}}\Bigg)$$

$$\times\Bigg(\prod_{s=1}^{n-1}(x_{i,\hat{k}_{m,s}} + \nu_i - \hat{\nu}_{\hat{k}_{m,s}} - s + 1)\Bigg)\|\Psi_{0,\bar{\nu}'^{m,n}}\|^2, \tag{3.60e}$$

$$\varepsilon'^{m}_{j} = \varepsilon_{j,1} + \cdots + \varepsilon_{j,N} - \delta_{ji} - \delta_{j,k_m},$$

$$\hat{\varepsilon}'_{\hat{m}} = \varepsilon_{1,\hat{m}} + \cdots + \varepsilon_{4,\hat{m}} - 2\delta_{\hat{m},\hat{\jmath}},$$

$$\nu'_j = \nu_j - \delta_{ij}, \quad \tilde{\nu}'^{m,n}_{\hat{m}} = \tilde{\nu}_{\hat{m}} - \delta_{\hat{m},\hat{k}_{m,n}},$$

$$\nu'^{m,n}_{i,\hat{k}_{m,1}} = \cdots = \nu'^{m,n}_{i,\hat{k}_{m,n}} = 0, \quad \nu'^{m,n}_{k_m,\hat{k}_{m,n}} = 1,$$

$$\textit{the rest of } \nu'^{m,n}_{k,\hat{\ell}} \textit{ are as } \nu_{k,\hat{\ell}}$$

$$C''^{m,n}_{\bar{\varepsilon},\bar{\nu}'} = \Bigg(\prod_{1\le j\le 4}\ \prod_{\substack{\hat{\jmath}\le\hat{m}\le N\\ (j,\hat{m})\neq(i,\hat{\jmath}),(i,\hat{\ell}_{m,n})}} (y_{j\hat{m}} + \varepsilon''^{m}_{j} - \hat{\varepsilon}''_{\hat{m}} + \nu''_j + \tilde{\nu}''^{m,n}_{\hat{m}})^{\varepsilon_{j\hat{m}}}\Bigg)$$

$$\times\Bigg(\prod_{s=1}^{n-1}(x_{\hat{\ell}_m,\hat{\ell}_{m,s}} + \nu_{\hat{\ell}_m} - \hat{\nu}_{\hat{\ell}_{m,s}} - s + 1)\Bigg)\|\Psi_{0,\bar{\nu}''^{m,n}}\|^2, \tag{3.60f}$$

$$\varepsilon''^{m}_{j} = \varepsilon_{j,1} + \cdots + \varepsilon_{j,N} - 2\delta_{ji},$$

$$\hat{\varepsilon}''_{\hat{m}} = \varepsilon_{1,\hat{m}} + \cdots + \varepsilon_{4,\hat{m}} - \delta_{\hat{m},\hat{\jmath}} - \delta_{\hat{m},\hat{\ell}_{m,n}},$$

$$\nu''_j = \nu_j - \delta_{j,\hat{\ell}_m}, \quad \tilde{\nu}''^{m,n}_{\hat{m}} = \tilde{\nu}_{\hat{m}} - \delta_{\hat{m},\hat{\ell}_{m,n}},$$

$$\nu''^{m,n}_{\ell_m,\hat{\ell}_{m,1}} = \cdots = \nu''^{m,n}_{\ell_m,\hat{\ell}_{m,n}} = 0, \quad \nu''^{m,n}_{\ell_m,\hat{\jmath}} = 1,$$

$$\textit{the rest of } \nu''^{m,n}_{k,\hat{\ell}} \textit{ are as } \nu_{k,\hat{\ell}}.$$

◇

The proof of this proposition is analogous to the one of Proposition 5, though more complicated since there are three possible mechanisms of transmutations corresponding to the three exceptional situations given. Thus, in a case described by (3.54),(3.55) the transmutation is

$$\Psi_{\bar{\varepsilon},\bar{\nu}} = Y^+_{i\hat{\jmath}}\dots Y^+_{j_m,\hat{\jmath}}\dots Y^+_{i,\hat{\jmath}_{m,n}}\dots v_0 \longrightarrow \dots Y^+_{j_m,\hat{\jmath}_{m,n}}\dots v_0. \tag{3.61}$$

In the case described by (3.56) and (3.57) the transmutation is

$$\Psi_{\bar{\varepsilon},\bar{\nu}} = Y^+_{i\hat{\jmath}}\dots Y^+_{k_m,\hat{\jmath}}\dots X^+_{i,\hat{k}_{m,n}}\dots v_0 \longrightarrow \dots X^+_{k_m,\hat{k}_{m,n}}\dots v_0. \tag{3.62}$$

In the case described by (3.58) and (3.59) the transmutation is

$$\Psi_{\bar{\varepsilon},\bar{\nu}} = Y^+_{i\hat{\jmath}}\dots Y^+_{i,\hat{\ell}_{m,n}}\dots X^+_{\ell_m,\hat{\ell}_{m,n}}\dots v_0 \longrightarrow \dots X^+_{\ell_m,\hat{\jmath}}\dots v_0. \tag{3.63}$$

Note that for $N = 1$ only the cases described by (3.56),(3.57) are possible. Furthermore, we proceed as for Proposition 5. □

Our final main result on the norms is

Proposition 7. *If a state is not fully factorizable then the general expression of its norm is*

$$\|\Psi_{\bar{\varepsilon}\bar{\nu}}\|^2 = \mathcal{N}_{\bar{\varepsilon}\bar{\nu}} - \mathcal{C}_{\bar{\varepsilon}\bar{\nu}}, \tag{3.64}$$

where $\mathcal{C}_{\bar{\varepsilon}\bar{\nu}}$ designates the possible counterterms. ◇

Proof. This follows from Propositions 5 and 6. Consider first $\Psi_{0,\bar{\nu}'}$ from Proposition 5. If it is fully factorizable, then (3.64) follows at once. If it is not fully factorizable but factorizable we first apply (3.44) one or more times until we are left with an unfactorizable state and then we apply Proposition 5 to the latter. We get another state which plays the role of $\Psi_{0,\bar{\nu}'}$. Proceeding further like this we establish (3.64) at the end. Analogously we consider $\Psi_{\bar{\varepsilon}',\bar{\nu}}$ from Proposition 6 until we establish (3.64) for this case. □

The above enables us to show that the conditions of the theorem are *sufficient* for $d > d_{11}^-$. Indeed, in that case $\mathcal{N}_{\bar{\varepsilon}\bar{\nu}} > 0$ for all states. What turns out to be important for the unitarity is that all counterterms are polynomials in d of lower degrees than $\mathcal{N}_{\bar{\varepsilon}\bar{\nu}}$ and all positivity requirements are determined by the terms $\mathcal{N}_{\bar{\varepsilon}\bar{\nu}}$. □

Unitarity at the reduction points will be considered in the next section.

3.1.3 Unitarity at the reduction points

3.1.3.1 The first reduction point

In this section we consider the unitarity of the irreps at the reducibility points d_{i1}^-. Unitarity is established by noting that there are no negative norm states and by factoring out the zero norm states, which are a typical feature of the Verma modules V^Λ at the reducibility points. These zero norm states generate invariant submodules I_{i1} and are decoupled into the factor modules V^Λ/I_{i1}, which realize the UIRs at the points $d = d_{i1}^-$.

In this subsection $d = d_{11}^-$, i. e., $x_{11} = 3$. We have the following.

Proposition 8. *Let $d = d_{11}^-$. There are no negative norm states. The zero norm states are described as follows. In the case $a_{\hat{k}} \neq 0$, $\hat{k} = 1, \ldots, N$, the states of zero norm $\mathcal{F}_0^\Lambda$ from $\mathcal{F}^\Lambda$ are given by $\Psi_{\bar{\varepsilon}\bar{\nu}}$ with*

$$\varepsilon_{i\hat{j}} = 0, 1, \quad \nu_{i\hat{j}} = \begin{cases} 1 & \hat{j} = 1 \\ 0, 1 & \textit{otherwise.} \end{cases} \tag{3.65}$$

The number of such states is 2^{8N-4} and the number of oddly generated states in the reduced irrep $L^\Lambda \equiv \mathcal{F}^\Lambda/\mathcal{F}_0^\Lambda$ is $15 \times 2^{8N-4}$.

In the cases $a_1 = \cdots = a_{\hat{k}} = 0$, $\hat{k} = 1, \ldots, N-1$, in addition to those in (3.65) the following states have zero norm:

$$\varepsilon_{i\hat{j}} = 0, 1,$$

$$v_{i\hat{j}} = \begin{cases} 1 & \hat{j} = 2 \\ 0 & i = \hat{j} = 1 \\ 0,1 & \textit{otherwise}, \end{cases}$$

$$\text{and} \quad v_{i\hat{j}} = \begin{cases} 1 & \hat{j} = 3 \\ 0 & i = 1, \hat{j} = 1, 2 \\ 0,1 & \textit{otherwise}, \end{cases}$$

$$\dots$$

$$\text{and} \quad v_{i\hat{j}} = \begin{cases} 1 & \hat{j} = \hat{k} + 1 \\ 0 & i = 1, \hat{j} = 1, \dots, \hat{k} \\ 0,1 & \textit{otherwise}. \end{cases} \tag{3.66}$$

The number of states in the kth row of (3.66) *is* $2^{8N-5}, 2^{8N-6}, \dots, 2^{8N-4-\tilde{k}}$*, thus, the overall number of states in* (3.66) *is* $2^{8N-4-\tilde{k}}(2^{\tilde{k}} - 1)$*, the number of states in the reduced* L^{Λ}*— factoring out both* (3.65) *and* (3.66)*—is* $2^{8N-4-\tilde{k}}(2^{4+\tilde{k}} - 2^{\tilde{k}+1} + 1)$.

In the case $r_1 = 0$ *(R-symmetry scalars) in addition to those in* (3.65) *and* (3.66) *for* $\hat{k} = N - 1$*, the following states have zero norm:*

$$\varepsilon_{i\hat{j}} = \begin{cases} 1 & \hat{j} = 1 \\ 0 & i = 1, \hat{j} > 1 \\ 0,1 & \textit{otherwise}, \end{cases}$$

$$v_{i\hat{j}} = \begin{cases} 0 & \hat{j} = 1 \\ 0 & i = 1 \\ 0,1 & \textit{otherwise}, \end{cases} \tag{3.67a}$$

$$\text{and} \quad \varepsilon_{i\hat{j}} = \begin{cases} 1 & \hat{j} = 2 \\ 0 & i = 1, \hat{j} > 2 \\ 0,1 & \textit{otherwise}, \end{cases}$$

$$v_{i\hat{j}} = \begin{cases} 0 & \hat{j} = 2 \\ 0 & i = 1 \\ 0,1 & \textit{otherwise}, \end{cases} \tag{3.67b}$$

$$\dots$$

$$\text{and} \quad \varepsilon_{i\hat{j}} = \begin{cases} 1 & \hat{j} = N - 1 \\ 0 & i = 1, \hat{j} = N \\ 0,1 & \textit{otherwise}, \end{cases}$$

$$v_{i\hat{j}} = \begin{cases} 0 & \hat{j} = N - 1 \\ 0 & i = 1 \\ 0,1 & \textit{otherwise}, \end{cases} \tag{3.67c}$$

$$\text{and} \quad \varepsilon_{i\hat{j}} = \begin{cases} 1 & \hat{j} = N \\ 0,1 & \textit{otherwise}, \end{cases}$$

$$\nu_{i\hat{j}} = \begin{cases} 0 & \hat{j} = N \\ 0 & i = 1 \\ 0,1 & \textit{otherwise}. \end{cases} \tag{3.67d}$$

The number of states in the kth row of (3.67) *is* $2^{6N-6}, 2^{6N-5}, \ldots, 2^{7N-8}, 2^{7N-7}$, *thus, the overall number of states in* (3.67) *is* $2^{6N-6}(2^N-1)$, *the number of states in the reduced* L^Λ*— factoring out* (3.65), (3.66) *(for* $\tilde{k} = N-1$*) and* (3.67)*—is* $2^{6N-6}(2^{2N+6} - (2^{N+3}+1)(2^N-1))$.◇

Proof. There are no negative norm states if $d > d_{11}^-$ and thus there are no such states for $d = d_{11}^-$ by continuity. For the zero norm states we start with the case $a_{\hat{k}} \neq 0$, $\hat{k} = 1, \ldots, N$. Inspecting formula (3.49) we see that the fully factorized states of zero norm have the form (3.65). Indeed, the only factor in $\mathcal{N}_{\bar{\varepsilon}\bar{\nu}}$ that can be zero is $(x_{11} + \tilde{\nu}_{11}) = (3 + \tilde{\nu}_{11})$, (hence $\nu_{11} = 1$), which happens if $\tilde{\nu}_{11} = -3$ which happens if $\nu_{i1} = 1$, $i = 2, 3, 4$. (In general, $(x_{i\hat{j}} + \tilde{\nu}_{i\hat{j}}) \geq (3 + \tilde{\nu}_{i\hat{j}}) \geq (i + \hat{j} - 2)$.) Further, the problem is reduced to unfactorizable states. The main term of the norm squared is given again by $\mathcal{N}_{\bar{\varepsilon}\bar{\nu}}$ which is zero. For further use we note more explicitly that for the states from (3.65) we have

$$\mathcal{N}_{\bar{\varepsilon}\bar{\nu}} \sim (x_{11} - 3)(x_{21} - 2)(x_{31} - 1)x_{41}. \tag{3.68}$$

Now we shall show that also the counterterms are zero. For this it is enough to show that $\nu_{i1}^{m,n} = 1$, $i = 1, 2, 3, 4$ in all auxiliary states that happen in the counterterms. Consider first states starting with $X_{i\hat{j}}^+$ for which the norm is given in Proposition 6. The only way $\nu_{i1}^{m,n}$ could differ from ν_{i1} is if one of the pairs in (3.50) is of the form $(k_m, 1)$, more precisely, that could be only one of the pairs $(k_m, \hat{k}_{m,q(m)}) = (k_m, 1)$. But then according to (3.51) for any possible m we should have $\nu_{i,1} = 1$ and $\nu_{k_m,1} = 0$, which does not hold. Thus, all counterterms are also zero. Consider next states starting with $Y_{i\hat{j}}^+$ for which the norm is given in Proposition 7. Here only the counterterms in (3.60e,f) might be non-zero. For the counterterm in (3.60e) the only way $\nu'^{m,n}_{i1}$ could differ from ν_{i1} is if one of the pairs in (3.56) is of the form $(k_m, 1)$, more precisely, that could be only one of the pairs $(k_m, \tilde{\lambda}_{m,q'(m)}) = (k_m, 1)$. But then according to (3.57) for any possible m we should have $\nu_{i,1} = 1$ and $\nu_{k_m,1} = 0$, which does not hold. For the counterterm in (3.60f) the considerations are simpler since it is immediately seen from (3.58) that there is no pair that can affect $\nu_{i,1}$ since all $\tilde{\lambda}_{m,n} > \hat{j} \geq 1$, and if we consider $\hat{j} = 1$ then our state does not fulfil the condition in (3.59) $\nu_{\ell_m,1} = 0$. Thus, all possible counterterms are zero and thus all states in (3.65) have zero norm. We continue with the cases $a_{\hat{k}} \neq 0$, $\hat{k} = 1, \ldots, N-1$. Then $x_{1,\hat{k}+1} = \cdots = x_{12} = x_{11} = 3$. Under the hypothesis in (3.66) for $k = 1$ we have $\tilde{\nu}_{12} = -3$, hence $x_{12} + \tilde{\nu}_{12} = 0$ and the corresponding states have zero norm—the argument for unfactorizable states goes analogously to above. The same reasoning goes for all other cases in (3.66). For further use we note more explicitly that, for the states from (3.66) for $\hat{\ell} = 1, 2, \ldots, \hat{k}$, we have

$$\mathcal{N}_{\bar{\varepsilon}\bar{\nu}} \sim (x_{1,\hat{\ell}+1} - 3)(x_{2,\hat{\ell}+1} - 2)(x_{3,\hat{\ell}+1} - 1)x_{4,\hat{\ell}+1}. \tag{3.69}$$

We continue with the case $r_1 = 0$. Then $y_{1,N} = \cdots = y_{11} = x_{11} = 3$. Under the hypothesis in (3.67) for $k = 1$ we have $\tilde{\varepsilon}_{11} = -3$, hence $y_{11} + \tilde{\nu}_{11} = 0$ and the corresponding states have zero norm—the argument for unfactorizable states goes analogously to above. The same reasoning goes for all other cases in (3.67). For further use we note more explicitly that, for the states from (3.67) for $\hat{\ell} = 1, 2, \ldots, N$, we have

$$\mathcal{N}_{\bar{\varepsilon}\bar{\nu}} \sim (y_{1,\hat{\ell}} - 3)(y_{2,\hat{\ell}} - 2)(y_{3,\hat{\ell}} - 1)y_{4,\hat{\ell}}. \tag{3.70}$$

The counting of states is straightforward. □

3.1.3.2 The other reduction points

We first consider the case (3.36b) of the theorem: $d = d_{21}^-$ and $n_1 = 0$, i. e., $x_{11} = x_{21} = 2$. We have the following.

Proposition 9. *Let $d = d_{21}^-$ and $n_1 = 0$. There are no negative norm states. All states of zero norm which are described in Proposition 8 have zero norm also under the present hypothesis. There are further states of zero norm which are described as follows. In the case $a_{\hat{k}} \neq 0$, $\hat{k} = 1, \ldots, N$, the additional states of zero norm are given by $\Psi_{\bar{\varepsilon}\bar{\nu}}$ with:*

$$\varepsilon_{i\hat{j}} = 0, 1, \quad \nu_{11} + \nu_{21} + \nu_{31} + \nu_{41} = 3, \quad \nu_{i\hat{j}} = 0, 1, \quad \hat{j} \neq 1. \tag{3.71}$$

The number of states in (3.71) is 2^{8N-2}, and the number of states in the reduced L^Λ—factoring out both (3.65) and (3.71)—is $11 \times 2^{8N-4}$.

In the case $a_1 = 0$ in addition to those in (3.71) the following states have zero norm for $N = 1$:

$$\varepsilon_{i1} = 1, \quad \nu_{11} = 0, \quad \nu_{21} + \nu_{31} + \nu_{41} = 1, \tag{3.72a}$$

$$\text{and} \quad \varepsilon_{11} + \varepsilon_{21} + \varepsilon_{31} + \varepsilon_{41} = 3, \quad \nu_{i1} = 0. \tag{3.72b}$$

The numbers of states in (3.72a) and (3.72b) are 3 and 4, respectively the overall number of zero states—including (3.65), (3.67), (3.71), and (3.72)—is 88, and thus the number of states of the reduced L^Λ is 168. ◇

Proof. We first have to show that the states of zero norm from Proposition 8 have zero norm also here. With this we shall establish also that there no negative norm states since those are states are the only suspects for this. For the cases described by (3.65) this follows by inspecting (3.68) which is zero also here. In the cases described by (3.66) and (3.67) this follows by inspecting (3.69), (3.70), which are zero also here. Further, the proof is as of Proposition 8. In particular, for the states from (3.71) we have

$$\mathcal{N}_{\bar{\varepsilon}\bar{\nu}} \sim (x_{i_1,1} - 2)(x_{i_2,1} - 1)x_{i_3,1}, \tag{3.73}$$

where i_j are from the set 1, 2, 3, 4, and thus at least one of them is equal to 1 or 2, hence the RHS of (3.73) is zero. Analogously, for the states from (3.67a) holds (3.70) for $\hat{\ell} = 1$, hence $\mathcal{N}_{\bar{\varepsilon}\bar{\nu}} = 0$. For the states from (3.67b) holds:

$$\mathcal{N}_{\bar{\varepsilon}\bar{\nu}} \sim (y_{i_1,1} - 2)(y_{i_2,1} - 1)y_{i_3,1}, \tag{3.74}$$

which is zero as (3.73) since $y_{i1} = x_{i1}$ for $a_1 = 0$. □

Next we consider the case (3.36c) of the theorem: $d = d_{31}^-$ and $n_1 = n_2 = 0$, i. e., $x_{11} = x_{21} = x_{31} = 1$. We have the following.

Proposition 10. *Let $d = d_{31}^-$ and $n_1 = n_2 = 0$. There are no negative norm states. All states of zero norm which are described in Propositions 8 and 9 have zero norm also under the present hypothesis. There are further states of zero norm which are described as follows. In the case $a_{\hat{k}} \neq 0$, $\hat{k} = 1, \ldots, N$, the additional states of zero norm are given by $\Psi_{\bar{\varepsilon}\bar{\nu}}$ with:*

$$\begin{aligned} \varepsilon_{i\hat{j}} &= 0, 1, \\ \nu_{11} + \nu_{21} + \nu_{31} + \nu_{41} &= 2. \end{aligned} \tag{3.75}$$

The number of states in (3.75) is $3 \times 2^{8N-3}$, and the number of states in the reduced L^Λ—factoring out (3.65), (3.71) and (3.75)—is $5 \times 2^{8N-4}$.

In the case $a_1 = 0$ in addition to those in (3.75) the following states have zero norm for $N = 1$:

$$\varepsilon_{i1} = 1, \quad \nu_{i1} = \delta_{i1}, \tag{3.76a}$$

$$\begin{aligned} \text{and} \quad \varepsilon_{11} + \varepsilon_{21} + \varepsilon_{31} + \varepsilon_{41} = 3, &\quad \nu_{11} + \nu_{21} + \nu_{31} + \nu_{41} = 1, \\ \varepsilon_{11} = 1 \quad \Rightarrow &\quad \nu_{11} = 0, \\ \varepsilon_{11} = 0 \quad \Rightarrow &\quad \nu_{21} = 0, \end{aligned} \tag{3.76b}$$

$$\text{and} \quad \varepsilon_{11} + \varepsilon_{21} + \varepsilon_{31} + \varepsilon_{41} = 2, \quad \nu_{i1} = 0. \tag{3.76c}$$

The number of states in (3.76a,b,c) is 1, 12, 6, respectively, the overall number of zero states—including (3.65), (3.67), (3.71), (3.72), (3.75), (3.76)—is 203, and thus the number of states of the reduced L^Λ is 53. ◇

Proof. We first have to show that the states of zero norm from Propositions 8 and 9 have zero norm also here (establishing also the lack of negative norm states). For the cases described by (3.65), (3.66), (3.67$\hat{\ell}$), (3.71),(3.72), this follows by inspecting (3.68),(3.69),(3.70),(3.73),(3.74), which are zero also here. Further, the proof is as of Propositions 8 and 9. In particular, for the states from (3.75) we have

$$\mathcal{N}_{\bar{\varepsilon}\bar{\nu}} \sim (x_{i_1,1} - 1)x_{i_2,1}, \tag{3.77}$$

where i_j are from the set 1, 2, 3, 4, and thus at least one of them is equal to 1 or 2 or 3, hence the RHS of (3.77) is zero. For the states from (3.67a), respectively, (3.67b) we have (3.70) for $\hat{\ell} = 1$, respectively, (3.74), hence $\mathcal{N}_{\bar{\varepsilon}\bar{\nu}} = 0$. For the states from (3.67c) we have

$$\mathcal{N}_{\bar{\varepsilon}\bar{\nu}} \sim (y_{i_1,1} - 1)y_{i_2,1}, \tag{3.78}$$

which is zero as (3.77) since $y_{i1} = x_{i1}$ for $a_1 = 0$. □

Finally, we consider case (3.36d) of the theorem: $d = d_{41}^-$ and $n_1 = n_2 = n_3 = 0$, i. e., $x_{11} = x_{21} = x_{31} = x_{41} = 0$. We have the following.

Proposition 11. *Let $d = d_{41}^-$ and $n_1 = n_2 = n_3 = 0$. There are no negative norm states. All states of zero norm which are described in Propositions 8, 9 and 10 have zero norm also under the present hypothesis. There are further states of zero norm which are described as follows. In the case $a_{\hat{k}} \neq 0$, $\hat{k} = 1, \ldots, N$, the additional states of zero norm are given by $\Psi_{\bar{\varepsilon}\bar{\nu}}$ with:*

$$\varepsilon_{i\hat{j}} = 0, 1, \quad \nu_{11} + \nu_{21} + \nu_{31} + \nu_{41} = 1, \quad \nu_{i\hat{j}} = 0, 1, \quad \hat{j} \neq 1. \tag{3.79}$$

The number of states in (3.79) is 2^{8N-2}, and the number of states in the reduced L^Λ—factoring out (3.65), (3.71), (3.75) and (3.79)—is 2^{8N-4}.

In the case $a_1 = 0$ in addition to those in (3.79) the following states have zero norm for $N = 1$:

$$\varepsilon_{11} + \varepsilon_{21} + \varepsilon_{31} + \varepsilon_{41} = 1, \quad \nu_{i1} = 0. \tag{3.80}$$

The number of states in (3.80) is 4, the overall number of zero states—including (3.65), (3.67), (3.71), (3.72), (3.75), (3.76), (3.79), (3.80)—is $2^8 - 1$ and thus the number of states of the reduced L^Λ is 1, i. e., this is the trivial representation. ◇

Proof. We first have to show that the states of zero norm from Propositions 8, 9 and 10 have zero norm also here (establishing also the lack of negative norm states). This is clear since in all cases the factor $\mathcal{N}_{\bar{\varepsilon}\bar{\nu}}$ contains as multiplicative factor some x_{i1} and hence is zero. The same holds for the states from (3.79). For $N = 1$ and $a_1 \neq 0$ there are 16 states which are of the form $\Psi_{\bar{\varepsilon},0}$. For $a_1 = 0$ all these, beside the vacuum state, are of zero norm since the factor $\mathcal{N}_{\bar{\varepsilon},0}$ contains as multiplicative factor some $y_{i1} = x_{y1} = 0$. For the counting of states we have to note that the 16 states in (3.72a) and (3.76a,b) are contained also in (3.79) if $a_1 = 0$. □

In the last case we gave the counting of states in the cases $a_{\hat{k}} \neq 0$, $\hat{k} = 1, \ldots, N$, only for $N = 1$. That would have taken many more pages due to the complicated combinatorics for $N > 1$ when $a_{\hat{k}} = 0$, and is left to the reader.

3.2 Positive energy UIRS of osp(1|2n, $\mathbb{R}$)

In the previous sections following the restrictions of [218] and [306] we considered conformal superalgebras for $D \le 6$. On the other hand the applications in string theory require the knowledge of the UIRs of the conformal superalgebras for $D > 6$. Most prominent role play the superalgebras osp(1|2n), cf. some of their applications in, e. g., [370, 231, 214, 26, 50, 97]. Initially, the superalgebra osp(1|32) was put forward for $D = 10$ [370]. Later it was realized that osp(1|2n) would fit any dimension, though they are minimal only for $D = 3, 9, 10, 11$ (for $n = 2, 16, 16, 32$, respectively) [97]. In all cases we need to find first the UIRs of osp(1|2n, $\mathbb{R}$). This can be done for general n. Thus, in the first subsections we treat the UIRs of osp(1|2n, $\mathbb{R}$) (for $n > 1$), while the implications for conformal supersymmetry for $D = 9, 10, 11$ are treated in the last subsection.

3.2.1 Representations of osp(1|2n) and osp(1|2n, $\mathbb{R}$)

3.2.1.1 The setting

In the beginning of this section we follow [148]. Our basic references for Lie superalgebras are [241, 243]. The even subalgebra of osp(1|2n, $\mathbb{R}$) is the algebra sp(2n, $\mathbb{R}$) (of rank n) with maximal compact subalgebra $\mathcal{K} = u(n) \cong \mathrm{su}(n) \oplus u(1)$. The algebra sp(2$n$, $\mathbb{R}$) contains the conformal algebra $\mathcal{C} = \mathrm{so}(D, 2)$, while $\mathcal{K}$ contains the maximal compact subalgebra $\mathrm{so}(D) \oplus \mathrm{so}(2)$ of $\mathcal{C}$, so(2) being identified with the $u(1)$ factor of $\mathcal{K}$.

Note that this setting trivializes for $n = 1$ since then the subalgebra sp(2, $\mathbb{R}$) coincides with the conformal algebra $\mathcal{C} = \mathrm{so}(1, 2)$, that is why we consider $n > 1$.

We label the relevant representations of $\mathcal{G}$ by the signature:

$$\chi = [d;\, a_1, \dots, a_{n-1}] \tag{3.81}$$

where d is the conformal weight, and $a_1, \dots, a_{n-1}$ are non-negative integers which are Dynkin labels of the finite-dimensional UIRs of the subalgebra su(n) (the simple part of $\mathcal{K}$).

Our aim is to classify the UIRs of $\mathcal{G}$ following the methods used for the $D = 4, 6$ conformal superalgebras, cf. [138, 140, 139, 141, 142], Section 2.1, [122], 3.1, respectively The main tool is an adaptation of the Shapovalov form on the Verma modules V^χ over the complexification $\mathcal{G}^{\mathbb{C}} = \mathrm{osp}(1|2n)$ of $\mathcal{G}$.

3.2.1.2 Verma modules

To introduce Verma modules we use the standard triangular decomposition:

$$\mathcal{G}^{\mathbb{C}} = \mathcal{G}^+ \oplus \mathcal{H} \oplus \mathcal{G}^- \tag{3.82}$$

where $\mathcal{G}^+$, $\mathcal{G}^-$, respectively, are the subalgebras corresponding to the positive, negative, roots, respectively, and $\mathcal{H}$ denotes the Cartan subalgebra.

We consider lowest weight Verma modules, so that $V^\Lambda \cong U(\mathcal{G}^+) \otimes v_0$, where $U(\mathcal{G}^+)$ is the universal enveloping algebra of $\mathcal{G}^+$, and v_0 is a lowest weight vector v_0 such that

$$\begin{aligned} Zv_0 &= 0, \quad Z \in \mathcal{G}^-, \\ Hv_0 &= \Lambda(H)v_0, \quad H \in \mathcal{H}. \end{aligned} \tag{3.83}$$

Furthermore, for simplicity we omit the sign $\otimes$, i. e., we write $pv_0 \in V^\Lambda$ with $p \in U(\mathcal{G}^+)$.

The lowest weight Λ is characterized by its values on the simple roots of the superalgebra. In the next subsection we describe the root system.

3.2.1.3 Root systems

We recall some facts about $\mathcal{G}^{\mathbb{C}} = \mathrm{osp}(1|2n)$ (denoted $B(0,n)$ in [241]). Their root systems are given in terms of $\delta_1 \dots, \delta_n$, $(\delta_i, \delta_j) = \delta_{ij}$, $i,j = 1,\dots,n$. The even and odd roots systems are [241]:

$$\begin{aligned} \Delta_{\bar{0}} &= \{\pm\delta_i \pm \delta_j, 1 \le i < j \le n, \pm 2\delta_i, 1 \le i \le n\}, \\ \Delta_{\bar{1}} &= \{\pm\delta_i, 1 \le i \le n\} \end{aligned} \tag{3.84}$$

(we recall that the signs $\pm$ are not correlated). We shall use the following distinguished simple root system [241]:

$$\Pi = \{\delta_1 - \delta_2, \dots, \delta_{n-1} - \delta_n, \delta_n\}, \tag{3.85}$$

or introducing standard notation for the simple roots:

$$\begin{aligned} \Pi &= \{\alpha_1, \dots, \alpha_n\}, \\ &\alpha_j = \delta_j - \delta_{j+1}, \quad j = 1,\dots,n-1, \quad \alpha_n = \delta_n. \end{aligned} \tag{3.86}$$

The root $\alpha_n = \delta_n$ is odd, the other simple roots are even. The Dynkin diagram is

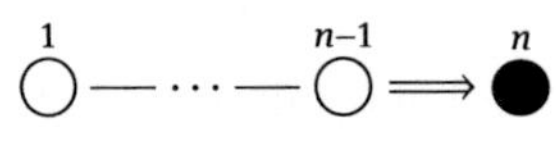

(3.87)

The black dot is used to signify that the simple odd root is not nilpotent, otherwise a gray dot would be used (cf. [241] and Section 1.3.2). In fact, the superalgebras $B(0,n) = \mathrm{osp}(1|2n)$ have no nilpotent generators unlike all other types of basic classical Lie superalgebras [241].

The corresponding to Π positive root system is

$$\Delta^+_{\bar{0}} = \{\delta_i \pm \delta_j, 1 \le i < j \le n, 2\delta_i, 1 \le i \le n\}, \quad \Delta^+_{\bar{1}} = \{\delta_i, 1 \le i \le n\}. \tag{3.88}$$

We record how the elementary functionals are expressed through the simple roots:

$$\delta_k = \alpha_k + \dots + \alpha_n. \tag{3.89}$$

The even root system $\Delta_{\bar{0}}$ is the root system of the rank n complex simple Lie algebra sp($2n$), with $\Delta_{\bar{0}}^+$ being its positive roots. The simple roots are

$$\Pi_0 = \{\delta_1 - \delta_2, \dots, \delta_{n-1} - \delta_n, 2\delta_n\} = \{\alpha_1^0, \dots, \alpha_n^0\}, \qquad (3.90)$$
$$\alpha_j^0 = \delta_j - \delta_{j+1}, \quad j = 1, \dots, n-1, \quad \alpha_n^0 = 2\delta_n.$$

The Dynkin diagram is

$$\overset{1}{\bigcirc} - \cdots - \overset{n-1}{\bigcirc} \Leftarrow \overset{n}{\bigcirc} \qquad (3.91)$$

The superalgebra $\mathcal{G}$ = osp(1|2n, ℝ) is a split real form of osp(1|2n) and has the same root system.

3.2.1.4 Lowest weight through the signature

Since we use a Dynkin labeling, we have the following relation with the signature χ from (3.81):

$$(\Lambda, \alpha_k^\vee) = \begin{cases} -a_k, & k < n, \\ \tilde{d}, & k = n, \end{cases} \qquad (3.92)$$

where $\alpha_k^\vee \equiv 2\alpha_k/(\alpha_k, \alpha_k)$, and $\tilde{d}$ differs from the conformal weight d as explained below. The minus signs in the first row are related to the fact that we work with lowest weight Verma modules (instead of the highest weight modules used in [243]) and to Verma module reducibility w.r.t. the roots α_k (this is explained in detail in [139]). The value of $\tilde{d}$ is a matter of normalization so as to correspond to some well-known cases. Thus, our choice is

$$\tilde{d} = 2d + a_1 + \cdots + a_{n-1}. \qquad (3.93)$$

Having in hand the values of Λ on the basis we can recover them for any element of $\mathcal{H}^*$. In particular, for the values on the elementary functionals we have, using (3.89), (3.92), and (3.93),

$$(\Lambda, \delta_j) = d + \frac{1}{2}(a_1 + \cdots + a_{j-1} - a_j - \cdots - a_{n-1}). \qquad (3.94)$$

Using (3.92) and (3.93) one can write easily $\Lambda = \Lambda(\chi)$ as a linear combination of the simple roots or of the elementary functionals δ_j, but this is not necessary in what follows. We shall need only $(\Lambda, \beta^\vee)$ for all positive roots β, and from (3.94) we have

$$\begin{aligned} (\Lambda, (\delta_i - \delta_j)^\vee) &= (\Lambda, \delta_i - \delta_j) = -a_i - \cdots - a_{j-1}, \\ (\Lambda, (\delta_i + \delta_j)^\vee) &= (\Lambda, \delta_i + \delta_j) = 2d + a_1 + \cdots + a_{i-1} - a_j - \cdots - a_{n-1}, \\ (\Lambda, \delta_i^\vee) &= (\Lambda, 2\delta_i) = 2d + a_1 + \cdots + a_{i-1} - a_i - \cdots - a_{n-1}, \\ (\Lambda, (2\delta_i)^\vee) &= (\Lambda, \delta_i) = d + \frac{1}{2}(a_1 + \cdots + a_{i-1} - a_i - \cdots - a_{n-1}). \end{aligned} \qquad (3.95)$$

3.2.1.5 Reducibility of Verma modules

Having established the relation between χ and Λ we turn our attention to the question of reducibility. A Verma module V^Λ is reducible w.r.t. the positive root β iff the following holds [243]:

$$(\rho - \Lambda, \beta^\vee) = m_\beta, \quad \beta \in \Delta^+,\ m_\beta \in \mathbb{N}, \tag{3.96}$$

where $\rho \in \mathcal{H}^*$ is the very important in representation theory element given by the difference of the half-sums $\rho_{\bar{0}}, \rho_{\bar{1}}$ of the even, odd, respectively, positive roots (cf. (3.88)):

$$\begin{aligned} \rho \doteq \rho_{\bar{0}} - \rho_{\bar{1}} &= \left(n - \frac{1}{2}\right)\delta_1 + \left(n - \frac{3}{2}\right)\delta_2 + \cdots + \frac{3}{2}\delta_{n-1} + \frac{1}{2}\delta_n, \\ \rho_{\bar{0}} &= n\delta_1 + (n-1)\delta_2 + \cdots + 2\delta_{n-1} + \delta_n, \\ \rho_{\bar{1}} &= \frac{1}{2}(\delta_1 + \cdots + \delta_n). \end{aligned} \tag{3.97}$$

To make (3.96) explicit we need first the values of ρ on the positive odd roots:

$$(\rho, \delta_i) = n - i + \frac{1}{2}. \tag{3.98}$$

Then for $(\rho, \beta^\vee)$ we have

$$\begin{aligned} (\rho, (\delta_i - \delta_j)^\vee) &= j - i, \\ (\rho, (\delta_i + \delta_j)^\vee) &= 2n - i - j + 1, \\ (\rho, \delta_i^\vee) &= 2n - 2i + 1, \\ (\rho, (2\delta_i)^\vee) &= n - i + \frac{1}{2}. \end{aligned} \tag{3.99}$$

Naturally, the value of ρ on the simple roots is 1: $(\rho, \alpha_i^\vee) = 1$, $i = 1, \ldots, n$.

Consequently we find that the Verma module $V^{\Lambda(\chi)}$ is reducible if one of the following relations holds (following the order of (3.95) and (3.99)):

$$\mathbb{N} \ni m_{ij}^- = j - i + a_i + \cdots + a_{j-1}, \tag{3.100a}$$

$$\begin{aligned} \mathbb{N} \ni m_{ij}^+ = 2n - i - j + 1 + a_j + \cdots + a_{n-1} - a_1 - \cdots \\ \cdots - a_{i-1} - 2d, \end{aligned} \tag{3.100b}$$

$$\begin{aligned} \mathbb{N} \ni m_i = 2n - 2i + 1 + a_i + \cdots + a_{n-1} - a_1 - \cdots \\ \cdots - a_{i-1} - 2d, \end{aligned} \tag{3.100c}$$

$$\begin{aligned} \mathbb{N} \ni m_{ii} = n - i + \frac{1}{2}(1 + a_i + \cdots + a_{n-1} - a_1 - \cdots \\ \cdots - a_{i-1}) - d. \end{aligned} \tag{3.100d}$$

Note that $m_i = 2m_{ii}$, thus, whenever (3.100d) is fulfilled, also (3.100c) is fulfilled.

If a condition from (3.100) is fulfilled then V^Λ contains a submodule which is a Verma module $V^{\Lambda'}$ with shifted weight given by the pair m, β: $\Lambda' = \Lambda + m\beta$. The embedding of $V^{\Lambda'}$ in V^Λ is provided by mapping the lowest weight vector v_0' of $V^{\Lambda'}$ to the singular vector $v_s^{m,\beta}$ in V^Λ which is completely determined by the conditions

$$\begin{aligned} &Xv_s^{m,\beta} = 0, \quad X \in \mathcal{G}^-, \\ &Hv_s^{m,\beta} = \Lambda'(H)v_0, \quad H \in \mathcal{H},\ \Lambda' = \Lambda + m\beta. \end{aligned} \tag{3.101}$$

Explicitly, $v_s^{m,\beta}$ is given by a polynomial in the positive root generators:

$$v_s^{m,\beta} = P^{m,\beta} v_0, \quad P^{m,\beta} \in U(\mathcal{G}^+). \tag{3.102}$$

Thus, the submodule of V^Λ which is isomorphic to $V^{\Lambda'}$ is given by $U(\mathcal{G}^+)P^{m,\beta}v_0$.

Here we should note that we may eliminate the reducibilities and embeddings related to the roots $2\delta_i$. Indeed, let (3.100d) hold, then the corresponding singular vector $v_s^{m_{ii},2\delta_i}$ has the properties prescribed by (3.101) with $\Lambda' = \Lambda + m_{ii}2\delta_i$. But as we mentioned above in this situation also (3.100c) holds and the corresponding singular vector $v_s^{m_i,\delta_i}$ has the properties prescribed by (3.101) with $\Lambda'' = \Lambda + m_i\delta_i$. But clearly, due to the fact that $m_i = 2m_{ii}$ we have $\Lambda'' = \Lambda'$, which means that the singular vectors $v_s^{m_{ii},2\delta_i}$ and $v_s^{m_i,\delta_i}$ coincide (up to nonzero multiplicative constant). On the other hand if (3.100c) holds with m_i being an odd number, then (3.100d) does not hold (since $m_{ii} = m_i/2$ is not integer).

We notice that all reducibility conditions in (3.100a) are fulfilled. In particular, for the simple roots from those condition (3.100a) is fulfilled with $\beta \to \alpha_i = \delta_i - \delta_{i+1}$, $i = 1, \ldots, n-1$ and $m_i^- \equiv m_{i,i+1}^- = 1 + a_i$. The corresponding submodules $I_i^\Lambda = U(\mathcal{G}^+)v_s^i$, where $\Lambda_i = \Lambda + m_i^-\alpha_i$ and $v_s^i = (X_i^+)^{1+a_i}v_0$, where X_i^+ are the root vectors of the these simple roots. These submodules generate an invariant submodule which we denote by I_c^Λ. Since these submodules are nontrivial for all our signatures instead of V^Λ we shall consider the factor modules:

$$F^\Lambda = V^\Lambda / I_c^\Lambda. \tag{3.103}$$

We shall denote the lowest weight vector of F^Λ by $\widetilde{|\Lambda\rangle}$ and the singular vectors above become null conditions in F^Λ:

$$(X_i^+)^{1+a_i}\widetilde{|\Lambda\rangle} = 0. \quad i = 1, \ldots, n-1. \tag{3.104}$$

If the Verma module V^Λ is not reducible w.r.t. the other roots, i. e., (3.100b,c,d) are not fulfilled, then F^Λ is irreducible and is isomorphic to the irrep L_Λ with this weight.

Other situations shall be discussed below in the context of unitarity.

3.2.1.6 Realization of osp(1|2n) and osp(1|2n, ℝ)

The superalgebras $\mathrm{osp}(m|2n) = \mathrm{osp}(m|2n)_{\bar{0}} + \mathrm{osp}(m|2n)_{\bar{1}}$ are defined as follows [241]:

$$\mathrm{osp}(m|2n)_s = \{X \in \mathrm{gl}(m/2n;\mathbb{C})_s : XW + i^s W^t X = 0\}, \quad s = \bar{0}, \bar{1},$$

where W is a matrix of order $m + 2n$:

$$W = \begin{pmatrix} iI_m & 0 & 0 \\ 0 & 0 & I_n \\ 0 & -I_n & 0 \end{pmatrix}$$

The even part $\mathrm{osp}(m|2n)_{\bar{0}}$ consists of matrices X such that

$$X = \begin{pmatrix} S & 0 & 0 \\ 0 & B & C \\ 0 & D & -{}^tB \end{pmatrix}, \qquad (3.105)$$
$${}^tS = -S,\ {}^tC = C,\ {}^tD = D.$$

In our case $m = 1$ and $S = 0$. The Cartan subalgebra $\mathcal{H}$ consists of diagonal matrices H such that

$$H = \begin{pmatrix} 0 & 0 & 0 \\ 0 & B & 0 \\ 0 & 0 & -B \end{pmatrix}$$

We take the following basis for the Cartan subalgebra:

$$H_i = \begin{pmatrix} 0 & 0 & 0 \\ 0 & B_i & 0 \\ 0 & 0 & -B_i \end{pmatrix}, \quad i < n,$$
$$H_n = \begin{pmatrix} 0 & 0 & 0 \\ 0 & I_n & 0 \\ 0 & 0 & -I_n \end{pmatrix}, \qquad (3.106)$$

where

$$B_i = \mathrm{diag}(0,\ldots 0,1,-1,0,\ldots,0)$$

the first non-zero entry being on the ith place. This basis shall be used also for the real form $\mathrm{osp}(1|2n,\mathbb{R})$ and is chosen to be consistent with the fact that the even subalgebra $\mathrm{sp}(2n,\mathbb{R})$ of the latter has as maximal noncompact subalgebra the algebra $\mathrm{sl}(n,\mathbb{R}) \oplus \mathbb{R}$. Via the Weyl unitary trick this related to the structure of $\mathrm{sp}(2n,\mathbb{R})$ as a hermitean symmetric space with maximal compact subalgebra $u(n) \cong \mathrm{su}(n) \oplus u(1)$.

The root vectors of the roots $\delta_i - \delta_j$, $(i \neq j)$, $\delta_i + \delta_j$, $(i \leq j)$, $-(\delta_i + \delta_j)$, $(i \leq j)$, respectively, are denoted X_{ij}, X^+_{ij}, X^-_{ij}, respectively. The latter are given by matrices of the type (3.105) with $S = 0$, given (up to multiplicative normalization) by $B = E_{ij}$, $C = E_{ij} + E_{ji}$, $D = E_{ij} + E_{ji}$, respectively, where E_{ij} is $n \times n$ matrix which has only one non-zero entry equal to 1 on the intersection of the ith row and jth column. Explicitly (including some choice of normalization), this is

$$X_{ij} = \begin{pmatrix} 0 & 0 & 0 \\ 0 & E_{ij} & 0 \\ 0 & 0 & -E_{ji} \end{pmatrix}, \quad i \neq j, \tag{3.107}$$

$$X^+_{ij} = \begin{pmatrix} 0 & 0 & 0 \\ 0 & 0 & -E_{ij} - E_{ji} \\ 0 & 0 & 0 \end{pmatrix}, \quad i < j, \quad X^+_{ii} = \begin{pmatrix} 0 & 0 & 0 \\ 0 & 0 & -E_{ii} \\ 0 & 0 & 0 \end{pmatrix},$$

$$X^-_{ij} = \begin{pmatrix} 0 & 0 & 0 \\ 0 & 0 & 0 \\ 0 & E_{ij} + E_{ji} & 0 \end{pmatrix}, \quad i < j, \quad X^-_{ii} = \begin{pmatrix} 0 & 0 & 0 \\ 0 & 0 & 0 \\ 0 & E_{ii} & 0 \end{pmatrix}.$$

The odd part $\mathrm{osp}(m|2n)_{\bar{1}}$ consists of matrices X such that

$$X = \begin{pmatrix} 0 & \xi & -\eta \\ {}^t\eta & 0 & 0 \\ {}^t\xi & 0 & 0 \end{pmatrix}.$$

The root vectors Y^+_i, Y^-_i, of the roots δ_i, $-\delta_i$ correspond to η, ξ, respectively, with only non-zero ith entry. Explicitly this is

$$Y^+_i = \begin{pmatrix} 0 & 0 & -E_{1i} \\ E_{i1} & 0 & 0 \\ 0 & 0 & 0 \end{pmatrix},$$

$$Y^-_i = \begin{pmatrix} 0 & E_{1i} & 0 \\ 0 & 0 & 0 \\ E_{i1} & 0 & 0 \end{pmatrix}. \tag{3.108}$$

In the calculations we need all commutators of the kind $[X_\beta, X_{-\beta}] = H_\beta$, $\beta \in \Delta^+_{\bar{0}}$. Explicitly, we have

$$[X_{ij}, X_{ji}] = H_{ij} = H_i + H_{i+1} + \cdots + H_{j-1}, \quad 1 \leq i < j \leq n, \tag{3.109a}$$

$$[Y^+_i, Y^-_i]_+ = H'_i \equiv \begin{pmatrix} 0 & 0 & 0 \\ 0 & E_{ii} & 0 \\ 0 & 0 & -E_{ii} \end{pmatrix}, \quad 1 \leq i \leq n, \tag{3.109b}$$

$$[X^+_{ij}, X^-_{ij}] = H'_{ij} = -H'_i - H'_j, \quad 1 \leq i < j \leq n, \tag{3.109c}$$

$$[X^+_{ii}, X^-_{ii}] = -H'_i, \quad 1 \leq i \leq n. \tag{3.109d}$$

The minus sign in (3.109d) is consistent with the relations:

$$\frac{1}{2}[Y^\pm_i, Y^\pm_i]_+ = (Y^\pm_i)^2 = X^\pm_{ii}. \tag{3.110}$$

We note also the following relations:

$$[Y_i^+, Y_j^-]_+ = X_{ij}, \quad i \neq j,$$
$$[Y_i^\pm, Y_j^\pm]_+ = X_{ij}^\pm, \quad i \neq j,$$
$$H_n = H_1' + \cdots + H_n'. \tag{3.111}$$

We shall use also the abstract defining relations of osp(1|2n) through the Chevalley basis. Let $\hat{H}_i$, $i = 1, \ldots, n$, be the basis of the Cartan subalgebra $\mathcal{H}$ associated with the simple roots, and $X_i^\pm$, $i = 1, \ldots, n$, be the simple root vectors (the Chevalley generators). The connection with the basis above is

$$\hat{H}_i = H_i, \quad i < n, \quad \hat{H}_n = H_n',$$
$$X_i^+ = X_{i,i+1}^+, \quad i < n, \quad X_n^+ = Y_n^+. \tag{3.112}$$

Let $A = (a_{ij})$ be the Cartan matrix [241]:

$$A = (a_{ij}) = \begin{pmatrix} 2 & -1 & 0 & \ldots & 0 & 0 & 0 \\ -1 & 2 & -1 & \ldots & 0 & 0 & 0 \\ 0 & -1 & 2 & \ldots & 0 & 0 & 0 \\ \ldots & \ldots & \ldots & \ldots & \ldots & \ldots & \ldots \\ 0 & 0 & 0 & \ldots & 2 & -1 & 0 \\ 0 & 0 & 0 & \ldots & -1 & 2 & -1 \\ 0 & 0 & 0 & \ldots & 0 & -2 & 2 \end{pmatrix}. \tag{3.113}$$

We shall also use the decomposition $A = A^d A^s$, where $A^d = \mathrm{diag}(1, \ldots, 1, 2)$, and A^s is a symmetric matrix:

$$A^s = (a_{ij}^s) = \begin{pmatrix} 2 & -1 & 0 & \ldots & 0 & 0 & 0 \\ -1 & 2 & -1 & \ldots & 0 & 0 & 0 \\ 0 & -1 & 2 & \ldots & 0 & 0 & 0 \\ \ldots & \ldots & \ldots & \ldots & \ldots & \ldots & \ldots \\ 0 & 0 & 0 & \ldots & 2 & -1 & 0 \\ 0 & 0 & 0 & \ldots & -1 & 2 & -1 \\ 0 & 0 & 0 & \ldots & 0 & -1 & 1 \end{pmatrix}. \tag{3.114}$$

Then the defining relations of osp(1|2n) are

$$[\hat{H}_i, \hat{H}_j] = 0, \quad [\hat{H}_i, X_j^\pm] = \pm a_{ij}^s X_j^\pm,$$
$$[X_i^+, X_j^-] = \delta_{ij} \hat{H}_i, \tag{3.115}$$
$$(\mathrm{Ad}\, X_j^\pm)^{n_{jk}} (X_k^\pm) = 0, \quad j \neq k, \ n_{jk} = 1 - a_{jk},$$

where in (3.115) one uses the supercommutator:

$$(\mathrm{Ad}\, X_j^\pm)(X_k^\pm) = [X_j^\pm, X_k^\pm]$$
$$\equiv X_j^\pm X_k^\pm - (-1)^{\deg X_j^\pm \deg X_k^\pm} X_k^\pm X_j^\pm. \tag{3.116}$$

3.2.1.7 Shapovalov form and unitarity

The Shapovalov form is a bilinear ℂ-valued form on $U(\mathcal{G}^+)$ [353], which we extend in the obvious way to Verma modules; cf. e. g., [141, 142]. We need also the involutive antiautomorphism ω of $U(\mathcal{G})$ which will provide the real form we are interested in. Since this is the split real form osp(1|2n, ℝ) we use

$$\omega(X_\beta) = X_{-\beta}, \quad \omega(H) = H, \tag{3.117}$$

where X_β is the root vector corresponding to the root β, $H \in \mathcal{H}$.

Thus, an adaptation of the Shapovalov form suitable for our purposes is defined as follows:

$$\begin{aligned}(u, u') &= (pv_0, p'v_0) \equiv (v_0, \omega(p)p'v_0) \\ &= (\omega(p')pv_0, v_0), \\ &\quad u = pv_0,\ u' = p'v_0,\ p, p' \in U(\mathcal{G}^+),\ u, u' \in V^\Lambda,\end{aligned} \tag{3.118}$$

supplemented by the normalization condition $(v_0, v_0) = 1$. The norms squared of the states would be denoted by

$$\|u\|^2 \equiv (u, u). \tag{3.119}$$

Now we need to introduce a PBW basis of $U(\mathcal{G}^+)$. We use the so-called normal ordering, namely, if we have the relation

$$\beta = \beta' + \beta'', \quad \beta, \beta', \beta'' \in \Delta^+$$

then the corresponding root vectors are ordered in the PBW basis as follows:

$$\ldots (X^+_{\beta'})^{k'} \ldots (X^+_\beta)^k \ldots (X^+_{\beta''})^{k''} \ldots, \quad k, k', k'' \in \mathbb{Z}_+. \tag{3.120}$$

We have also to take into account the relation (3.110) between the root vectors corresponding to the roots δ_i and $2\delta_i$. Because of this relation and consistently with (3.120) the generators X^+_{ii}, $i = 1, \ldots, n$, are not present in the PBW basis. On the other hand the PBW basis of the even subalgebra of $U(\mathcal{G}^+)$ would differ from the above only in the fact that the powers of X^+_i, $i = 1, \ldots, n$, are only even representing powers of the even generators X^+_{ii}, $i = 1, \ldots, n$.

3.2.2 Unitarity

3.2.2.1 Calculation of some norms

In this subsection we show how to use the form (3.118) to calculate the norms of the states. We shall use the isomorphism between the Cartan subalgebra $\mathcal{H}$ and its dual

$\mathcal{H}^*$. This is given by the following correspondence: corresponding to every element $\beta \in \mathcal{H}^*$ there is a unique element $H_\beta \in \mathcal{H}$, so that

$$\mu(H_\beta) = (\mu, \beta^\vee), \tag{3.121}$$

for every $\mu \in \mathcal{H}^*, \mu \neq 0$. Applying this to the positive roots we have to $\beta = \delta_i - \delta_j, \delta_i, \delta_i + \delta_j$, respectively, correspond: $H_\beta = H_{ij}, H'_i, H'_i + H'_j$, respectively

We give now explicitly the norms of the one-particle states introducing also notation for future use:

$$\begin{aligned}
x^-_{ij} &\equiv \|X_{ij}v_0\|^2 = (X_{ij}v_0, X_{ij}v_0) \\
&= (v_0, X_{ji}X_{ij}v_0) = (v_0, (X_{ij}X_{ji} - H_{ij})v_0) \\
&= -\Lambda(H_{ij}) = -(\Lambda, (\delta_i - \delta_j)^\vee) = a_i + \cdots + a_{j-1}, \quad i < j, \\
x^+_{ij} &\equiv \|X^+_{ij}v_0\|^2 = (X^+_{ij}v_0, X^+_{ij}v_0) \\
&= (v_0, X^-_{ij}X^+_{ij}v_0) = (v_0, (X^+_{ij}X^-_{ji} - H'_{ij})v_0) \\
&= -\Lambda(H'_{ij}) = \Lambda(H'_i + H'_j) = (\Lambda, (\delta_i + \delta_j)^\vee) \\
&= 2d + a_1 + \cdots + a_{i-1} - a_j - \cdots - a_{n-1}, \\
x_i &\equiv \|X^+_i v_0\|^2 = (X^+_i v_0, X^+_i v_0) \\
&= (v_0, X^-_i X^+_i v_0) = (v_0, (-X^+_i X^-_i + H'_i)v_0) = \Lambda(H'_i) \\
&= (\Lambda, \delta^\vee_i) = 2d + a_1 + \cdots + a_{i-1} - a_i - \cdots - a_{n-1}.
\end{aligned} \tag{3.122}$$

Positivity of all these norms gives the following necessary conditions for unitarity:

$$\begin{aligned}
&a_i \geq 0, \quad i = 1, \ldots, n-1, \\
&d \geq \frac{1}{2}(a_1 + \cdots + a_{n-1}).
\end{aligned} \tag{3.123}$$

In fact, the boundary values are possible due to factoring out of the corresponding null states when passing from the Verma module to the unitary irreducible factor module.

Below, we shall discuss only norms which involve the conformal weight since the others are related to unitarity of the irrep restricted to the maximal simple compact subalgebra su(n). The norms that we are going to consider can be written in terms of factors $(d - \cdots)$, and the leading term in d has a positive coefficient. Thus, for d large enough all norms will be positive. When d is decreasing there is a critical point at which one (or more) norm(s) will become zero. This critical point (called the 'first reduction point' in [182]) can be read off from the reducibility conditions, since at that point the Verma module is reducible (and it is the corresponding submodule that has zero norm states).

The maximal d coming from the different possibilities in (3.100b) are obtained for $m^+_{ij} = 1$ and they are, denoting also the corresponding root:

$$d_{ij} \equiv n + \frac{1}{2}(a_j + \cdots + a_{n-1} - a_1 - \cdots - a_{i-1} - i - j), \tag{3.124}$$

the corresponding root being $\delta_i + \delta_j$. The maximal d coming from the different possibilities in (3.100c,d), respectively, are obtained for $m_i = 1$, $m_{ii} = 1$, respectively, and they are

$$\begin{aligned} d_i &\equiv n - i + \frac{1}{2}(a_i + \cdots + a_{n-1} - a_1 - \cdots - a_{i-1}), \\ d_{ii} &= d_i - \frac{1}{2}, \end{aligned} \tag{3.125}$$

the corresponding roots being δ_i, $2\delta_j$, respectively These are some orderings between these maximal reduction points:

$$\begin{aligned} d_1 &> d_2 > \cdots > d_n, \\ d_{i,i+1} &> d_{i,i+2} > \cdots > d_{in}, \\ d_{1,j} &> d_{2,j} > \cdots > d_{j-1,j}, \\ d_i &> d_{jk} > d_\ell, \quad i \leq j < k \leq \ell. \end{aligned} \tag{3.126}$$

Obviously the first reduction point is

$$d_1 = n - 1 + \frac{1}{2}(a_1 + \cdots + a_{n-1}). \tag{3.127}$$

3.2.2.2 Main result

The first results on the unitarity were given in [148], and then improved in [143]. Thus, the statement below is called the *Dobrev–Zhang–Salom theorem*:

Theorem (Dobrev–Zhang–Salom). *All positive energy unitary irreducible representations of the superalgebras* osp(1|2n, ℝ) *($n > 1$) characterized by the signature χ in (3.81) are obtained for real d and are given as follows:*

$$\begin{aligned} d &\geq n - 1 + \frac{1}{2}(a_1 + \cdots + a_{n-1}) = d_1, \quad a_1 \neq 0, \\ d &\geq n - \frac{3}{2} + \frac{1}{2}(a_2 + \cdots + a_{n-1}) = d_{12}, \\ &\quad a_1 = 0, a_2 \neq 0, \\ d &= n - 2 + \frac{1}{2}(a_2 + \cdots + a_{n-1}) = d_2 > d_{13}, \\ &\quad a_1 = 0, a_2 \neq 0, \\ d &\geq n - 2 + \frac{1}{2}(a_3 + \cdots + a_{n-1}) = d_2 = d_{13}, \\ &\quad a_1 = a_2 = 0, a_3 \neq 0, \\ d &= n - \frac{5}{2} + \frac{1}{2}(a_3 + \cdots + a_{n-1}) = d_{23} > d_{14}, \\ &\quad a_1 = a_2 = 0, a_3 \neq 0, \end{aligned}$$

$$\begin{aligned}
&d = n-3+\frac{1}{2}(a_3+\cdots+a_{n-1}) = d_3 = d_{24} > d_{15},\\
&\quad a_1 = a_2 = 0, a_3 \neq 0,\\
&\dots \\
&\dots \\
&d \geq n-1-\kappa+\frac{1}{2}(a_{2\kappa+1}+\cdots+a_{n-1}),\\
&\quad a_1 = \cdots = a_{2\kappa} = 0, a_{2\kappa+1} \neq 0,\\
&\quad \kappa = \frac{1}{2}, 1, \dots, \frac{1}{2}(n-1),\\
&d = n-\frac{3}{2}-\kappa+\frac{1}{2}(a_{2\kappa+1}+\cdots+a_{n-1}),\\
&\quad a_1 = \cdots = a_{2\kappa} = 0, a_{2\kappa+1} \neq 0,\\
&\dots \\
&d = n-1-2\kappa+\frac{1}{2}(a_{2\kappa+1}+\cdots+a_{n-1}),\\
&\quad a_1 = \cdots = a_{2\kappa} = 0, a_{2\kappa+1} \neq 0,\\
&\dots \\
&\dots \\
&d \geq \frac{1}{2}(n-1), \quad a_1 = \cdots = a_{n-1} = 0,\\
&d = \frac{1}{2}(n-2), \quad a_1 = \cdots = a_{n-1} = 0,\\
&\dots \\
&d = \frac{1}{2}, \quad a_1 = \cdots = a_{n-1} = 0,\\
&d = 0, \quad a_1 = \cdots = a_{n-1} = 0.
\end{aligned} \tag{3.128}$$

◇

Remark. Note that for $n = 2$ the result of the theorem was shown first in [224] though by a different method. ◇

Proof. The statement of the theorem for $d > d_1$ is clear from the general considerations since this is the First reduction point. For $d = d_1$ we have the first zero norm state which is naturally given by the corresponding singular vector $v^1_{\delta_1} = \mathcal{P}^{1,\delta_1} v_0$. In fact, all states of the embedded submodule $V^{\Lambda+\delta_1}$ built on $v^1_{\delta_1}$ have zero norms. Due to the above singular vector we have the following additional null condition in F^Λ_c:

$$\mathcal{P}^{1,\delta_1}\widetilde{|\Lambda\rangle} = 0. \tag{3.129}$$

The above condition factorizes the submodule built on $v^1_{\delta_1}$. There are no other vectors with zero norm at $d = d_1$ since by a general result [243], the elementary embeddings between Verma modules are one dimensional. Thus, F^Λ is the UIR $L_\Lambda = F^\Lambda$. Below $d < d_1$ there is no unitarity for $a_1 \neq 0$. On the other hand for $a_1 = 0$ the singular vector $v^1_{\delta_1}$ is descendant of the compact root singular vector $X^+_1 v_0$ which is already factored out for $a_1 = 0$. Thus, below we set $a_1 = 0$.

The next reducibility point is $d = d_{12} = n - \frac{3}{2} + \frac{1}{2}(a_2 + \cdots + a_{n-1})$. The corresponding root is $\delta_1 + \delta_2 = \alpha_1 + 2\alpha_2 + \cdots + 2\alpha_n$. The corresponding singular vector is $v^1_{\delta_1+\delta_2} = \mathcal{P}^{1,\delta_1+\delta_2} v_0$. All states of the embedded submodule $V^{\Lambda+\delta_1+\delta_2}$ built on $v^1_{\delta_1+\delta_2}$ have zero norms for $d = d_{12}$. Due to the above singular vector we have the following additional null condition in F_c^Λ:

$$\mathcal{P}^{1,\delta_1+\delta_2} \widetilde{|\Lambda\rangle} = 0, \quad d = d_{12}. \tag{3.130}$$

The above conditions factorizes the submodule built on $v^1_{\delta_1+\delta_2}$. Thus, F_c^Λ is the UIR $L_\Lambda = F_c^\Lambda$.

Below $d < d_{12}$ there is no unitarity for $a_2 \neq 0$, except at the isolated point: $d_2 = n - 2 + \frac{1}{2}(a_2 + \cdots + a_{n-1})$. At the latter point there is a singular vector $v^1_{\delta_2}$ which must be factored for unitarity. In addition, the previous singular vector is descendant of $v^1_{\delta_2}$ and the compact root singular vector $X_1^+ v_0$.

Furthermore, for $a_2 = 0$ the singular vectors $v^1_{\delta_1+\delta_2}$ and $v^1_{\delta_2}$ are descendants of the compact root singular vectors $X_1^+ v_0$ and $X_2^+ v_0$ which are factored out for $a_1 = a_2 = 0$. Thus, below we set also $a_2 = 0$ and there would be no obstacles for unitarity until the next reducibility points (coinciding due $a_2 = 0$): $d_2 = d_{13} = n - 2 + \frac{1}{2}(a_3 + \cdots + a_{n-1})$. The singular vector for $d = d_{13}$ and $m = 1$ has weight $\delta_1 + \delta_3 = \alpha_1 + \alpha_2 + 2\alpha_3 + \cdots 2\alpha_n$ and for $a_1 = 0$ it is a descendant of the compact root singular vector $X_1 v_0$ [115]. However, at $d_2 = d_{13}$ there is a subsingular vector which must be factored for unitarity. For $d < d_2 = d_{13}$ and $a_3 \neq 0$ the norm of that subsingular vector is negative, and there will not be unitarity except at some lower reducibility points.

For $d_{23} = n - \frac{5}{2} + \frac{1}{2}(a_3 + \cdots + a_{n-1})$ there is singular vector $v^1_{\delta_2+\delta_3}$ of weight $\delta_2 + \delta_3 = \alpha_2 + 2\alpha_3 + \cdots 2\alpha_n$ [115] which must be factored for unitarity. The previous subsingular vector is also factored out since it is descendant of $v^1_{\delta_2+\delta_3}$ and compact root singular vectors.

Further, the proof goes along similar lines. We list the points at which there are subsingular vectors—these happen when reducibility points coincide due the zero values of some a_i:

$$\begin{aligned}
& d_2 = d_{13} = n - 2 + \frac{1}{2}(a_3 + \cdots + a_{n-1}), \quad a_1 = a_2 = 0, \\
& d_{23} = d_{14} = n - 5/2 + \frac{1}{2}(a_4 + \cdots + a_{n-1}), \quad a_1 = a_2 = a_3 = 0,\ n > 3, \\
& d_3 = d_{24} = d_{15} = n - 3 + \frac{1}{2}(a_5 + \cdots + a_{n-1}), \quad a_1 = a_2 = a_3 = a_4 = 0,\ n > 3, \\
& \ldots \\
& d_j = d_{1,2j-1} = d_{2,2j-2} = \cdots = d_{j-1,j+1} = n - j + \frac{1}{2}(a_{2j-1} + \cdots + a_{n-1}), \\
& \qquad a_1 = \cdots = a_{2j-2} = 0, \quad j < n, \\
& d_{j,j+1} = d_{1,2j} = d_{2,2j-1} = \cdots = d_{j-1,j+2} = n - j - \frac{1}{2}
\end{aligned} \tag{3.131}$$

$$+\frac{1}{2}(a_{2j} + \cdots + a_{n-1}),$$
$$a_1 = \cdots = a_{2j-1} = 0, \quad j < n-1.$$

Above it is understood that $a_j \equiv 0$ for $j \geq n$.

At the points of the subsingular vectors the associated singular vectors are factored out automatically. This happens also when the subsingular vectors are inside a continuous part of the unitarity spectrum. □

The proof above is not as explicit as one may like it to be, but the interested reader may study the explicit examples for $n = 3$ in [143] and $n = 4$ in [144].

3.2.3 Character formulas

3.2.3.1 Character formulas: generalities

In the beginning of this subsection we follow Dixmier [110]. Let $\hat{\mathcal{G}}$ be a simple Lie algebra of rank ℓ with Cartan subalgebra $\hat{\mathcal{H}}$, root system $\hat{\mathcal{D}}$, simple root system $\hat{\pi}$. Let Γ (respectively, Γ_+) be the set of all integral (respectively, integral dominant) elements of $\hat{\mathcal{H}}^*$, i. e., $\lambda \in \hat{\mathcal{H}}^*$ such that $(\lambda, \alpha_i^\vee) \in \mathbb{Z}$ (respectively $\mathbb{Z}_+$), for all simple roots α_i ($\alpha_i^\vee \equiv 2\alpha_i/(\alpha_i, \alpha_i)$). Let V be a lowest weight module with lowest weight Λ and lowest weight vector v_0. It has the following decomposition:

$$V = \underset{\mu\in\Gamma_+}{\oplus} V_\mu, \quad V_\mu = \{u \in V \mid Hu = (\lambda + \mu)(H)u, \ \forall H \in \mathcal{H}\}. \tag{3.132}$$

(Note that $V_0 = \mathbb{C}v_0$.) Let $E(\mathcal{H}^*)$ be the associative abelian algebra consisting of the series $\sum_{\mu\in\mathcal{H}^*} c_\mu e(\mu)$, where $c_\mu \in \mathbb{C}, c_\mu = 0$ for μ outside the union of a finite number of sets of the form $D(\lambda) = \{\mu \in \mathcal{H}^* \mid \mu \geq \lambda\}$, using some ordering of $\mathcal{H}^*$, e. g., the lexicographic one; the formal exponents $e(\mu)$ have the properties: $e(0) = 1$, $e(\mu)e(\nu) = e(\mu + \nu)$.

Then the (formal) character of V is defined by

$$\operatorname{ch}_0 V = \sum_{\mu\in\Gamma_+} (\dim V_\mu) e(\Lambda + \mu) = e(\Lambda) \sum_{\mu\in\Gamma_+} (\dim V_\mu) e(\mu). \tag{3.133}$$

(We shall use subscript '0' for the even case.)

For a Verma module, i. e., $V = V^\Lambda$ one has $\dim V_\mu = P(\mu)$, where $P(\mu)$ is a generalized partition function, $P(\mu) = \#$ of ways μ can be presented as a sum of positive roots β, each root taken with its multiplicity $\dim \mathcal{G}_\beta$ (= 1 here), $P(0) \equiv 1$. Thus, the character formula for Verma modules is

$$\operatorname{ch}_0 V^\Lambda = e(\Lambda) \sum_{\mu\in\Gamma_+} P(\mu) e(\mu) = e(\Lambda) \prod_{\alpha\in\Delta^+} (1 - e(\alpha))^{-1}. \tag{3.134}$$

Furthermore, we recall the standard reflections in $\hat{\mathcal{H}}^*$:

$$s_\alpha(\lambda) = \lambda - (\lambda, \alpha^\vee)\alpha, \quad \lambda \in \hat{\mathcal{H}}^*, \ \alpha \in \hat{\mathcal{D}}. \tag{3.135}$$

The Weyl group W is generated by the simple reflections $s_i \equiv s_{\alpha_i}$, $\alpha_i \in \hat{\pi}$. Thus every element $w \in W$ can be written as the product of simple reflections. It is said that w is written in a reduced form if it is written with the minimal possible number of simple reflections; the number of reflections of a reduced form of w is called the length of w, denoted by $\ell(w)$.

The Weyl character formula for the finite-dimensional irreducible LWM L_Λ over $\hat{\mathcal{G}}$, i. e., when $\Lambda \in -\Gamma_+$, has the form

$$\mathrm{ch}_0\, L_\Lambda = \sum_{w\in W} (-1)^{\ell(w)}\, \mathrm{ch}_0\, V^{w\cdot\Lambda}, \quad \Lambda \in -\Gamma_+ \tag{3.136}$$

where the dot $\cdot$ action is defined by $w \cdot \lambda = w(\lambda - \rho) + \rho$. For future reference we note that

$$s_\alpha \cdot \Lambda = \Lambda + n_\alpha \alpha \tag{3.137}$$

where

$$n_\alpha = n_\alpha(\Lambda) \doteq (\rho - \Lambda, \alpha^\vee) = (\rho - \Lambda)(H_\alpha), \quad \alpha \in \Delta^+. \tag{3.138}$$

In the case of basic classical Lie superalgebras the first character formulas were given by Kac [243, 242]. For all such superalgebras—except osp(1/2n)—the character formula for Verma modules is (2.72) (taken from [243, 242]). We are, however, interested exactly in the osp(1/2n) when the Verma module character formula is [243, 242]

$$\mathrm{ch}\, V^\Lambda = e(\Lambda)\Big(\prod_{\alpha\in\bar{\Delta}^+} (1 - e(\alpha))^{-1}\Big). \tag{3.139}$$

Naturally, the character formula for the finite-dimensional irreducible LWM L_Λ is again (3.136) using the Weyl group W_n of B_n.

3.2.3.2 Multiplets

A Verma module V^Λ may be reducible w.r.t. to many positive roots, and thus there maybe many Verma modules isomorphic to its submodules. They themselves may be reducible, and so on.

One main ingredient of the approach of [114] is as follows. We group the (reducible) Verma modules with the same Casimir operators in sets called *multiplets* [112]. The multiplet corresponding to fixed values of the Casimir operators may be depicted as a connected graph, the vertices of which correspond to the reducible Verma modules and the lines between the vertices correspond to embeddings between them. The explicit parametrization of the multiplets and of their Verma modules is important for understanding of the situation.

If a Verma module V^Λ is reducible w.r.t. to all simple roots (and thus w.r.t. all positive roots), i. e., $m_k \in \mathbb{N}$ for all k, then the irreducible submodules are isomorphic to the finite-dimensional irreps of $\mathcal{G}^{\mathbb{C}}$ [114]. (Actually, this is a condition only for m_n since $m_k \in \mathbb{N}$ for $k = 1, \ldots, n-1$.) In these cases we have the *main multiplets* which are isomorphic to the Weyl group of $\mathcal{G}^{\mathbb{C}}$ [114].

In the cases of non-dominant weight Λ the character formula for the irreducible LWM is [253]:

$$\operatorname{ch} L_\Lambda = \sum_{\substack{w \in W \\ w \leq w_\Lambda}} (-1)^{\ell(w_\Lambda w)} P_{w,w_\Lambda}(1) \operatorname{ch} V^{w\cdot(w_\Lambda^{-1}\cdot\Lambda)}, \quad \Lambda \in \Gamma, \tag{3.140}$$

where $P_{y,w}(u)$ are the Kazhdan–Lusztig polynomials $y, w \in W$ [253] (for an easier exposition see [121]), w_Λ is a unique element of W with minimal length such that the signature of $\Lambda_0 = w_\Lambda^{-1} \cdot \Lambda$ is anti-dominant or semi-anti-dominant:

$$\chi_0 = (m'_1, \ldots, m'_n), \quad m'_k = 1 - \Lambda_0(H_k) \in \mathbb{Z}_-. \tag{3.141}$$

Note that $P_{y,w}(1) \in \mathbb{N}$ for $y \leq w$.

When Λ_0 is semi-anti-dominant, i. e., at least one $m'_k = 0$, then in fact W is replaced by a reduced Weyl group W_R.

Most often the value of $P_{y,w}(1)$ is equal to 1 (as in the character formula for the finite-dimensional irreps), while the cases $P_{y,w}(1) > 1$ are related to the appearance of subsingular vectors, though the situation is more subtle; see [121].

It is interesting to see how the reducible points relevant for unitarity fit in the multiplets. In the case of d_{ij} (3.124) and using (3.93) we have

$$m_n(d_{ij}) = 1 - 2m_j - \cdots - 2m_{n-1} - m_i - \cdots - m_{j-1}. \tag{3.142}$$

In the case of d_i (3.125) we have

$$m_n(d_i) = 1 - 2m_i - \cdots - 2m_{n-1}. \tag{3.143}$$

As expected the weights related to positive energy d are not dominant ($m_n(d_{ij}) \in \mathbb{Z}_-$, $m_n(d_i) \in -\mathbb{N}$ $(i < n)$), since the positive energy UIRs are infinite-dimensional. (Naturally, $m_n(d_n) = 1$ falls out of the picture since $d_n < 0$.)

Thus, the Verma modules with weights related to positive energy would be somewhere in the main multiplet (or in a reduction of the main multiplet), and the first task for calculating the character is to find the w_Λ in the character formula (3.140). This we do in the next subsection in the case $n = 3$.

3.2.4 The case $n = 3$

Here we follow [143, 145]. In order to illustrate the main ideas we consider the first non-trivial example $n = 3$, i. e., osp(1/6) actually using the Weyl group of B_3. The Weyl group W_n of B_n has $2^n n!$ elements, i. e., 48 for B_3. Let $S = (s_1, s_2, s_3)$, $s_i \equiv s_{\alpha_i}$, be the simple reflections. They fulfill the following relations:

$$s_1^2 = s_2^2 = s_3^2 = e, (s_1 s_2)^3 = e, (s_2 s_3)^4 = e, s_1 s_3 = s_3 s_1, \tag{3.144}$$

e being the identity of W_3. The 48 elements may be listed as

$$\begin{aligned}
&e,\ s_1,\ s_2,\ s_3 \\
&s_1s_2,\ s_1s_3,\ s_2s_1,\ s_2s_3,\ s_3s_2, \\
&s_1s_2s_1,\ s_1s_2s_3,\ s_1s_3s_2,\ s_2s_1s_3,\ s_2s_3s_2,\ s_3s_2s_1,\ s_3s_2s_3, \\
&s_1s_2s_1s_3,\ s_1s_2s_3s_2,\ s_1s_3s_2s_1,\ s_1s_3s_2s_3, \\
&s_2s_3s_2s_1,\ s_2s_1s_3s_2,\ s_3s_2s_3s_1,\ s_3s_2s_3s_2, \\
&s_1s_2s_3s_2s_1,\ s_1s_3s_2s_1s_3,\ s_1s_2s_1s_3s_2,\ s_1s_3s_2s_3s_2, \\
&s_2s_1s_3s_2s_1,\ s_2s_1s_3s_2s_3,\ s_3s_2s_3s_1s_2,\ s_3s_2s_3s_2s_1, \\
&s_1s_3s_2s_3s_2s_1,\ s_1s_3s_2s_1s_3s_2,\ s_1s_2s_1s_3s_2s_1,\ s_2s_1s_3s_2s_1s_3, \\
&s_2s_1s_3s_2s_3s_2,\ s_3s_2s_3s_1s_2s_1,\ s_3s_2s_3s_1s_2s_3, \\
&s_2s_1s_3s_2s_3s_2s_1,\ s_2s_1s_3s_2s_3s_1s_2,\ s_3s_2s_1s_2s_3s_2s_1, \\
&s_3s_2s_3s_1s_2s_1s_3,\ s_3s_2s_3s_1s_2s_3s_2, \\
&s_2s_3s_2s_1s_2s_3s_2s_1,\ s_3s_2s_1s_3s_2s_3s_2s_1,\ s_3s_2s_1s_3s_2s_3s_1s_2, \\
&s_2s_3s_2s_1s_3s_2s_3s_2s_1.
\end{aligned} \tag{3.145}$$

This Weyl group may be pictorially represented on a cube as in the figure, where we have given only the simple root reflections, namely, continuous (red) arrows represent action of reflection s_1, dashed (blue) arrows represent action of reflection s_2, dotted (green) arrows represent action of reflection s_3. Each face of the cube contains eight elements related by blue and green arrows representing the Weyl group of B_2 generated by s_2 and s_3. The figure contains also eight sextets (around the eight corners of the cube). Each sextet is related by red and green arrows representing the Weyl group of A_2 generated by s_1 and s_2. Finally, there are 12 quartets (straddling the edges of the cube). Each quartet is formed by red and blue arrows representing the Weyl group of $A_1 \times A_1$ generated by the commuting reflections s_1 and s_3.

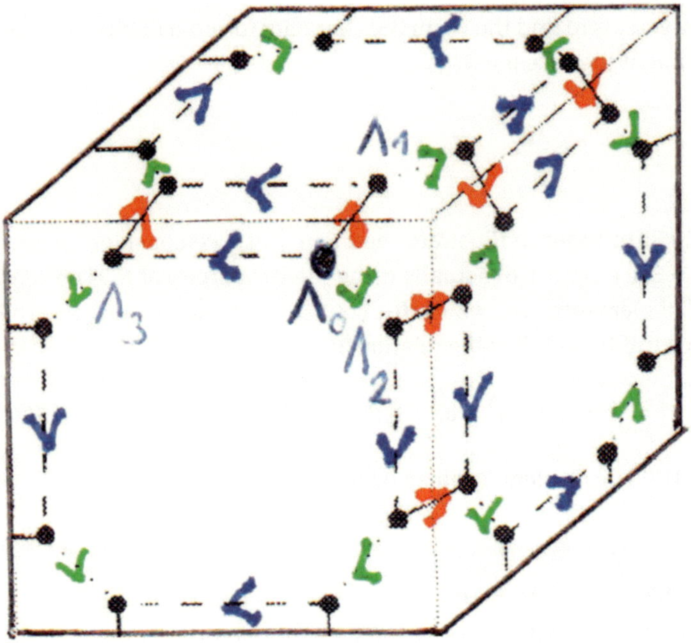

We use the same diagram to depict the main multiplets containing the Verma modules V^{Λ_0} which contain (as factor module) the finite-dimensional irreps of B_3, i. e., with dominant weights Λ_0, i. e., with Dynkin labels (m_1, m_2, m_3), $m_k \in \mathbb{N}$. We may do this since these multiplets are isomorphic to the Weyl group, W_3 in our case. On the picture we have indicated the modules, Λ_0 and $\Lambda_k = s_k \cdot \Lambda_0$, $k = 1, 2, 3$. The mentioned isomorphism is fixed by assigning to Λ_0 the identity element e of W_3, and to Λ_k the reflections s_k.

The character formula for the Verma modules in our case is given explicitly by

$$\begin{aligned} \operatorname{ch} V^{\Lambda} &= \frac{e(\Lambda)}{(1-t_1)(1-t_2)(1-t_1t_2)} \\ &\times \frac{1}{(1-t_3)(1-t_2t_3)(1-t_1t_2t_3)(1-t_2t_3^2)(1-t_1t_2t_3^2)(1-t_1t_2^2t_3^2)} \end{aligned} \tag{3.146}$$

where $t_j \equiv e(\alpha_j)$.

Now we give the character formulas of the five boundary or isolated unitarity cases. Below we shall denote the signature of the dominant weight Λ_0 which determines the main multiplet by (m'_1, m'_2, m'_3), $m'_k \in \mathbb{N}$, using primes to distinguish from the signatures of the weights we are interested. We shall use also reductions of the main multiplet when the weights are semi-dominant, i. e., when some $m'_k = 0$.

- In the case of $d = d_1 = 2 + \frac{1}{2}(a_1 + a_2)$ there are 12 members of the multiplet which is a submultiplet of a main multiplet. (Remember that $m_1 > 1$ since $a_1 \neq 0$.) They are grouped into two standard sl(3) submultiplets of six members. The first submultiplet starts from $V^{\Lambda_0^{d_1}}$, where $\Lambda_0^{d_1} = w \cdot \Lambda_0$, $w = w_{\Lambda_0^{d_1}} = s_2 s_1 s_3 s_2 s_3$, with signature:

$$\Lambda_0^{d_1} : (m_1, m_2, m_3' = 1 - 2m_{12}), \quad m_1, m_2 \in \mathbb{N},\ m_{12} \equiv m_1 + m_2. \tag{3.147}$$

The other submultiplet starts from $V^{\Lambda_0'}$ with $\Lambda_0' = \Lambda_0^{d_1} + \delta_1 = \Lambda_0^{d_1} + \alpha_1 + \alpha_2 + \alpha_3$, with signature: $\Lambda_0' : (m_1 - 1, m_2, m_3' = 1 - 2m_{12})$, $m_1 > 1$. The character formula is (3.140) with $w_\Lambda = w_{\Lambda_0^{d_1}}$:

$$\begin{aligned}\operatorname{ch}\Lambda_0^{d_1} &= \frac{e(\Lambda_0^{d_1})}{(1-t_3)(1-t_2t_3)(1-t_1t_2t_3)(1-t_2t_3^2)(1-t_1t_2t_3^2)(1-t_1t_2^2t_3^2)} \\ &\quad \times \{\operatorname{ch}\Lambda_{m_1,m_2}(t_1,t_2) - t_1t_2t_3 \operatorname{ch}\Lambda_{m_1-1,m_2}(t_1,t_2)\}, \quad m_1 > 1,\end{aligned} \tag{3.148}$$

where $\operatorname{ch}\Lambda_{m_1,m_2}(t_1,t_2)$ is the normalized character of the finite-dimensional sl(3) irrep with Dynkin labels (m_1, m_2) (and dimension $m_1m_2(m_1+m_2)/2$):

$$\operatorname{ch}\Lambda_{m_1,m_2}(t_1,t_2) = \frac{1 - t_1^{m_1} - t_2^{m_2} + t_1^{m_1}t_2^{m_{12}} + t_1^{m_{12}}t_2^{m_2} - t_1^{m_{12}}t_2^{m_{12}}}{(1-t_1)(1-t_2)(1-t_1t_2)}. \tag{3.149}$$

Naturally, the latter formula is a polynomial in t_1, t_2, e. g., $\operatorname{ch}\Lambda_{1,1}(t_1,t_2) = 1$, $\operatorname{ch}\Lambda_{2,1}(t_1,t_2) = 1 + t_1 + t_1t_2$. Note that (3.148) trivializes for $m_1 = 1$ since the second term disappears by the formal substitution: $\operatorname{ch}\Lambda_{0,m_2}(t_1,t_2) = 0$.

In the case $m_1 = 2$, $m_2 = 1$ the character formula (3.148) simplifies to

$$\begin{aligned}\operatorname{ch}\Lambda_0^{d_1} &= \frac{e(\Lambda_0^{d_1})}{(1-t_3)(1-t_2t_3)(1-t_2t_3^2)(1-t_1t_2t_3^2)(1-t_1t_2^2t_3^2)} \\ &\quad \times \left(1 + \frac{t_1(1+t_2)}{1-t_1t_2t_3}\right), \quad m_1 = 2,\ m_2 = 1.\end{aligned} \tag{3.150}$$

- In the case of $d = d_{12} = \frac{1}{2}(3 + a_2)$, which is relevant for unitarity, i. e., $m_1 = 1$, there are again 12 members of the multiplet. The corresponding signature is

$$\Lambda_0^{d_{12}} : (1, m_2, m_3' = -2m_2), \quad m_2 \in \mathbb{N}. \tag{3.151}$$

The multiplet is submultiplet of a reduced multiplet with 24 members obtained from a main multiplet for $m_3' = 0$. As above our multiplet consists of two standard sl(3) submultiplets of six members. The first submultiplet starts from $V^{\Lambda_0^{d_{12}}}$, where $\Lambda_0^{d_{12}} = w \cdot \Lambda_0$, $w = w_{\Lambda_0^{d_{12}}} = s_3 s_2 s_1$. The other submultiplet starts from $V^{\Lambda_0'}$ with $\Lambda_0' = \Lambda_0^{d_{12}} + m_2(\alpha_1 + 2\alpha_2 + 2\alpha_3) = \Lambda_0^{d_{12}} + m_2(\delta_1 + \delta_2)$ with signature: $\Lambda_0' : (1, m_2 - 1, -2m_2)$. The character formula is (3.140), with $W \mapsto W_R$ (where W_R is a reduced 24-member Weyl group) and with $w_\Lambda = w_{\Lambda_0^{d_{12}}}$:

$$\begin{aligned}\operatorname{ch}\Lambda_0^{d_{12}} &= \frac{e(\Lambda_0^{d_{12}})}{(1-t_3)(1-t_2t_3)(1-t_1t_2t_3)(1-t_2t_3^2)(1-t_1t_2t_3^2)(1-t_1t_2^2t_3^2)} \\ &\quad \times \{\operatorname{ch}\Lambda_{1,m_2}(t_1,t_2) - (t_1t_2^2t_3^2)^{m_2} \operatorname{ch}\Lambda_{1,m_2-1}(t_1,t_2)\}, \quad m_2 > 1,\end{aligned} \tag{3.152}$$

where $\operatorname{ch}\Lambda_{m_1,m_2}$ are the sl(3) characters defined in (3.149).

In the case $m_2 = 2$ it simplifies to

$$\operatorname{ch} \Lambda_0^{d_{12}} = \frac{e(\Lambda_0^{d_{12}})}{(1-t_3)(1-t_2t_3)(1-t_1t_2t_3)(1-t_2t_3^2)(1-t_1t_2t_3^2)} \times \left\{1 + t_1t_2^2t_3^2 + \frac{t_2(1+t_1)}{1-t_1t_2^2t_3^2}\right\}. \tag{3.153}$$

- In the case of $d = d_2 = 1 + \frac{1}{2}a_2 \geq d_{13}$, i.e., $m_3' = 1 - 2m_2$, the corresponding signature is

$$\Lambda_0^{d_2} : (m_1, m_2, m_3' = 1 - 2m_2). \tag{3.154}$$

We should consider two subcases

$$1 + m_1 - m_2 > 0 \quad \text{or} \quad 1 + m_1 - m_2 \leq 0$$

We start with the *first subcase* which is relevant when $d = d_2 = d_{13} = 1$ and $a_1 = a_2 = 0$, then $m_1 = m_2 = 1$, and the signature is

$$\Lambda_0^{d_2=d_{13}} : (1, 1, -1). \tag{3.155}$$

Our multiplet is a submultiplet of a 12-member reduced multiplet obtained when the signature of Λ_0 is $(m_1', m_2', m_3') = (1, 0, 1)$, and then $\Lambda_0^{d_2=d_{13}}$ is a submodule with signature (3.155). Thus, we have $\Lambda_0^{d_2=d_{13}} = s_3 \cdot \Lambda_0$, i. e., $w_{\Lambda_0^{d_2=d_{13}}} = s_3$.

Explicitly, our 12-member multiplet has two sl(3) submultiplets. First we take into account a sl(3) sextet starting from $\Lambda_0^{d_2=d_{13}}$ with parameters (1, 1). Then there is a sl(3) sextet starting from $\Lambda_0^{d_2=d_{13}}+\alpha_1+2\alpha_2+3\alpha_3$ with parameters (1, 1). Note that $\alpha_1+2\alpha_2+3\alpha_3 = \delta_1 + \delta_2 + \delta_3$ is the weight of the subsingular vector

$$v_{ss} = (X_{\delta_1}X_{\delta_2}X_{\delta_3} - X_{\delta_3}X_{\delta_2}X_{\delta_1})v_0 \tag{3.156}$$

(with norm $16d(d-1)(2d-1)$), yet the corresponding KL polynomial $P_{y,w}(1)$ is equal to 1.

The character formula is (3.140), with $W \mapsto W_R$ (where W_R is a reduced 12-member Weyl group) and $w_\Lambda = s_3$:

$$\operatorname{ch} \Lambda_0^{d_2=d_{13}} = \frac{e(\Lambda_0^{d_2=d_{13}})}{(1-t_3)(1-t_2t_3)(1-t_1t_2t_3)(1-t_2t_3^2)(1-t_1t_2t_3^2)(1-t_1t_2^2t_3^2)} \times \{1 - t_1t_2^2t_3^3\}. \tag{3.157}$$

Note that the above formula may be rewritten as

$$\operatorname{ch} \Lambda_0^{d_2=d_{13}} = \frac{e(\Lambda_0^{d_2=d_{13}})}{(1-t_2t_3)(1-t_1t_2t_3)(1-t_2t_3^2)(1-t_1t_2t_3^2)} \times \left(\frac{1}{1-t_1t_2^2t_3^2} + \frac{t_3}{1-t_3}\right). \tag{3.158}$$

- In the case of $d = d_2 = 1 + \frac{1}{2}a_2 > d_{13} = 1$, i. e., $m_1 = 1$, $m_2 = 1 + a_2 > 1$, thus, this is the subcase $1 + m_1 - m_2 = m_{13} \le 0$. The multiplet has 24 members for $m_2 > 2$ ($m_{13} < 0$) and starts with $\Lambda_0^{'d_2} == s_3 s_2 s_1 \cdot \Lambda_0$, with signatures:

$$\Lambda_0 : (m_2 - 2, 1, 1),$$
$$\Lambda_0^{'d_2} : (1, m_2, m_3' = 1 - 2m_2), \quad m_2 \in 1 + \mathbb{N}. \tag{3.159}$$

It has four sl(3) submultiplets. First we take into account a sl(3) sextet starting from $\Lambda_0^{'d_2}$ with parameters $(1, m_2)$. Then there is a sl(3) sextet starting from $\Lambda_0^{'d_2} + \alpha_{23}$ with parameters $(2, m_2 - 1)$. Then there is a sl(3) sextet with parameters $(2, m_2 - 2)$ starting from a Verma module $V^{\Lambda''}$, $\Lambda'' = \Lambda_0^{'d_2} + \alpha_1 + 3\alpha_{23}$. Finally, there is a sl(3) sextet with parameters $(1, m_2 - 2)$, starting from a Verma module $V^{\Lambda'''}$, $\Lambda''' = \Lambda_0^{'d_2} + 2(\alpha_1 + 2\alpha_2 + 2\alpha_3)$. Thus, the character formula is (3.140) with $w_\Lambda = s_3 s_2 s_1$:

$$\begin{aligned}\operatorname{ch} \Lambda_0^{'d_2} &= \frac{e(\Lambda_0^{'d_2})}{(1 - t_3)(1 - t_2 t_3)(1 - t_1 t_2 t_3)(1 - t_2 t_3^2)(1 - t_1 t_2 t_3^2)(1 - t_1 t_2^2 t_3^2)} \\ &\quad \times \{\operatorname{ch} \Lambda_{1,m_2}(t_1, t_2) - t_2 t_3 \operatorname{ch} \Lambda_{2,m_2-1}(t_1, t_2) \\ &\quad + t_1 t_2^3 t_3^3 \operatorname{ch} \Lambda_{2,m_2-2}(t_1, t_2) - t_1^2 t_2^4 t_3^4 \operatorname{ch} \Lambda_{1,m_2-2}(t_1, t_2)\}.\end{aligned} \tag{3.160}$$

When $m_2 = 2$ ($a_2 = 1$, $m_{13} = 0$) the weight Λ_0 is semi dominant, the main multiplet reduces to 24 members, our multiplet reduces to only 12 members, consisting of the first two sl(3) submultiplets mentioned above. The character formula takes this into account by construction since for $m_2 = 2$ the terms in the second row are automatically zero (due to the fact that the sl(3) character formula gives zero: $\operatorname{ch} \Lambda_{1,0}(t_1, t_2) = 0$). Thus, the character formula simplifies to

$$\begin{aligned}\operatorname{ch} \Lambda_0^{'d_2} &= \frac{e(\Lambda_0^{'d_2})}{(1 - t_2 t_3^2)(1 - t_1 t_2 t_3^2)(1 - t_1 t_2^2 t_3^2)} \\ &\quad \times \left\{\frac{1}{(1 - t_3)(1 - t_1 t_2 t_3)} + \frac{t_2}{(1 - t_3)(1 - t_2 t_3)} \right. \\ &\quad \left. + \frac{t_1 t_2}{(1 - t_2 t_3)(1 - t_1 t_2 t_3)}\right\}.\end{aligned} \tag{3.161}$$

- In the case of $d = d_{23} = \frac{1}{2}$, $a_1 = a_2 = 0$, i. e., $m_1 = m_2 = 1$, and the signature is

$$\Lambda_0^{d_{23}} : (1, 1, 0). \tag{3.162}$$

This is in fact a multiplet with 24 members which is reduction of the main multiplet starting with the semi dominant weight (3.162).

The multiplet consists of four sl(3) submultiplets. First there is a sl(3) sextet starting from $\Lambda_0^{d_{23}}$ with parameters $(1, 1)$. Then a sl(3) sextet starting from $\Lambda_0^{d_{23}} + \alpha_2 + 2\alpha_3$ with parameters $(2, 1)$. Then a sl(3) sextet starting from $\Lambda_0^{d_{23}} + \alpha_1 + 2\alpha_2 + 4\alpha_3$ with parameters $(1, 2)$. Then a sl(3) sextet starting from $\Lambda_0^{d_{23}} + 2\alpha_1 + 4\alpha_2 + 6\alpha_3$ with parameters $(1, 1)$.

The character formula is (3.140), however, with $W \mapsto W_R$, where W_R is the reduced 24-member Weyl group (generated by s_1, s_2, $s_3s_2s_3$) and $w_\Lambda = 1$:

$$\begin{aligned}\operatorname{ch}\Lambda_0^{d_{23}} &= \frac{e(\Lambda_0^{d_{23}})}{(1-t_3)(1-t_2t_3)(1-t_1t_2t_3)} \\ &\quad\times\frac{1}{(1-t_2t_3^2)(1-t_1t_2t_3^2)(1-t_1t_2^2t_3^2)} \\ &\quad\times\{1-t_2t_3^2\operatorname{ch}\Lambda_{2,1}(t_1,t_2)+t_1t_2^2t_3^4\operatorname{ch}\Lambda_{1,2}(t_1,t_2)-t_1^2t_2^4t_3^6\} \\ &= \frac{e(\Lambda_0^{d_{23}})}{(1-t_3)(1-t_2t_3)(1-t_1t_2t_3)(1-t_2t_3^2)(1-t_1t_2t_3^2)(1-t_1t_2^2t_3^2)} \\ &\quad\times\{1-t_2t_3^2(1+t_1+t_1t_2)+t_1t_2^2t_3^4(1+t_2+t_1t_2)-t_1^2t_2^4t_3^6\}.\end{aligned} \tag{3.163}$$

Finally, note that the above formula may be rewritten as

$$\operatorname{ch}\Lambda_0^{d_{23}} = \frac{e(\Lambda_0^{d_{23}})}{(1-t_3)(1-t_2t_3)(1-t_1t_2t_3)}. \tag{3.164}$$

3.2.5 Conformal superalgebras for $D = 9, 10, 11$

In this subsection we follow [135]. The conformal superalgebras in $D = 9, 10, 11$ are $\mathcal{G} = \operatorname{osp}(1|2n, \mathbb{R})$, $n = 16, 16, 32$, respectively, cf. [370, 97]. The even subalgebra of $\operatorname{osp}(1|2n, \mathbb{R})$ is the algebra $\operatorname{sp}(2n, \mathbb{R})$ with maximal compact subalgebra $\mathcal{K} = u(n) \cong \operatorname{su}(n) \oplus u(1)$.

From the prescription of [97] follows that the even subalgebra $\operatorname{sp}(2n, \mathbb{R})$, $n = 16, 16, 32$, respectively, contains the conformal algebra $\mathcal{C} = \operatorname{so}(D, 2)$, $D = 9, 10, 11$. Then $\mathcal{K}$ contains the maximal compact subalgebra $\operatorname{so}(D) \oplus \operatorname{so}(2)$ of $\mathcal{C}$, $\operatorname{so}(2)$ being identified with the $u(1)$ factor of $\mathcal{K}$, and $\operatorname{su}(n)$ contains the algebra $\operatorname{so}(D)$. The easiest way to describe the embeddings is via the root systems. The superalgebra $\mathcal{G}$ is the split real form of the basic classical superalgebra $\operatorname{osp}(1|2n)$ and has the same root system. The root system of $\operatorname{su}(n)$, actually of $\operatorname{sl}(n)$, is comprised of the even simple roots of $\mathcal{G}$: α_i, $i = 1, \dots, n-1$, with standard non-zero products: $(\alpha_i, \alpha_i) = 2$ $(i = 1, \dots, n-1)$, $(\alpha_i, \alpha_{i+1}) = -1$ $(i = 1, \dots, n-2)$. The root system of $\operatorname{so}(D)$ is comprised of simple roots γ_j, $j = 1, \dots, \ell \equiv [D/2]$.

For even D the non-zero scalar products are $(\gamma_j, \gamma_j) = 2\kappa$ $(j = 1, \dots, \ell)$, $(\gamma_j, \gamma_{j+1}) = -\kappa$ $(j = 1, \dots, \ell-2)$, $(\gamma_{\ell-2}, \gamma_\ell) = -\kappa$, where κ is a non-zero common multiple (it is inessential since it cancels in the Cartan matrix elements). In the case of interest $D = 10$, $\ell = 5$, $n = 16$:

$$\begin{aligned}\gamma_1 &= \alpha_4 + \alpha_7 + \alpha_9 + \alpha_{12}, \\ \gamma_2 &= \alpha_3 + \alpha_6 + \alpha_{10} + \alpha_{13},\end{aligned}$$

$$\gamma_3 = \alpha_2 + \alpha_4 + \alpha_5 + \alpha_6 + \alpha_7 + 2\alpha_8 + \alpha_9 + \alpha_{10} + \alpha_{11} + \alpha_{12} + \alpha_{14},$$
$$\gamma_4 = \alpha_3 + \alpha_5 + \alpha_7 + \alpha_{15},$$
$$\gamma_5 = \alpha_1 + \alpha_9 + \alpha_{11} + \alpha_{13}. \qquad (3.165)$$

(The roots γ_i satisfy the prescribed products with $\kappa = 4$.) Correspondingly, the so(10) Dynkin labels $r_k \equiv -\kappa(\Lambda, \gamma_k^\vee) = -(\Lambda, \gamma_k)$ are

$$r_1 = a_4 + a_7 + a_9 + a_{12},$$
$$r_2 = a_3 + a_6 + a_{10} + a_{13},$$
$$r_3 = a_2 + a_4 + a_5 + a_6 + a_7 + 2a_8 + a_9 + a_{10} + a_{11} + a_{12} + a_{14},$$
$$r_4 = a_3 + a_5 + a_7 + a_{15},$$
$$r_5 = a_1 + a_9 + a_{11} + a_{13}. \qquad (3.166)$$

The dimensions of the so(10) UIRs is

$$\dim L(r_1, \ldots, r_5) = \prod_{1\le s<t\le 5} \frac{n_t^2 - n_s^2}{(t-s)(t+s-2)}, \qquad (3.167)$$

where we use the additional parametrization:

$$n_1 = \frac{1}{2}(r_5 - r_4), \quad n_2 = \frac{1}{2}(r_5 + r_4) + 1$$
$$n_s = \frac{1}{2}(r_5 + r_4) + r_3 + \cdots + r_{6-s} + s - 1, \quad s = 3, \ldots, 5. \qquad (3.168)$$

The parameters n_s are either all integer or all half-integer obeying:

$$n_5 > n_4 > n_3 > n_2 > |n_1| \ge 0. \qquad (3.169)$$

It is known that the unitarity restrictions for a conformal superalgebra $\mathcal{A}$ are stronger than those for the even subalgebra of $\mathcal{A}$ (for $D = 4, 6$ cf. [137]). Here, in addition the conformal algebra so$(D, 2)$ is smaller than the even subalgebra sp$(2n)$. Thus, the unitarity conditions for so$(D, 2)$ are given only in terms of r_i. Thus, so$(D, 2)$ unitarity would not require that all parameters a_k are non-negative integers—that would be required only for their combinations r_i. The unrestricted parameters a_k are combined in so-called tensorial charges [370]. We shall leave this together with a more detailed analysis for a follow-up publication.

Here, due to the lack of space, we only consider briefly the reduction of the fundamental irreps of su(n) to so(D) irreps. We list only the main so(D) component with signature following directly from the embedding formulas, e. g., (3.166) for so(10).

Of course, the one-dimensional irreps, when $a_i = 0 = r_s$ for all i, s, coincide. In the Dynkin labeling of the $n-1$ fundamental irreps Λ_k of su(n) are characterized for fixed k by $a_i = \delta_{ik}$. The fundamental irreps Λ_t^o of so(D) are characterized for fixed k by $r_s =$

δ_{st}. The 16-dimensional fundamental su(16) UIR with $a_1 = 1$ gives the fundamental 16-dimensional so(10) spinor when $r_5 = 1$, while the conjugated 16-dimensional fundamental su(16) UIR with $a_{15} = 1$ gives the conjugated 16-dimensional so(10) spinor with $r_4 = 1$. We summarize the results in the table included in equation (3.170):

Λ_i	$\dim(\Lambda_i)$	$\chi = [r_1, r_2, r_3, r_4, r_5]$	$\dim(\chi)$
Λ_1	16	$[0,0,0,0,1]$	16
Λ_{15}	16	$[0,0,0,1,0]$	16
Λ_2, Λ_{14}	120	$[0,0,1,0,0]$	120
Λ_3	560	$[0,1,0,1,0]$	560
Λ_{13}	560	$[0,1,0,0,1]$	560
Λ_4, Λ_{12}	1820	$[1,0,1,0,0]$	945
Λ_5	4368	$[0,0,1,1,0]$	1200
Λ_{11}	4368	$[0,0,1,0,1]$	1200
Λ_6, Λ_{10}	8008	$[0,1,1,0,0]$	2970
Λ_7	11440	$[1,0,1,1,0]$	8800
Λ_9	11440	$[1,0,1,0,1]$	8800
Λ_8	12870	$[0,0,2,0,0]$	4125

(3.170)

Part of the above analysis is done in the oscillator approach in [214].

For odd D the non-zero scalar products are $(\gamma_j, \gamma_j) = 2\kappa$ $(j = 1, \ldots, \ell - 1)$, $(\gamma_\ell, \gamma_\ell) = \kappa$, $(\gamma_j, \gamma_{j+1}) = -\kappa$ $(j = 1, \ldots, \ell - 1)$.

In the case $D = 9$, $\ell = 4$, $n = 16$, we have

$$\begin{aligned}
\gamma_1 &= \alpha_5 + 2\alpha_6 + \alpha_7 + \alpha_9 + 2\alpha_{10} + \alpha_{11},\\
\gamma_2 &= \alpha_3 + 2\alpha_4 + \alpha_5 + \alpha_{11} + 2\alpha_{12} + \alpha_{13},\\
\gamma_3 &= \alpha_2 + \alpha_6 + \alpha_7 + 2\alpha_8 + \alpha_9 + \alpha_{10} + \alpha_{14},\\
\gamma_4 &= \frac{1}{2}(\alpha_1 + \alpha_3 + \alpha_5 + \alpha_7 + \alpha_9 + \alpha_{11} + \alpha_{13} + \alpha_{15})
\end{aligned} \tag{3.171}$$

(with $\kappa = 4$). Then the so(9) Dynkin labels and dimensions are

$$\begin{aligned}
r_1 &= a_5 + 2a_6 + a_7 + a_9 + 2a_{10} + a_{11},\\
r_2 &= a_3 + 2a_4 + a_5 + a_{11} + 2a_{12} + a_{13},\\
r_3 &= a_2 + a_6 + a_7 + 2a_8 + a_9 + a_{10} + a_{14},\\
r_4 &= a_1 + a_3 + a_5 + a_7 + a_9 + a_{11} + a_{13} + a_{15},
\end{aligned} \tag{3.172}$$

$$\dim L(r_1, \ldots, r_4) = \frac{2^4 n_1 n_2 n_3 n_4}{7.5.3} \prod_{1\le s<t\le 4} \frac{n_t^2 - n_s^2}{(t-s)(t+s-1)}, \tag{3.173}$$

$$\begin{aligned}
&n_1 = \frac{1}{2}(r_4 + 1), \quad n_s = \frac{1}{2}(r_4 - 1) + r_3 + \cdots + r_{5-s} + s, \quad s = 2,3,4,\\
&n_4 > n_3 > n_2 > n_1 > 0.
\end{aligned} \tag{3.174}$$

The fundamental 16-dimensional so(9) spinor—obtained for $r_4 = 1$—is contained without reduction in both conjugated 16-dimensional fundamental su(16) UIRs with $a_1 = 1$ and $a_{15} = 1$. We summarize the results in the table included in equation (3.175):

$$\begin{array}{lrcr}
\Lambda_i & \dim(\Lambda_i) & \chi = [r_1, r_2, r_3, r_4] & \dim(\chi) \\
\Lambda_1, \Lambda_{15} & 16 & [0,0,0,1] & 16 \\
\Lambda_2, \Lambda_{14} & 120 & [0,0,1,0] & 84 \\
\Lambda_3, \Lambda_{13} & 560 & [0,1,0,1] & 432 \\
\Lambda_4, \Lambda_{12} & 1820 & [0,2,0,0] & 495 \\
\Lambda_5, \Lambda_{11} & 4368 & [1,1,0,1] & 2560 \\
\Lambda_6, \Lambda_{10} & 8008 & [2,0,1,0] & 2457 \\
\Lambda_7, \Lambda_9 & 11440 & [1,0,1,1] & 5040 \\
\Lambda_8 & 12870 & [0,0,2,0] & 1980
\end{array} \tag{3.175}$$

In the case $D = 11$, $\ell = 5$, $n = 32$, we have

$$\begin{aligned}
y_1 &= \alpha_5 + \alpha_8 + \alpha_{10} + \alpha_{13} + \alpha_{14} + \alpha_{15} \\
&\quad + \alpha_{17} + \alpha_{18} + \alpha_{19} + \alpha_{22} + \alpha_{24} + \alpha_{27}, \\
y_2 &= \alpha_4 + \alpha_7 + \alpha_9 + \alpha_{10} + \alpha_{11} + \alpha_{12} + \alpha_{14} + 2\alpha_{15} + 2\alpha_{16} \\
&\quad + 2\alpha_{17} + \alpha_{18} + \alpha_{20} + \alpha_{21} + \alpha_{22} + \alpha_{23} + \alpha_{25} + \alpha_{28}, \\
y_3 &= \alpha_3 + \alpha_5 + \alpha_6 + \alpha_8 + \alpha_9 + \alpha_{12} + \alpha_{13} + \alpha_{14} + \alpha_{15} + 2\alpha_{16} \\
&\quad + a_{17} + \alpha_{18} + \alpha_{19} + \alpha_{20} + \alpha_{23} + \alpha_{24} + \alpha_{26} + \alpha_{27} + \alpha_{29}, \\
y_4 &= \alpha_2 + \alpha_4 + \alpha_6 + \alpha_7 + \alpha_8 + \alpha_{10} + \alpha_{11} + \alpha_{12} + \alpha_{13} + \alpha_{14} \\
&\quad + \alpha_{18} + \alpha_{19} + a_{20} + \alpha_{21} + \alpha_{22} + \alpha_{24} + a_{25} + \alpha_{26} + \alpha_{28} + \alpha_{30}, \\
y_5 &= \frac{1}{2}(\alpha_1 + \alpha_3 + \alpha_5 + \alpha_7 + \alpha_9 + \alpha_{11} + \alpha_{13} + \alpha_{15} \\
&\quad + \alpha_{17} + \alpha_{19} + \alpha_{21} + \alpha_{23} + \alpha_{25} + \alpha_{27} + \alpha_{29} + \alpha_{31})
\end{aligned} \tag{3.176}$$

(with $\kappa = 8$). Then the so(11) Dynkin labels and dimensions are

$$\begin{aligned}
r_1 &= a_5 + a_8 + a_{10} + a_{13} + a_{14} + a_{15} \\
&\quad + a_{17} + a_{18} + a_{19} + a_{22} + a_{24} + a_{27}, \\
r_2 &= a_4 + a_7 + a_9 + a_{10} + a_{11} + a_{12} + a_{14} + 2a_{15} + 2a_{16} \\
&\quad + 2a_{17} + a_{18} + a_{20} + a_{21} + a_{22} + a_{23} + a_{25} + a_{28}, \\
r_3 &= a_3 + a_5 + a_6 + a_8 + a_9 + a_{12} + a_{13} + a_{14} + a_{15} + 2a_{16} \\
&\quad + a_{17} + a_{18} + a_{19} + a_{20} + a_{23} + a_{24} + a_{26} + a_{27} + a_{29}, \\
r_4 &= a_2 + a_4 + a_6 + a_7 + a_8 + a_{10} + a_{11} + a_{12} + a_{13} + a_{14} \\
&\quad + a_{18} + a_{19} + a_{20} + a_{21} + a_{22} + a_{24} + a_{25} + a_{26} + a_{28} + a_{30}, \\
r_5 &= a_1 + a_3 + a_5 + a_7 + a_9 + a_{11} + a_{13} + a_{15}
\end{aligned}$$

$$+ a_{17} + a_{19} + a_{21} + a_{23} + a_{25} + a_{27} + a_{29} + a_{31}, \tag{3.177}$$

$$\dim L(r_1,\dots,r_5) = \frac{2^5 n_1 n_2 n_3 n_4 n_5}{9.7.5.3} \prod_{1\le s<t\le 5} \frac{n_t^2 - n_s^2}{(t-s)(t+s-1)}, \tag{3.178}$$

$$n_1 = \frac{1}{2}(r_5+1), \quad n_s = \frac{1}{2}(r_5-1) + r_4 + \cdots + r_{6-s} + s, \quad s = 2,\dots,\ell,$$
$$n_5 > n_4 > n_3 > n_2 > n_1 > 0. \tag{3.179}$$

The fundamental 32-dimensional so(11) spinor—obtained for $r_5 = 1$—is contained without reduction in the two 32-dimensional fundamental su(32) UIRs with $a_1 = 1$ and $a_{31} = 1$. We summarize the results in the table included in equation (3.180):

Λ_i	$\dim(\Lambda_i)$	$\chi = [r_1, r_2, r_3, r_4, r_5]$	$\dim(\chi)$
Λ_1, Λ_{31}	32	[0, 0, 0, 0, 1]	32
Λ_2, Λ_{30}	496	[0, 0, 0, 1, 0]	330
Λ_3, Λ_{29}	4960	[0, 0, 1, 0, 1]	3520
Λ_4, Λ_{28}	35960	[0, 1, 0, 1, 0]	11583
Λ_5, Λ_{27}	201376	[1, 0, 1, 0, 1]	28512
Λ_6, Λ_{26}	906192	[0, 0, 1, 1, 0]	23595
Λ_7, Λ_{25}	3365856	[0, 1, 0, 1, 1]	160160
Λ_8, Λ_{24}	10518300	[1, 0, 1, 1, 0]	178750
Λ_9, Λ_{23}	28048800	[0, 1, 1, 0, 1]	91520
$\Lambda_{10}, \Lambda_{22}$	64512240	[1, 1, 0, 1, 0]	78650
$\Lambda_{11}, \Lambda_{21}$	129024480	[0, 1, 0, 1, 1]	160160
$\Lambda_{12}, \Lambda_{20}$	225792840	[0, 1, 1, 1, 0]	525525
$\Lambda_{13}, \Lambda_{19}$	347373600	[1, 0, 1, 1, 1]	2114112
$\Lambda_{14}, \Lambda_{18}$	471435600	[1, 1, 1, 1, 0]	3128697
$\Lambda_{15}, \Lambda_{17}$	565722720	[1, 2, 1, 0, 1]	6040320
Λ_{16}	601080390	[0, 2, 2, 0, 0]	1718496

(3.180)

4 Quantum superalgebras

This chapter is in a field which is the intersection of two major developments in physics starting in the seventies and in the eighties: supersymmetry and quantum groups, respectively. The extension of the activity on quantum groups to the field of supersymmetry was started with the paper of Manin [293], where the standard multiparametric quantum deformation of the supergroup $GL(m/n)$ was introduced. These deformations of $GL(m/n)$ were extensively studied in the physics literature; cf., e. g., [87, 345, 95, 365, 71, 262]. In the case of one-parametric deformation the superalgebra $U_q(gl(m/n))$ in duality with $GL_q(m/n)$ and its quantum subsuperalgebra $U_q(sl(m/n))$ were studied in, e. g., [78, 64, 184, 251, 263, 344, 183, 254, 319, 117, 80, 133, 318, 363, 394]. In the study of the multiparameter deformations of $U(gl(m/n))$ and $U(sl(m/n))$ and their interrelations, early results for two-parameter deformations were obtained for $m = n = 1$ in [229, 95, 71, 94, 366, 187], and multiparameter deformations of $U(sl(m/n))$ were obtained in [249]. However, until recently the superalgebra in duality with the standard multiparameter deformation $GL_{u\mathbf{q}}(m/n)$ was not known. This dual Hopf superalgebra, which we denote as $\mathcal{U} \equiv \mathcal{U}_{u\mathbf{q}}(gl(m/n))$, was found in [146], and is presented in Section 4.2 below. The induced representations of $\mathcal{U}$ were constructed in [147] and are presented in Section 4.3.

In the next section we give the general picture by spelling out the definition of q-deformed superalgebras. Then we give some important examples. After that we discuss in detail the multiparameter case.

4.1 Defining relations for q-deformed superalgebras

Let $\mathcal{G}$ be a complex Lie superalgebra with a symmetrizable Cartan matrix $A = (a_{jk}) = A^d \hat{A}^s$, where $\hat{A}^s = (\hat{a}^s_{jk})$ is a symmetric matrix, and $A^d = \mathrm{diag}(\hat{d}_1, \dots, \hat{d}_\ell)$, $\hat{d}_k \neq 0$. (Note that $\hat{a}^s_{jk} = d_j \hat{d}_j a^s_{jk}$, cf. (1.41).) Then the q-deformation $U_q(\mathcal{G})$ of the universal enveloping algebras $U(\mathcal{G})$ is defined [368, 78, 254] as the associative algebra over $\mathbb{C}$ with generators $X^\pm_j$, H_j, $j \in J = \{1, \dots, \ell\}$ and with relations similar to the even case:

$$[H_i, H_j] = 0, \quad [H_i, X^\pm_j] = \pm \hat{a}^s_{ij} X^\pm_j,$$
$$[X^+_i, X^-_j] = \delta_{ij} [H_i]_{q_i}, \quad q_i = q^{\hat{d}_i} \tag{4.1}$$

([,] being the supercommutator),

$$(\mathrm{Ad}_{q^\kappa} X^\pm_j)^{n_{jk}} (X^\pm_k) = 0, \quad \text{for } j \neq k, \ \kappa = \pm; \tag{4.2}$$

and for every three simple roots, say, α_j, $\alpha_{j\pm1}$, such that $(\alpha_j, \alpha_j) = 0$, $(\alpha_{j\pm1}, \alpha_{j\pm1}) \neq 0$, $(\alpha_{j+1}, \alpha_{j-1}) = 0$, $(\alpha_j, \alpha_{j+1} + \alpha_{j-1}) = 0$, we also have

$$[[X^\pm_j, X^\pm_{j-1}]_{q^\kappa}, [X^\pm_j, X^\pm_{j+1}]_{q^\kappa}] = 0, \tag{4.3}$$

https://doi.org/10.1515/9783110527490-004

where

$$n_{jk} = \begin{cases} 1 & \text{if } \hat{a}^s_{jj} = \hat{a}^s_{jk} = 0 \\ 2 & \text{if } \hat{a}^s_{jj} = 0,\ \hat{a}^s_{jk} \neq 0 \\ 1 - 2\hat{a}^s_{jk}/\hat{a}^s_{jj} & \text{if } \hat{a}^s_{jj} \neq 0 \end{cases}$$

and in (4.1) and (4.2) one uses the deformed supercommutator:

$$(\mathrm{Ad}_{q^\kappa} X^\pm_j)(X^\pm_k) = [X^\pm_j, X^\pm_k]_{q^\kappa} \equiv X^\pm_j X^\pm_k - (-1)^{p(X^\pm_j)p(X^\pm_k)} q^{\kappa(\alpha_j,\alpha_k)/2} X^\pm_k X^\pm_j . \tag{4.4}$$

The above is applicable to the even case; then equations (4.1) for $\kappa = 1$ are the same as for $\kappa = -1$ and coincide with the usual even Serre relations.

The Hopf algebra structure is given by the formulas of the even case, however, with $\rho = \rho_{\bar{0}} - \rho_{\bar{1}}$.

The necessity of the extra equations (4.3) (including the classical $q = 1$ case) was communicated to the author in May 1991 independently by Scheunert [344], Kac [245], and Leites [183]. (Later the author became aware that at the same time these relations were found by Yamane [389].) In the q-deformed case these relations were written first for $U_q(\mathrm{sl}(M/N))$ in [344] (cf. also [183]); then for $U_q(\mathrm{osp}(M/2N))$ in [183]. Here they are given as in [254], except that in [254] also the external supercommutator is given as deformed, while in fact it is not. The reason is that the would-be deformation is given by the factor $q^{\kappa(\alpha_j+\alpha_{j-1},\alpha_j+\alpha_{j+1})}$, and as can easily be checked the scalar product in the q-exponent it is actually zero.

Remark. Later extra Serre relations were written also for the affine (q-deformed) Kac–Moody superalgebras [255, 390] and for any Lie superalgebra with Cartan matrix [210]. See also recent papers on the q-deformed affine superalgebra $D(2,1;\bar{\sigma})$ [222] and the su(2/2) Yangian [360] relevant in our current context and related to [36]. ◇

4.1.1 Example of $U_q(\mathrm{sl}(M/N;\mathbb{C}))$

Let $\mathcal{G} = \mathrm{sl}(M/N;\mathbb{C})$, $\ell = M+N$. We choose a Cartan matrix with elements $a_{jj} = \hat{a}^s_{jj} = 2(1-\delta_{jM})$, $a_{jj\pm1} = \hat{a}^s_{j\pm1} = -1$ except for $a_{MM+1} = 1$, all other elements are zero; $d_j = 1, j \leq M$, $d_j = -1, j > M$. Consistently the products between the simple roots are $(\alpha_j,\alpha_j) = 2, 0, -2$ for $j < M, j = 0, j > M$, respectively, $(\alpha_j,\alpha_{j+1}) = -1, 1$ for $j \leq M, j > M$, respectively, all other products are zero. The root system is given by $\Delta^\pm = \{\pm\alpha_{jk} = \pm(\alpha_j + \alpha_{j+1} + \cdots + \alpha_k) \mid 1 \leq j < k \leq \ell, \alpha_{jj+1} = \alpha_j\}$. The roots $\pm\alpha_{jk}$ with $1 \leq j \leq M$, $M < k \leq \ell$ are odd, the rest are even. The Cartan–Weyl generators corresponding to nonsimple roots are defined inductively in analogy to (1.48) of Volume 2:

$$X^+_{jk} = (\mathrm{Ad}_q X^+_j) X^+_{j+1,k} \equiv X^+_j X^+_{j+1,k} - (-1)^{p(X^+_j)p(X^+_{j+1,k})} q^{(\alpha_j,\alpha_{j+1,k+1})/2} X^+_{j+1,k} X^+_j$$

$$
\begin{aligned}
&= X_j^+ X_{j+1,k}^+ - q^{(\alpha_j,\alpha_{j+1,k+1})/2} X_{j+1,k}^+ X_j^+ \\
&= (\mathrm{Ad}_q X_{j,k-1}^+) X_k^+ \\
&\equiv X_{j,k-1}^+ X_k^+ - (-1)^{p(X_k^+)p(X_{j,k-1}^+)} q^{(\alpha_k,\alpha_{jk})/2} X_k^+ X_{j,k-1}^+ \\
&= X_{j,k-1}^+ X_k^+ - q^{(\alpha_k,\alpha_{jk})/2} X_k^+ X_{j,k-1}^+, \quad j < k, \\
X_{kj}^- &= (\mathrm{Ad}_{q^{-1}} X_k^-) X_{k-1,j}^- \\
&\equiv X_k^- X_{k-1,j}^- - (-1)^{p(X_k^-)p(X_{k-1,j}^-)} q^{-(\alpha_k,\alpha_{jk})/2} X_{k-1,j}^- X_k^- \\
&= X_k^- X_{k-1,j}^- - q^{-(\alpha_k,\alpha_{jk})/2} X_{k-1,j}^- X_k^- \\
&= (\mathrm{Ad}_{q^{-1}} X_{k,j+1}^-) X_j^- \\
&\equiv X_{k,j+1}^- X_j^- - (-1)^{p(X_j^-)p(X_{k,j+1}^-)} q^{-(\alpha_j,\alpha_{j+1,k+1})/2} X_j^- X_{k,j+1}^- \\
&= X_{k,j+1}^- X_j^- - q^{-(\alpha_j,\alpha_{j+1,k+1})/2} X_j^- X_{k,j+1}^-, \quad j < k,
\end{aligned} \tag{4.5}
$$

where $X_j^\pm \equiv X_{jj}^\pm$, and we have also stressed the fact that there are no sign factors since all expressions bilinear in the generators involve at most one odd generator. Indeed, at most one of the two roots α_j, $\alpha_{j+1,k+1}$, or of the two roots α_k, α_{jk}, can be odd.

The central element of $\mathrm{sl}(M/M;\mathbb{C})$ may be written explicitly as

$$Z_M = H_1 - H_{2M-1} + 2(H_2 - H_{2M-2}) + \cdots + (M-1)(H_{M-1} - H_{M+1}) + MH_M. \tag{4.6}$$

4.1.2 The cases of $\mathrm{sl}(2/2;\mathbb{C})$ and $D(2,1;\bar{\sigma})$

In this subsection we are in the setting of the choice of deformation parameter $q = 1$ and it could have been placed earlier in the book. On the other hand, for the algebra $\mathrm{sl}(2/2;\mathbb{C})$ the new phenomenon displayed in (4.3) plays an important role, which we explain here.

We first recall that recently the superalgebra su(2/2) (which is a real form of $\mathrm{sl}(2/2;\mathbb{C})$) played an important role in the study of planar $N = 4$ super-Yang–Mills theory and in the construction of S-matrix for dynamic spin chains; cf. [35]. In [35], the fact was exploited that the superalgebra $\mathrm{sl}(2/2;\mathbb{C})$ has an extraordinary feature, namely, it is the only basic classical Lie superalgebra that has nontrivial central extension [236] (besides the trivial extension of $\mathrm{psl}(n/n;\mathbb{C})$ to $\mathrm{sl}(n/n;\mathbb{C})$). This feature is related to the fact that the nontrivial central extension of $\mathrm{sl}(2/2;\mathbb{C})$ may be obtained by contraction from the algebra $D(2,1;\bar{\sigma})$. This fact was noted in [126] and is exposed below.

Pictorially, one case of the situation with the three roots in (4.3) is given by the following (part of) a Dynkin diagram:

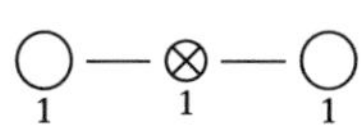

(4.7)

The three roots in (4.3) and three nodes in (4.7) by themselves determine the root system and the Dynkin diagram of the superalgebra $\mathrm{sl}(2/2;\mathbb{C})$. We recall that this superalgebra is 15 dimensional (over $\mathbb{C}$), the even subalgebra being seven dimensional:

$$\mathrm{sl}(2/2;\mathbb{C})_{\bar{0}} \cong \mathrm{sl}(2,\mathbb{C}) \oplus \mathrm{sl}(2,\mathbb{C}) \oplus \mathbb{C},$$

while the odd part $\mathrm{sl}(2/2;\mathbb{C})_{\bar{1}}$ is eight dimensional. We choose a distinguished root system (with one odd simple root) corresponding to (4.7), the simple roots being denoted as $\alpha_1, \alpha_2, \alpha_3$, of which α_2 is odd, the other even. The positive roots are

$$\begin{aligned}\Delta^+ &= \Delta^+_{\bar{0}} \cup \Delta^+_{\bar{1}}, \quad \Delta^+_{\bar{0}} = \{\alpha_1, \alpha_3\},\\ \Delta^+_{\bar{1}} &= \{\alpha_2, \alpha_1 + \alpha_2, \alpha_2 + \alpha_3, \alpha_1 + \alpha_2 + \alpha_3\}.\end{aligned} \tag{4.8}$$

(Note that $\alpha_1 + \alpha_2 + \alpha_3$ is the highest root.)

We denote the Chevalley generators as $X_j^{\pm}, H_j, j = 1,2,3$. Then the Cartan subalgebra of $\mathrm{sl}(2/2;\mathbb{C})$ is spanned by $H_j, j = 1,2,3$, though often instead of H_2 (related to the odd root α_2), the central generator $Z_4 = H_1 + 2H_2 - H_3$, see (4.6), is used. (In a matrix realization of 4×4 matrices, $K \sim I_4$, the unit 4×4 matrix.)

The defining equations (4.1) and (4.2) (for $q = 1$) are clear. We are interested in the new equations (4.3), which we write out explicitly for $q = 1$:

$$[[X_2^{\pm}, X_1^{\pm}], [X_2^{\pm}, X_3^{\pm}]]_+ = 0. \tag{4.9}$$

We now pass to the superalgebra $D(2,1;\bar{\sigma})$. We recall that this superalgebra is 17 dimensional, the even subalgebra being nine dimensional:

$$\mathrm{sl}(2/2;\mathbb{C})_{\bar{0}} = \mathrm{sl}(2,\mathbb{C}) \oplus \mathrm{sl}(2,\mathbb{C}) \oplus \mathrm{sl}(2,\mathbb{C}),$$

while the odd part $\mathrm{sl}(2/2;\mathbb{C})_{\bar{1}}$ is eight dimensional [244]. We choose a distinguished root system, the simple roots being denoted as $\alpha_1, \alpha_2, \alpha_3$, of which α_2 is odd, the other even. The positive roots are

$$\begin{aligned}\Delta^+ &= \Delta^+_{\bar{0}} \cup \Delta^+_{\bar{1}}, \quad \Delta^+_{\bar{0}} = \{\alpha_1, \alpha_3, \alpha_1 + 2\alpha_2 + \alpha_3\},\\ \Delta^+_{\bar{1}} &= \{\alpha_2, \alpha_1 + \alpha_2, \alpha_2 + \alpha_3, \alpha_1 + \alpha_2 + \alpha_3\}.\end{aligned} \tag{4.10}$$

We note that $\tilde{\alpha} = \alpha_1 + 2\alpha_2 + \alpha_3$ is the highest root, and the root system corresponds to the following Dynkin diagram:

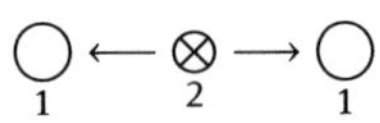

(4.11)

For this superalgebra the defining relations are only (4.1) and (4.2) for $q = 1$, since the Dynkin diagram is not of the type (4.7). Our interest is to give an explicit expression for the Cartan–Weyl generators $X^{\pm}_{\tilde{\alpha}}$ corresponding to the highest root $\tilde{\alpha}$. After a simple calculation we obtain

$$X^{\pm}_{\tilde{\alpha}} = [[X^{\pm}_2, X^{\pm}_1], [X^{\pm}_2, X^{\pm}_3]]_+. \tag{4.12}$$

Identifying the simple roots in (4.8) and (4.10) and comparing (4.9) and (4.12) we see that the superalgebras sl(2/2; $\mathbb{C}$), $D(2,1;\bar{\sigma})$ differ by the extra Serre equations (4.9) which are needed for sl(2/2; $\mathbb{C}$) but are absent for $D(2,1;\bar{\sigma})$. Consequently, we can pass from $D(2,1;\bar{\sigma})$ to sl(2/2; $\mathbb{C}$) by a contracting procedure in which the root vectors $X^{\pm}_{\tilde{\alpha}}$ (4.12) are sent to zero. However, we may use a more general contraction procedure, as in [35], in which the root vectors are replaced by central elements (not necessarily zero), i. e., setting

$$X^{\pm}_{\tilde{\alpha}} \longrightarrow C^{\pm} \in \mathbb{C}, \tag{4.13}$$

we obtain the central extension of sl(2/2; $\mathbb{C}$) found in [236].

4.1.3 q-deformed conformal superalgebras U_q(su(2, 2/N))

The Lie superalgebra $\mathcal{G}^S \equiv \mathrm{su}(2,2/N)$ [241] is a real non-compact form of $\mathcal{G}^{\mathbb{C}} = \mathrm{sl}(4/N;\mathbb{C})$ with Cartan decomposition and splitting into even and odd parts: $\mathcal{G}^S = \mathcal{K}^S + \mathcal{P}^S = \mathcal{G}^S_{(0)} + \mathcal{G}^S_{(1)}$ such that $\mathcal{G}^S_{(0)} \cong \mathrm{su}(2,2) \oplus u(1) \oplus \mathrm{su}(N)$, $\mathcal{K}^S_{(0)} \cong u(2) \oplus u(2) \oplus \mathrm{su}(N)$, $\dim_R \mathcal{P}^S_{(0)} = 8$, $\dim_R \mathcal{K}^S_{(1)} = \dim_R \mathcal{P}^S_{(1)} = 4N$.

The parabolic subalgebras of $\mathcal{G}^S$ are determined by the parabolic subalgebras of the non-compact subalgebra su(2, 2) of the even part $\mathcal{G}^S_{(0)}$. Here we consider only a q-deformation of $\mathcal{G}$ consistent with the maximal parabolic subalgebra $\mathcal{P}^S_{\max} = \mathcal{M}^S \oplus \mathcal{A}^S \oplus \mathcal{N}^S$, where

$$\begin{aligned}
\mathcal{A}^S &= \mathcal{A}^S_{(0)} = l.s.\{D\} \cong \mathcal{A}', \\
\mathcal{M}^S &= \mathcal{M}^S_{(0)} \cong \mathcal{M}' \oplus u(1) \oplus \mathrm{su}(N), \quad \mathcal{M}' \cong \mathrm{so}(3,1), \\
\mathcal{N}^S &= \mathcal{G}^-_1 \oplus \mathcal{G}^-_2, \quad \mathcal{G}^-_k \equiv \mathcal{G}_{-\lambda_k}, \quad \lambda_1(D) = 1/2, \quad \lambda_2 = 2\lambda_1, \\
\dim \mathcal{G}^-_1 &= 4N, \quad \mathcal{N}^S_{(0)} = \mathcal{G}^-_2 \cong \mathcal{N}', \\
\tilde{\mathcal{N}}^S &= \mathcal{G}^+_1 \oplus \mathcal{G}^+_2, \quad \mathcal{G}^+_k \equiv \mathcal{G}_{\lambda_k} = \theta\mathcal{G}^-_k, \quad \tilde{\mathcal{N}}^S_{(0)} = \mathcal{G}^+_2 \cong \tilde{\mathcal{N}}',
\end{aligned} \tag{4.14}$$

where the primed objects are su(2, 2) subalgebras. The Cartan subalgebra $\mathcal{H}^S \subset \mathcal{G}^S_{(0)}$ is chosen as follows:

$$\mathcal{H}^S = \mathcal{H} \oplus u(1) \oplus \mathcal{H}_N, \tag{4.15}$$

where $\mathcal{H}$ is the Cartan subalgebra of su(2, 2), $\mathcal{H}_N$ is the Cartan subalgebra of su(N).

Now we express the generators of $U_q(\mathcal{G}^S)$ in terms of those of $U_q(\mathcal{G}^{\mathbb{C}})$.

For $U_q(\mathrm{su}(2,2))$ we use equations (28)–(31) from [117]. In detail, for the Lorentz algebra generators we use

$$H = -Y_{30} = (1/2)(H_1 + H_3), \quad M^{\pm} = -iY_{13} \pm iY_{10} = X_1^{\pm} + X_3^{\pm}, \qquad (4.16)$$
$$\tilde{D} = -Y_{12} = (i/2)(H_1 - H_3), \quad N^{\pm} = -iY_{20} \pm iY_{23} = i(X_1^{\pm} - X_3^{\pm}),$$

while for the dilatations, translations and special conformal transformations we have

$$D = (1/2)(H_1 + H_3) + H_2, \qquad (4.17)$$
$$P_0 = i(X_{13}^+ + X_2^+), \quad P_1 = i(X_{12}^+ + X_{23}^+),$$
$$P_2 = X_{12}^+ - X_{23}^+, \quad P_3 = i(X_2^+ - X_{13}^+),$$
$$K_0 = -i(X_{13}^- + X_2^-), \quad K_1 = i(X_{12}^- + X_{23}^-),$$
$$K_2 = X_{23}^- - X_{12}^-, \quad K_3 = i(X_2^- - X_{13}^-).$$

Then for $U_q(\mathrm{su}(N))$ we use the formulas for $U_q(\mathrm{sl}(N))$ adapted to the compact case. Therefore, we note that $\{\pm i\alpha_{jk} \mid 5 \le j \le k \le N+3\}$ form the root system of $\mathrm{su}(N)$.

Finally, for the generator of the $u(1)$ subalgebra in $\mathcal{G}^S_{(0)}$, $\mathcal{M}^S_{(0)}$, and $\mathcal{H}^S$ we have

$$e_N = \sum_{k=1}^{4} kH_k + \frac{4}{N}\sum_{k=5}^{N+3}(k-4-N)H_k. \qquad (4.18)$$

Note that e_4 coincides with Z_4 described above.

Next we have to express the $8N$ generators of $\mathcal{G}^S_{(1)}$. Let us denote the generators of $\tilde{\mathcal{N}}^S_{(1)} = \mathcal{G}_1^+$ by $P^{\pm}_{ak}$, and of $\mathcal{N}^S_{(1)} = \mathcal{G}_1^-$ by $K^{\pm}_{ak}$. Then we have

$$P^+_{ak} = iX^+_{a,k+4} - X^+_{a+2,k+4}, \quad P^-_{ak} = X^+_{a,k+4} - iX^+_{a+2,k+4}, \qquad (4.19)$$
$$K^+_{ak} = iX^-_{a+2,k+4} - X^-_{a,k+4}, \quad K^-_{ak} = X^-_{a+2,k+4} - iX^-_{a,k+4},$$

where $a = 1,2$, $k = 1,\dots,N$. The commutation and Hopf algebra relations of $U_q(\mathrm{su}(2,2/N))$ can be explicitly written now using equations (4a), (7), (11), (28)–(31), (38), (41), (44), and (45) from [117].

4.2 Duality for multiparameter quantum deformation of the supergroup GL(m/n)

4.2.1 Multiparameter quantum deformation of GL(m/n)

The multiparameter quantum deformation $\mathcal{A} = \mathrm{GL}_{\mathbf{uq}}(m/n)$ of the supergroup $\mathrm{GL}(m/n)$ was introduced first in [293], and later, in a slightly different form, in [262]. It is generated by the elements of a quantum supermatrix M [262]:

$$M = (T_{IJ}) = \begin{pmatrix} A_{ij} & B_{i\alpha} \\ C_{\beta j} & D_{\beta\alpha} \end{pmatrix} \qquad (4.20)$$

where $I, J = 1, \dots, m+n$; $i, j = 1, \dots, m$, and $\alpha, \beta = m+1, \dots, m+n$, which obey the following commutation relations:

$$
\begin{aligned}
T_{IK}T_{IL} &= (-1)^{\widehat{IK}+\widehat{IL}+\widehat{KL}}(-u^2)^{\widehat{I}}pT_{IL}T_{IK}, \quad \text{for } K < L, \\
T_{IK}T_{JK} &= (-1)^{\widehat{KI}+\widehat{KJ}+\widehat{IJ}}(-u^2)^{\widehat{K}}qT_{JK}T_{IK}, \quad \text{for } I < J, \\
pT_{IL}T_{JK} &= (-1)^{(\widehat{I}+\widehat{L})(\widehat{J}+\widehat{K})}qT_{JK}T_{IL}, \quad \text{for } I < J,\ K < L, \\
(\ 1)^{(\widehat{I}+\widehat{K})(\widehat{J}+\widehat{L})}&uqT_{JL}T_{IK} - (up)^{-1}T_{IK}T_{JL} \\
&= (-1)^{\widehat{JK}+\widehat{JL}+\widehat{KL}}(u-u^{-1})T_{IL}T_{JK}, \quad \text{for } I < J,\ K < L, \\
(T_{IK})^2 &= 0, \quad \text{for } \widehat{I}+\widehat{K} = 1, \\
p &= \frac{q_{KL}}{u^2}, \quad q = \frac{1}{q_{IJ}},
\end{aligned}
\tag{4.21}
$$

where ^ denotes the parity, which for the indices is defined by $\widehat{I} = 0$ if $I = i = 1, \dots, m$ and $\widehat{I} = 1$ if $I = \alpha = m+1, \dots, m+n$. Further, we define the parity $\widehat{T_{IJ}}$ of the generators T_{IJ} through the parity of the indices, namely we set $\widehat{T_{IK}} = (\widehat{I} + \widehat{K}) \pmod 2$. Thus, the supermatrix M is in the so-called standard form, so that the elements of A and D are even and those of B and C are odd.[1] We note also that the number of parameters is equal to $1 + (m+n)(m+n-1)/2$.

Considered as a superbialgebra, $\mathcal{A}$ has the following comultiplication $\delta_{\mathcal{A}}$ and counit $\varepsilon_{\mathcal{A}}$ [293]:

$$\delta_{\mathcal{A}}(T_{IJ}) = \sum_{N=1}^{m+n} T_{IN} \otimes T_{NJ} = (T_{IJ})_{(1)} \otimes (T_{IJ})_{(2)}, \tag{4.22a}$$

$$\varepsilon_{\mathcal{A}}(T_{IJ}) = \delta_{IJ}, \tag{4.22b}$$

where in (4.22a) we have used Sweedler's notation for the coproduct of an element a: $\delta_{\mathcal{A}}(a) = a_{(1)} \otimes a_{(2)}$.

We note that for a superbialgebra $\mathcal{A} = \mathcal{A}_0 \otimes \mathcal{A}_1$, where $\mathcal{A}_0, \mathcal{A}_1$, denotes the even and odd subalgebras, respectively, the coproduct preserves the parity, namely, one has

$$
\begin{aligned}
\delta_{\mathcal{A}} &: \mathcal{A} \longrightarrow \mathcal{A} \otimes \mathcal{A} \\
\delta_{\mathcal{A}} &: \mathcal{A}_0 \longrightarrow \mathcal{A}_0 \otimes \mathcal{A}_0 + \mathcal{A}_1 \otimes \mathcal{A}_1 \\
\delta_{\mathcal{A}} &: \mathcal{A}_1 \longrightarrow \mathcal{A}_0 \otimes \mathcal{A}_1 + \mathcal{A}_1 \otimes \mathcal{A}_0
\end{aligned}
\tag{4.23}
$$

where we define the parity of the tensor product standardly by $\widehat{X \otimes Y} \equiv (\widehat{X} + \widehat{Y}) \pmod 2$.

We are looking for the superalgebra which is in duality with $GL_{u\mathbf{q}}(m/n)$. We recall that two bialgebras $\mathcal{U}$, $\mathcal{A}$ are said to be *bialgebras in duality* if there exists a doubly

1 Note that we have changed X and q_{IJ} in [262] to $\frac{1}{u^2}$ and $(-1)^{\widehat{IJ}}\frac{1}{q_{IJ}}$, respectively.

nondegenerate bilinear form

$$\langle , \rangle : \mathcal{U} \times \mathcal{A} \to \mathbb{C}, \quad \langle , \rangle : (u,a) \mapsto \langle u,a \rangle, \quad u \in \mathcal{U},\ a \in \mathcal{A}, \tag{4.24}$$

such that, for $u, v \in \mathcal{U}$, $a, b \in \mathcal{A}$,

$$\langle u, ab \rangle = \langle \delta_{\mathcal{U}}(u), a \otimes b \rangle, \quad \langle uv, a \rangle = \langle u \otimes v, \delta_{\mathcal{A}}(a) \rangle, \tag{4.25}$$

$$\langle 1_{\mathcal{U}}, a \rangle = \varepsilon_{\mathcal{A}}(a), \quad \langle u, 1_{\mathcal{A}} \rangle = \varepsilon_{\mathcal{U}}(u). \tag{4.26}$$

All this extends to superbialgebras [293]. The only subtlety is that the tensor product is also graded, and, if (using Sweedler's notation) $\delta_{\mathcal{U}}(u) = \sum u_{(1)} \otimes u_{(2)}$, $\delta_{\mathcal{A}}(a) = \sum a_{(1)} \otimes a_{(2)}$, then

$$\begin{aligned} \langle u, ab \rangle &= \langle \delta_{\mathcal{U}}(u), a \otimes b \rangle \\ &= (-1)^{\widehat{u_{(2)}} \hat{a}} \sum \langle u_{(1)}, a \rangle \langle u_{(2)}, b \rangle, \\ \langle uv, a \rangle &= \langle u \otimes v, \delta_{\mathcal{A}}(a) \rangle \\ &= (-1)^{\widehat{a_{(1)}} \hat{v}} \sum \langle u, a_{(1)} \rangle \langle v, a_{(2)} \rangle. \end{aligned} \tag{4.27}$$

It is enough to define the pairing (4.24) between the generating elements of the two algebras. The pairing between any other elements of $\mathcal{U}$, $\mathcal{A}$ follows then from equations (4.25), (4.26), and (4.27).

4.2.1.1 Commutation relations for the superalgebra $U_{u\mathbf{q}}(m/n)$

The duality between bialgebras may be used also to obtain the unknown algebra from a known one if the two are in duality. For that it is enough to give the pairing between the generating elements of the unknown algebra with arbitrary elements of the PBW basis of the known algebra. Using these initial pairings and the duality properties one may find the unknown algebra. Following the method of [116] many examples were treated in Volume 2 of our trilogy. In [146] the method of [116] was used to find the superalgebra in duality with $\mathrm{GL}_{u\mathbf{q}}(m/n)$.

In order to find the superalgebra in duality with $\mathrm{GL}_{u\mathbf{q}}(m/n)$ following [116] we first need to fix a PBW basis of $\mathcal{A}$. This basis consists of monomials:

$$f = T_{11}^{k_1} \dots T_{m+n,m+n}^{k_{m+n}} T_{12}^{v_{12}} \dots T_{m+n-1,m+n}^{v_{m+n-1,m+n}} T_{m+n,m+n-1}^{w_{m+n,m+n-1}} \dots T_{21}^{w_{21}} \tag{4.28}$$

where $k_I, v_{ij}, v_{\alpha\beta}, w_{ij}, w_{\alpha\beta} \in \mathbb{Z}_+$, $v_{i\alpha}, w_{\alpha i} = 0, 1$. We use here the so-called normal ordering of the elements T_{IJ}. Namely, we first put the elements T_{II}; then we put the elements T_{IJ} with $I < J$ in lexicographic order, i. e., if $I < K$ then T_{IJ} is before T_{KL}, and T_{MI} is before T_{MK}; finally, we put the elements T_{IJ} with $I > J$ in antilexicographic order, i. e., if $I > K$ then T_{IJ} is before T_{KL}, and T_{MI} is before T_{MK}.

Let us now denote the generating elements of the unknown superalgebra in duality by D_I, $1 \le I \le m+n$, E_{IJ}, $1 \le I < J \le m+n$, F_{IJ}, $1 \le J < I \le m+n$. Following

[116] we shall *postulate* the pairing $\langle Z, f\rangle$, $Z = D_I, E_{IJ}, F_{IJ}, f$ from (4.28), as if we use the classical tangent vector at the identity:

$$\langle D_I, f\rangle \equiv k_I \delta_{\mathbf{M}0}\delta_{\mathbf{N}0}, \quad 1 \le I \le m+n, \tag{4.29a}$$

$$\langle E_{IJ}, f\rangle \equiv \delta_{v_{IJ}1}\delta^{IJ}_{\mathbf{M}0}\delta_{\mathbf{N}0}, \quad 1 \le I < J \le m+n, \tag{4.29b}$$

$$\langle F_{IJ}, f\rangle \equiv \delta_{w_{IJ}1}\delta_{\mathbf{M}0}\delta^{IJ}_{\mathbf{N}0}, \quad 1 \le J < I \le m+n, \tag{4.29c}$$

$$\delta_{\mathbf{M}0} = \prod_{1\le J<K\le m+n} \delta_{v_{JK}0}, \quad \delta_{\mathbf{N}0} = \prod_{1\le K<J\le m+n} \delta_{w_{JK}0}, \tag{4.29d}$$

$$\delta^{IJ}_{\mathbf{M}0} = \prod_{\substack{1\le K<L\le m+n \\ (K,L)\neq(I,J)}} \delta_{v_{KL}0}, \quad \delta^{IJ}_{\mathbf{N}0} = \prod_{\substack{1\le L<K\le m+n \\ (K,L)\neq(I,J)}} \delta_{w_{KL}0}. \tag{4.29e}$$

Note that in (4.29b,c) the corresponding classical derivative would be a superderivative if the parity $\hat{I} + \hat{J}$ were odd, however, there are no additional signs arising (from commuting of odd variables) since only the terms of exactly first degree in the odd variables give a non-zero result. We note also that from the duality properties follows

$$\langle 1_{\mathcal{U}}, f\rangle = \varepsilon(f) = \delta_{\mathbf{M}0}\delta_{\mathbf{N}0}. \tag{4.30}$$

To obtain the commutation relations between the generators D_I, E_{IJ}, and F_{IJ} we first need to evaluate the action of their bilinear products on the elements of $GL_{u\mathbf{q}}(m/n)$. We shall show that it is enough to do this for the Chevalley-like basis generators D_I, $1 \le I \le m+n$, $E_I \equiv E_{I,I+1}$, $F_I \equiv F_{I+1,I}$, $1 \le I \le m+n-1$. Then through them we shall express the rest of the generators E_{IJ} and F_{IJ}. For the action of the bilinear products on the elements of $GL_{u\mathbf{q}}(m/n)$ we refer to [146], equations (3.4)–(3.14).

Thus, we have the following commutation relations:

$$\begin{aligned}
[D_I, D_J] &= 0, \\
[D_I, E_J] &= (\delta_{IJ} - \delta_{I,J+1})E_J, \\
[D_I, F_J] &= -(\delta_{IJ} - \delta_{I,J+1})F_J, \\
uE_iF_i - u^{-1}F_iE_i &= \frac{u^{2(D_i - D_{i+1})} - 1}{u - u^{-1}}, \\
E_mF_m + F_mE_m &= -\frac{u^{2(D_m + D_{m+1})} - 1}{u^2 - 1}, \\
u'E_\alpha F_\alpha - u'^{-1}F_\alpha E_\alpha &= \frac{u'^{2(D_\alpha - D_{\alpha+1})} - 1}{u' - u'^{-1}}, \\
E_IF_J &= g_{IJ}F_JE_I, \quad I \neq J, \\
E_IE_J &= g_{IJ}^{-1}E_JE_I, \quad I < J-1, \\
F_IF_J &= g_{IJ}^{-1}F_JF_I, \quad I > J+1,
\end{aligned} \tag{4.31}$$

where

$$g_{IJ} = \frac{e_{IJ}}{f_{IJ}} = \begin{cases} \frac{q_{I+1,J+1}q_{IJ}}{q_{I,J+1}q_{I+1,J}}, & I < J-1 \\ \frac{q_{i+1,i+2}q_{i,i+1}}{q_{i,i+2}}, & I = J-1 = i < m \\ \frac{q'_{m+1,m+2}q'_{m,m+1}}{q'_{m,m+2}}, & I = J-1 = m \\ \frac{q'_{\alpha+1,\alpha+2}q'_{\alpha,\alpha+1}}{q'_{\alpha,\alpha+2}}, & I = J-1 = \alpha \\ \frac{u^2 q_{i-1,i+1}}{q_{i-1,i}q_{i,i+1}}, & I = J+1 = i < m \\ \frac{u^2 q_{m-1,m+1}}{q_{m-1,m}q_{m,m+1}}, & I = J+1 = m \\ \frac{u'^2 q'_{\alpha-1,\alpha+1}}{q'_{\alpha-1,\alpha}q'_{\alpha,\alpha+1}}, & I = J+1 = \alpha, \\ \frac{q_{J+1,I}q_{J,I+1}}{q_{J+1,I+1}q_{JI}}, & I > J+1. \end{cases} \tag{4.32}$$

We also obtain the expected nilpotency of the odd generators:

$$E_m^2 = F_m^2 = 0. \tag{4.33}$$

Similarly we derive the analogue of the cubic Serre relations:

$$\begin{aligned} &p_i^{\pm} E_i^2 E_{i\pm1} - (u + u^{-1}) E_i E_{i\pm1} E_i + (p_i^{\pm})^{-1} E_{i\pm1} E_i^2 = 0, \quad i < m, \\ &p_\alpha'^{\pm} E_\alpha^2 E_{\alpha\pm1} - (u' + u'^{-1}) E_\alpha E_{\alpha\pm1} E_\alpha + (p_\alpha'^{\pm})^{-1} E_{\alpha\pm1} E_\alpha^2 = 0, \\ &p_i^{\pm} F_i^2 F_{i\pm1} - (u + u^{-1}) F_i F_{i\pm1} F_i + (p_i^{\pm})^{-1} F_{i\pm1} F_i^2 = 0, \quad i < m, \\ &p_\alpha'^{\pm} F_\alpha^2 F_{\alpha\pm1} - (u' + u'^{-1}) F_\alpha F_{\alpha\pm1} F_\alpha + (p_\alpha'^{\pm})^{-1} F_{\alpha\pm1} F_\alpha^2 = 0, \\ &p_i^{+} = \frac{q_{i,i+1}q_{i+1,i+2}}{u q_{i,i+2}}, \quad p_i^{-} = \frac{u q_{i-1,i+1}}{q_{i-1,i}q_{i,i+1}}, \\ &p_\alpha'^{+} = \frac{q'_{\alpha,\alpha+1}q'_{\alpha+1,\alpha+2}}{u' q'_{\alpha,\alpha+2}}, \quad p_\alpha'^{-} = \frac{u' q'_{\alpha-1,\alpha+1}}{q'_{\alpha-1,\alpha}q'_{\alpha,\alpha+1}}. \end{aligned} \tag{4.34}$$

The full set of Serre relations will be checked in Section 4.3. Here we complete the construction of $\mathcal{U}_{u\mathbf{q}}$ in an alternative way, namely, by defining the rest of the generators E_{IJ}, F_{IJ} through the Chevalley-like ones, analogously to one-parameter deformations (cf., e. g., [117]):

$$\begin{aligned} &E_{IJ} \equiv E_I E_{I+1,J} - p_{IJ}^{-1} E_{I+1,J} E_I, \quad I < J-1 \\ &F_{JI} \equiv F_{J,I+1} F_I - p_{IJ} F_I F_{J,I+1}, \quad I < J-1, \\ &p_{IJ} = \begin{cases} p_{iJ} & I = i < J-1,\ i < m \\ p'_{m\alpha} & I = m,\ J = \alpha > m+1, \\ p'_{\alpha\beta} & I = \alpha < \beta - 1 = J-1, \end{cases} \\ &p_{IJ}^{(\prime)} \equiv \frac{q_{I,I+1}^{(\prime)} q_{I+1,J}^{(\prime)}}{q_{IJ}^{(\prime)}}. \end{aligned} \tag{4.35}$$

The above definitions for the generators E_{IJ}, F_{IJ} when $|I - J| \neq 1$ should be checked for consistency with equations (4.29); cf. [146]. We refer to [146] also for checking the nilpotency of the odd generators:

$$\langle E_{i\alpha}^2, f\rangle = \langle E_{i\alpha} \otimes E_{i\alpha}, \delta_{\mathcal{A}}(f)\rangle = 0. \tag{4.36}$$

For further use let us introduce besides the generators D_I also the generators

$$\begin{aligned} K &= D_1 + \cdots + D_{m+n} \\ H_I &= D_I - \frac{d_I}{d_{I+1}} D_{I+1}, \quad 1 \le I \le m+n-1 \\ d_1 &= \cdots = d_m = -d_{m+1} = \cdots = -d_{m+n} = 1. \end{aligned} \tag{4.37}$$

First we note that the generator K is central:

$$[K, D_J] = 0, \quad [K, E_J] = 0, \quad [K, F_J] = 0, \tag{4.38}$$

while the generators H_I, E_I, F_I form a commutation subsuperalgebra, namely, instead of equations (4.31) we have

$$[H_I, H_J] = 0, \tag{4.39a}$$

$$[H_I, E_J] = c_{IJ} E_J, \tag{4.39b}$$

$$[H_I, F_J] = -c_{IJ} F_J, \tag{4.39c}$$

$$uE_iF_i - u^{-1}F_iE_i = \frac{u^{2H_i} - 1}{u - u^{-1}}, \quad i < m, \tag{4.39d}$$

$$E_mF_m + F_mE_m = -\frac{u^{2H_m} - 1}{u^2 - 1}, \tag{4.39e}$$

$$u'E_\alpha F_\alpha - u'^{-1}F_\alpha E_\alpha = \frac{u'^{2H_\alpha} - 1}{u' - u'^{-1}}, \tag{4.39f}$$

$$c_{IJ} = \left(1 + \frac{d_I}{d_{I+1}}\right)\delta_{IJ} - \delta_{I,J+1} - \frac{d_I}{d_{I+1}}\delta_{I+1,J}, \quad 1 \le I, J \le m+n-1, \tag{4.39g}$$

where the numbers c_{IJ} form the Cartan matrix of the superalgebra sl(m/n).

Further we recall the well-known peculiarity that for $m = n$ the generator K is a linear combination of the generators H_I:

$$K = \sum_{j=1}^{m} jH_j + \sum_{\beta=m+1}^{2m-1} (\beta - 2m)H_\beta, \tag{4.40}$$

in particular, for $m = n = 1$, $K = H_1 = D_1 + D_2$. Thus, for $m = n$ instead of K we have to use another linear combination of D_I which is not a linear combination of H_I, e. g.,

$$\tilde{K} \equiv \sum_{I=1}^{2m} d_I D_I = D_1 + \cdots + D_m - D_{m+1} - \cdots - D_{2m}, \quad m = n, \tag{4.41}$$

in particular, for $m = n = 1$, $\tilde{K} = D_1 - D_2 \neq H_1$.

4.2.2 Hopf structure of the superalgebra $\mathcal{U}_{u\mathbf{q}}$

Now we shall use the duality to derive the Hopf structure of $\mathcal{U}_{u\mathbf{q}}$. We start with the coproducts in $\mathcal{U}_{u\mathbf{q}}$, i. e., we shall use the following relations:

$$\langle Y, f\rangle = \langle \delta_{\mathcal{U}}(Y), f_1 \otimes f_2\rangle, \tag{4.42}$$

for every splitting $f = f_1 f_2$. Thus we derive

$$\delta_{\mathcal{U}}(D_I) = D_I \otimes 1_{\mathcal{U}} + 1_{\mathcal{U}} \otimes D_I, \tag{4.43a}$$
$$\delta_{\mathcal{U}}(H_I) = H_I \otimes 1_{\mathcal{U}} + 1_{\mathcal{U}} \otimes H_I, \tag{4.43b}$$
$$\delta_{\mathcal{U}}(K') = K' \otimes 1_{\mathcal{U}} + 1_{\mathcal{U}} \otimes K', \quad K' = K, \tilde{K}. \tag{4.43c}$$

Then we try the following Ansätze:

$$\begin{aligned} \delta_{\mathcal{U}}(E_I) &= E_I \otimes \mathcal{P}_I + 1_{\mathcal{U}} \otimes E_I, \\ \delta_{\mathcal{U}}(F_I) &= F_I \otimes \mathcal{Q}_I + 1_{\mathcal{U}} \otimes F_I. \end{aligned} \tag{4.44}$$

After some calculations done in [146] the result is

$$\begin{aligned}
\mathcal{P}_i &= \left(\prod_{s=1}^{i-1}\left(\frac{q_{si}}{q_{s,i+1}}\right)^{D_s}\right)\left(\frac{u^2}{q_{i,i+1}}\right)^{D_i} \\
&\quad \times \left(\frac{1}{q_{i,i+1}}\right)^{D_{i+1}}\left(\prod_{J=i+2}^{m+n}\left(\frac{q_{i+1,J}}{q_{iJ}}\right)^{D_J}\right), \\
\mathcal{P}_m &= \left(\prod_{s=1}^{m-1}\left(\frac{q_{sm}}{q_{s,m+1}}\right)^{D_s}\right)\left(\frac{u^2}{q_{m,m+1}}\right)^{D_m} \\
&\quad \times \left(\frac{1}{q'_{m,m+1}}\right)^{D_{m+1}}\left(\prod_{\beta=m+2}^{m+n}\left(\frac{q_{m+1,\beta}}{q_{m\beta}}\right)^{D_\beta}\right), \\
\mathcal{P}_\alpha &= \left(\prod_{I=1}^{\alpha-1}\left(\frac{q'_{I\alpha}}{q'_{I,\alpha+1}}\right)^{D_I}\right)\left(\frac{u'^2}{q'_{\alpha,\alpha+1}}\right)^{D_\alpha} \\
&\quad \times \left(\frac{1}{q'_{\alpha,\alpha+1}}\right)^{D_{\alpha+1}}\left(\prod_{\beta=\alpha+2}^{m+n}\left(\frac{q'_{\alpha+1,\beta}}{q'_{\alpha\beta}}\right)^{D_\beta}\right),
\end{aligned} \tag{4.45}$$

$$\mathcal{Q}_I = u^{2d_I H_I}\mathcal{P}_I^{-1}. \tag{4.46}$$

We obtain the coproducts of the rest of the generators using (4.35) and the coproducts of the generators E_I and F_I; cf. [146].

The counit relations in $\mathcal{U}_{u\mathbf{q}}$ are given by

$$\varepsilon_{\mathcal{U}}(Y) = 0, \quad Y = D_I, E_{IJ}, F_{IJ}, K, H_I, \tilde{K}, \tag{4.47}$$

which follows easily using

$$\varepsilon_{\mathcal{U}}(Y) = \langle Y, 1_A \rangle = 0. \tag{4.48}$$

The antipode in $\mathcal{U}_{u\mathbf{q}}$ is obtained using one of the basic axioms of Hopf (super-) algebras (cf. Abe [1] in general, and Manin [293] in the super-case):

$$m \circ (\mathrm{id}_{\mathcal{U}} \otimes \gamma_{\mathcal{U}}) \circ \delta_{\mathcal{U}} = i \circ \varepsilon_{\mathcal{U}}, \tag{4.49}$$

where both sides are maps $\mathcal{U} \to \mathcal{U}$, m is the usual product in the algebra: $m(Y \otimes Z) = YZ$, $Y, Z \in \mathcal{U}$, and i is the natural embedding of $\mathbb{C}$ into $\mathcal{U}$: $i(c) = c1_{\mathcal{U}}$, $c \in \mathbb{C}$. Applying this for the antipode map of the superalgebra $\mathcal{U}_{u\mathbf{q}}$ we obtain

$$\gamma_{\mathcal{U}}(D_I) = -D_I, \tag{4.50a}$$
$$\gamma_{\mathcal{U}}(H_I) = -H_I, \tag{4.50b}$$
$$\gamma_{\mathcal{U}}(K) = -K, \quad \gamma_{\mathcal{U}}(\tilde{K}) = -\tilde{K}, \tag{4.50c}$$
$$\gamma_{\mathcal{U}}(E_I) = -E_I \mathcal{P}_I^{-1}, \quad \gamma_{\mathcal{U}}(F_I) = -F_I \mathcal{Q}_I^{-1}. \tag{4.50d}$$

We obtain the antipode map for the rest of the generators E_{IJ}, F_{IJ} using (4.35) and (4.50).

4.2.3 Drinfeld–Jimbo form of the superalgebra $\mathcal{U}_{u\mathbf{q}}$

In this section we show how to transform the superalgebra $\mathcal{U}_{u\mathbf{q}}$ to a Drinfeld–Jimbo form, for which form we follow the exposition of [254, 117]. We first note that if we set all parameters equal $q_{IJ} = u$ for all I, J then we have $\mathcal{P}_I = u^{d_I H_I} = \mathcal{Q}_I$. Thus, we are prompted to try for the analogue of the transformation which was used for the GL(n) multiparameter case [136]:

$$E_I = u^{-\hat{I}/2} X_I^+ \mathcal{P}_I^{1/2}, \quad F_I = (-)^{\hat{I}} u^{-\hat{I}/2} X_I^- \mathcal{Q}_I^{1/2}, \tag{4.51}$$

where $\hat{I} = 1$ for $I = m$ and 0 otherwise. Further the bracket $[,]$ will denote the super-commutator:

$$[Y, Z] \equiv YZ - (-1)^{\hat{Y}\hat{Z}} ZY. \tag{4.52}$$

First we have to show the Chevalley commutation relations:

$$[H_I, H_J] = 0, \tag{4.53a}$$
$$[H_I, X_J^\pm] = \pm c_{IJ} X_J^\pm, \tag{4.53b}$$
$$[X_I^+, X_J^-] = \delta_{IJ} [H_I]_u, \tag{4.53c}$$

$$[H_I]_u \equiv \lambda^{-1}(u^{H_I} - u^{-H_I}). \tag{4.53d}$$

We note that (4.53a) is (4.39a), while for (4.53b) we have

$$\begin{aligned}[H_I, X_J^+] &= [H_I, u^{\hat{I}/2} E_J \mathcal{P}_I^{-1/2}] = c_{IJ} X_J^+ \\ [H_I, X_J^-] &= [H_I, (-)^{\hat{I}} u^{\hat{I}/2} F_J \mathcal{Q}_I^{-1/2}] = -c_{IJ} X_J^-,\end{aligned} \tag{4.54}$$

using (4.39b,c). For (4.53c) with $I = J$ we have

$$\begin{aligned}[X_I^+, X_I^-] &= (-)^{\hat{I}} u^{\hat{I}} [E_I \mathcal{P}_I^{-1/2}, F_I \mathcal{Q}_I^{-1/2}] \\ &= \begin{cases} (uE_iF_i - u^{-1}F_iE_i)u^{-H_i}, & I = i < m \\ -u(E_mF_m + F_mE_m)u^{-H_m}, & I = m \\ (u'E_\alpha F_\alpha - u'^{-1}F_\alpha E_\alpha)u'^{-H_\alpha}, & I = \alpha \end{cases} \\ &= \lambda^{-1}(u^{H_I} - u^{-H_I}) = [H_I]_u = [H_I]_{u'},\end{aligned} \tag{4.55}$$

where we have used (4.51), (4.39), (4.45), and (4.46). Furthermore, for $I \neq J$ we have

$$\begin{aligned}[X_I^+, X_J^-] &= (-)^{\hat{J}} u^{\hat{I}/2} u^{\hat{J}/2} [E_I \mathcal{P}_I^{-1/2}, F_J \mathcal{Q}_J^{-1/2}] \\ &= (-)^{\hat{J}} u^{\hat{I}/2} u^{\hat{J}/2} (E_I \mathcal{P}_I^{-1/2} F_J \mathcal{Q}_J^{-1/2} - (-1)^{\widehat{E_I}\widehat{F_J}} F_J \mathcal{Q}_J^{-1/2} E_I \mathcal{P}_I^{-1/2}) \\ &= (-)^{\hat{J}} u^{\hat{I}/2} u^{\hat{J}/2} \left(E_I F_J \left(\frac{\mathcal{P}_J}{g_{IJ}\mathcal{P}_I}\right)^{1/2} u^{-d_J H_J} - F_J E_I \left(\frac{\mathcal{P}_J}{g_{JI}\mathcal{P}_I}\right)^{1/2} u^{-d_J H_J} u^{-d_J c_{JI}} \right) \\ &= (-)^{\hat{J}} u^{\hat{I}/2} u^{\hat{J}/2} \left(F_J E_I \left(\frac{\mathcal{P}_J}{g_{JI}\mathcal{P}_I}\right)^{1/2} u^{-d_J H_J} (g_{IJ}^{1/2} g_{JI}^{1/2} - u^{-d_J c_{JI}}) \right) = 0,\end{aligned} \tag{4.56}$$

where we have used $(-1)^{\widehat{E_I}\widehat{F_J}} = 1$ if $I \neq J$, also (4.51), (4.45), (4.46), and

$$g_{IJ} g_{JI} = u^{-2d_J c_{JI}} = u^{-2d_I c_{IJ}}, \quad I \neq J, \tag{4.57}$$

the latter following from (4.32) and (4.39g).

Furthermore, we consider the Serre relations:

$$X_I^\pm (X_J^\pm) = X_J^\pm (X_I^\pm), \quad |I - J| > 1, \tag{4.58a}$$

$$(\mathrm{ad}_{u^\kappa} X_I^\pm)^2 X_J^\pm = 0, \quad |I - J| = 1, \quad \kappa = \pm, \tag{4.58b}$$

$$[[X_m^\pm, X_{m-1}^\pm]_{u^\kappa}, [X_m^\pm, X_{m+1}^\pm]_{u^\kappa}]_{u^\kappa} = 0, \quad \kappa = \pm, \tag{4.58c}$$

where the deformed adjoint action and deformed supercommutator are given as

$$\begin{aligned}(\mathrm{ad}_{u^\kappa} X_j^\pm)(X_k^\pm) &= [X_j^\pm, X_k^\pm]_{u^\kappa} \\ &\equiv X_j^\pm X_k^\pm - (-1)^{\deg X_j^\pm \deg X_k^\pm} u^{\kappa(\alpha_j, \alpha_k)} X_k^\pm X_j^\pm.\end{aligned} \tag{4.59}$$

The explicit check of (4.58) is done in [146].

Thus, as a **commutation superalgebra** we have the classical structure, namely, $\mathcal{U}_{\mathbf{uq}} \cong \mathcal{U}' \otimes \mathcal{K}$, where $\mathcal{U}'$ is isomorphic to the one-parametric deformation $U_u(\mathrm{sl}(m/n))$ discussed above (cf. [254, 117]) with Chevalley generators $H_I, X_I^{\pm}$ $(I = 1, \dots, m+n-1)$, and $\mathcal{K}$ is generated by $K', K' = K(m \neq n), \tilde{K}(m = n)$. This is a splitting of subalgebras, and $\mathcal{K}$ is central in $\mathcal{U}_{\mathbf{uq}}$ for $m \neq n$. (We recall that K is central for all m, n.)

We turn now to the co-algebra structure. For the generators H_I and K or $\tilde{K}$ the co-algebra relations are given in (4.43b,c), (4.47), and (4.50b,c). For the generators $X^{\pm}$ the coproducts are given by

$$\begin{aligned}
\delta_{\mathcal{U}}(X_I^+) &= u^{\hat{I}/2}\delta_{\mathcal{U}}(E_I)\delta_{\mathcal{U}}(\mathcal{P}_I^{-1/2}) \\
&= u^{\hat{I}/2}(E_I \otimes \mathcal{P}_I + 1_{\mathcal{U}} \otimes E_I)(\mathcal{P}_I^{-1/2} \otimes \mathcal{P}_I^{-1/2}) \\
&= X_I^+ \otimes \mathcal{P}_I^{1/2} + \mathcal{P}_I^{-1/2} \otimes X_I^+, \\
\delta_{\mathcal{U}}(X_I^-) &= (-)^{\hat{I}} u^{\hat{I}/2}\delta_{\mathcal{U}}(F_I)\delta_{\mathcal{U}}(\mathcal{Q}_I^{-1/2}) \\
&= (-)^{\hat{I}} u^{\hat{I}/2}(F_I \otimes \mathcal{Q}_I + 1_{\mathcal{U}} \otimes F_I)(\mathcal{Q}_I^{-1/2} \otimes \mathcal{Q}_I^{-1/2}) \\
&= X_I^- \otimes \mathcal{Q}_I^{1/2} + \mathcal{Q}_I^{-1/2} \otimes X_I^-,
\end{aligned} \tag{4.60}$$

while the counit and the antipode are given by

$$\begin{aligned}
\varepsilon_{\mathcal{U}}(X_I^+) &= u^{\hat{I}/2}\varepsilon_{\mathcal{U}}(E_I)\varepsilon_{\mathcal{U}}(\mathcal{P}_I^{-1/2}) = 0, \\
\varepsilon_{\mathcal{U}}(X_I^-) &= (-)^{\hat{I}} u^{\hat{I}/2}\varepsilon_{\mathcal{U}}(F_I)\varepsilon_{\mathcal{U}}(\mathcal{Q}_I^{-1/2}) = 0, \\
\gamma_{\mathcal{U}}(X_I^+) &= u^{\hat{I}/2}\gamma_{\mathcal{U}}(\mathcal{P}_I^{-1/2})\gamma_{\mathcal{U}}(E_I) = -u^{\hat{I}/2}\mathcal{P}_I^{1/2}E_I\mathcal{P}_I^{-1} \\
&= -u_I u^{\hat{I}/2}E_I\mathcal{P}_I^{-1/2} = -u_I X_I^+, \\
\gamma_{\mathcal{U}}(X_I^-) &= (-)^{\hat{I}} u^{\hat{I}/2}\gamma_{\mathcal{U}}(\mathcal{Q}_I^{-1/2})\gamma_{\mathcal{U}}(F_I) = -(-)^{\hat{I}} u^{\hat{I}/2}\mathcal{Q}_I^{1/2}F_I\mathcal{P}_I^{-1} \\
&= -u_I^{-1}(-)^{\hat{I}} u^{\hat{I}/2}F_I\mathcal{Q}_I^{-1/2} = -u_I^{-1}X_I^-, \\
u_I &= \begin{cases} u & I = i < m \\ 1 & I = m \\ u' & I = \alpha. \end{cases}
\end{aligned} \tag{4.61}$$

Thus, the splitting $\mathcal{U}_{\mathbf{uq}} \cong \mathcal{U}' \otimes \mathcal{K}$ is preserved also by the counit and the antipode, (cf. (4.47), (4.50b,c), and (4.61)), and also by the coproducts of $H_I, K, \tilde{K}$; cf. (4.43). Thus, $\mathcal{K}$ is a Hopf subalgebra of $\mathcal{U}_{\mathbf{uq}}$. However, the above splitting is not preserved in general by the coproducts of the Chevalley generators $X_I^{\pm}$, since these coproducts involve (through the operators $\mathcal{P}_I$, $\mathcal{Q}_I$) the generator K' of $\mathcal{K}$; cf. (4.60).

To make the last statement explicit we need to express the operators $\mathcal{P}_I$ (and through them $\mathcal{Q}_I$) in terms of the generators H_I and K'. For this purpose, we first express the generators D_I through H_I and K'. We first introduce the following notation:

$$\hat{H}_I \equiv \sum_{J=I}^{m+n-1} d_J H_J, \quad \hat{H}_{m+n} = 0. \tag{4.62}$$

Then for $m \neq n$ we have

$$D_I = d_I \hat{K} + d_I \hat{H}_I, \qquad \hat{K} \equiv \frac{1}{m-n}(K - K_0), \qquad K_0 \equiv \sum_{j=1}^{m} jH_j + \sum_{\beta=m+1}^{m+n-1} (\beta - 2m)H_\beta, \tag{4.63}$$

while for $m = n$ we have

$$D_I = d_I \check{K} + d_I \hat{H}_I, \qquad \check{K} \equiv \frac{1}{2m}(\tilde{K} - \tilde{K}_0), \qquad \tilde{K}_0 \equiv \sum_{I=1}^{2m-1} I d_I H_I. \tag{4.64}$$

(Note that multiplying with $m - n$ the second equation in (4.63) and setting $m = n$ we recover (4.40).)

Now we substitute (4.63) in (4.45) to obtain

$$\mathcal{P}_i = (\tilde{q}_i)^{\hat{K}'} \left(\prod_{s=1}^{i-1} \left(\frac{q_{si}}{q_{s,i+1}} \right)^{\hat{H}_s} \right) \left(\frac{u^2}{q_{i,i+1}} \right)^{\hat{H}_i} \times \left(\frac{1}{q_{i,i+1}} \right)^{\hat{H}_{i+1}} \left(\prod_{J=i+2}^{m+n} \left(\frac{q_{i+1,J}}{q_{iJ}} \right)^{d_J \hat{H}_J} \right), \tag{4.65a}$$

$$\mathcal{P}_m = (\tilde{q}_m)^{\hat{K}'} \left(\prod_{s=1}^{m-1} \left(\frac{q_{sm}}{q_{s,m+1}} \right)^{\hat{H}_s} \right) \left(\frac{u^2}{q_{m,m+1}} \right)^{\hat{H}_m} \times \left(\frac{1}{q'_{m,m+1}} \right)^{-\hat{H}_{m+1}} \left(\prod_{\beta=m+2}^{m+n} \left(\frac{q_{m+1,\beta}}{q_{m\beta}} \right)^{-\hat{H}_\beta} \right), \tag{4.65b}$$

$$\mathcal{P}_\alpha = (\tilde{q}_\alpha)^{\hat{K}'} \left(\prod_{I=1}^{\alpha-1} \left(\frac{q_{I\alpha}}{q_{I,\alpha+1}} \right)^{d_I \hat{H}_I} \right) \left(\frac{u'^2}{q'_{\alpha,\alpha+1}} \right)^{-\hat{H}_\alpha} \times \left(\frac{1}{q'_{\alpha,\alpha+1}} \right)^{-\hat{H}_{\alpha+1}} \left(\prod_{\beta=\alpha+2}^{m+n} \left(\frac{q_{\alpha+1,\beta}}{q_{\alpha\beta}} \right)^{-\hat{H}_\beta} \right) \tag{4.65c}$$

where $\hat{K}' = \hat{K}$ or $\check{K}$, and

$$\tilde{q}_i = \left(\prod_{s=1}^{i-1} \left(\frac{q_{si}}{q_{s,i+1}} \right) \right) \left(\frac{u^2}{q_{i,i+1}^2} \right) \left(\prod_{J=i+2}^{m+n} \left(\frac{q_{i+1,J}}{q_{iJ}} \right)^{d_J} \right), \tag{4.66a}$$

$$\tilde{q}_m = \left(\prod_{s=1}^{m-1} \left(\frac{q_{sm}}{q_{s,m+1}} \right) \right) \left(\prod_{\beta=m+2}^{m+n} \left(\frac{q_{m\beta}}{q_{m+1,\beta}} \right) \right), \tag{4.66b}$$

$$\tilde{q}_\alpha = \left(\prod_{I=1}^{\alpha-1}\left(\frac{q_{I\alpha}}{q_{I,\alpha+1}}\right)^{d_I}\right)\left(\frac{q^2_{\alpha,\alpha+1}}{u^2}\right)\left(\prod_{\beta=\alpha+2}^{m+n}\left(\frac{q_{\alpha\beta}}{q_{\alpha+1,\beta}}\right)\right). \tag{4.66c}$$

Thus, indeed for generic parameters the generator K' takes part in the coproducts of $X_I^\pm$, except for $m = n = 1$ when $\tilde{q}_1 = 1$.

Thus, in general the subsuperalgebra $\mathcal{U}'$ is not a Hopf subsuperalgebra of $\mathcal{U}_{u\mathbf{q}}$. The exceptions from this rule are discussed now.

4.2.4 Multiparameter superalgebra deformation of $U(\mathrm{sl}(m/n))$

Here we consider the cases when the deformations of $U(\mathrm{sl}(m/n))$ generated by the Chevalley generators $X_I^\pm$, H_I are Hopf subsuperalgebras of $\mathcal{U}_{u\mathbf{q}}$ which are deformations of $U(\mathrm{gl}(m/n))$. A sufficient condition for this is to set all parameters equal: $q_{IJ} = u$, $\forall I$, J, when we obtain the Hopf superalgebra $U_u(\mathrm{sl}(m/n))$ (cf. [254, 117]). But as we show it is not a necessary condition and below we obtain multiparameter Hopf superalgebra deformations of $U(\mathrm{sl}(m/n))$. For this purpose, we need to analyze only the coproducts of the Chevalley generators $X_I^\pm$ in (4.60), i. e., the operators $\mathcal{P}_I$ (and so $\mathcal{Q}_I$) in (4.65).

We start with the special case $m = n = 1$. In this case the deformation of GL(1/1) depends on two parameters: u and $q = q_{12}$. The superalgebra $\mathcal{U}_{u,q}$ has four generators, $X^\pm \equiv X_1^\pm$, $H \equiv H_1$, $\tilde{K}$, and is denoted by $U_{u,q}(\mathrm{gl}(1/1))$. The operators $\mathcal{P}_1$, $\mathcal{Q}_1$ depend only on H, cf. (4.65b), and so $X^\pm$, H generate a nilpotent Hopf subsuperalgebra of $\mathcal{U}_{u,q}$ depending on both parameters. We shall denote this subsuperalgebra by $U_{u,q}(\mathrm{sl}(1/1))$. For later reference we record its algebra and co-algebra relations, though they are just special cases of (4.53b,c), (4.43b), (4.60), (4.47), (4.61), and (4.50b):

$$\begin{aligned}
&[H, X^\pm] = 0, \quad [X^+, X^-] \equiv X^+X^- + X^-X^+ = [H]_u, \\
&\delta_{\mathcal{U}}(H) = H \otimes 1_{\mathcal{U}} + 1_{\mathcal{U}} \otimes H, \\
&\delta_{\mathcal{U}}(X^+) = X^+ \otimes \left(\frac{u^2}{q}\right)^{H/2} + \left(\frac{u^2}{q}\right)^{-H/2} \otimes X^+, \\
&\delta_{\mathcal{U}}(X^-) = X^- \otimes q^{H/2} + q^{-H/2} \otimes X^-, \\
&\varepsilon_{\mathcal{U}}(Y) = 0, \quad Y = X^\pm, H, \\
&\gamma_{\mathcal{U}}(H) = -H, \quad \gamma_{\mathcal{U}}(X^\pm) = -X^\pm.
\end{aligned} \tag{4.67}$$

The relations involving the generator $\tilde{K}$ are

$$\begin{aligned}
&[\tilde{K}, H] = 0, \quad [\tilde{K}, X^\pm] = \pm 2X^\pm, \\
&\delta_{\mathcal{U}}(\tilde{K}) = \tilde{K} \otimes 1_{\mathcal{U}} + 1_{\mathcal{U}} \otimes \tilde{K}, \quad \varepsilon_{\mathcal{U}}(\tilde{K}) = 0, \quad \gamma_{\mathcal{U}}(\tilde{K}) = -\tilde{K}.
\end{aligned} \tag{4.68}$$

Thus, equations (4.67) and (4.68) give the complete structure of $\mathcal{U}_{u,q}$. This superalgebra was obtained first in a slightly different but equivalent form in [229] from the analysis of the XY quantum chain in a magnetic field and a little later but independently in [71] using the approach of [162]. A two-parameter deformation of $U(\mathrm{gl}(1/1))$ without a Hopf subsuperalgebra deformation of $U(\mathrm{sl}(1/1))$ was given in [95].

In the cases when $m+n>1$ the subsuperalgebra $\mathcal{U}'$ may be a Hopf subsuperalgebra of $\mathcal{U}_{u\mathbf{q}}$ only when some of the parameters coincide. From the expressions (4.65) it is clear that in order to decouple $K' = K(m \neq n)$ or $K' = \tilde{K}(m = n)$, from the system the $m+n-1$ constants $\tilde{q}_I$ (cf. (4.66)) should become equal to unity. This brings about $m+n-1$ conditions on the deformation parameters q_{IJ} when $m \neq n$ and $m+n-2$ conditions when $m = n$. The reason for the missing condition is that for $m = n$ the constants $\tilde{q}_I$ are related as follows:

$$\tilde{q}_m^m \prod_{s=1}^{m-1} (\tilde{q}_s \tilde{q}_{2m-s})^s = 1, \quad m = n. \tag{4.69}$$

(For $m = n = 1$ (4.69) reduces to $\tilde{q}_1 = 1$ which is just a repetition of (4.66b).)

First we consider the case $m \neq n$. Then we may fix the $m+n-1$ next-to-main diagonal parameters $q_{I,I+1}$ as

$$q^0_{i,i+1} \equiv u \prod_{\substack{1\leq s\leq i,\ i+1\leq T\leq m+n\\ s<T-1}} \left(\frac{u}{q_{sT}}\right)^{d_T}, \quad i \leq m-1 \tag{4.70a}$$

$$q^0_{m,m+1} \equiv u \prod_{\substack{1\leq s\leq m,\ m+1\leq \alpha\leq m+n\\ s<\alpha-1}} \frac{u}{q_{s\alpha}}, \tag{4.70b}$$

$$q^0_{\alpha,\alpha+1} \equiv u \prod_{\substack{1\leq S\leq \alpha,\ \alpha+1\leq \beta\leq m+n\\ S<\beta-1}} \left(\frac{q_{S\beta}}{u}\right)^{d_S}, \tag{4.70c}$$

or in more compact form

$$q^0_{I,I+1} \equiv u\left(\prod_{\substack{1\leq S\leq I,\ I+1\leq T\leq m+n\\ S<T-1}} \left(\frac{u}{q_{ST}}\right)^{d_S d_T}\right)^{d_I d_{I+1}}, \quad 1 \leq I \leq m+n-1. \tag{4.70d}$$

Using this we obtain the desired result:

$$(\tilde{q}_I)_{q_{I,I+1}=q^0_{I,I+1}} = 1, \quad 1 \leq I \leq m+n-1, \tag{4.71}$$

by which the generator of $\mathcal{K}$ decouples from the rest.

It is useful to have the explicit form of the operators $\mathcal{P}_I$ with this choice of parameters, for which we refer to [146], f-lae (6.6).

We thus have the deformation of $U(\mathrm{sl}(m/n))$ depending on $2+(m+n)(m+n-3)/2$ parameters, cf. (4.53), (4.43b,c), (4.47), (4.50b,c), (4.60), and (4.61). Such a deformation using a different parametrization was derived using the FRT approach in [249]. Later, in [65] a Witten-Woronowicz type of deformation [387, 388] of $\mathrm{sl}(m/1)$ was constructed depending on $m(m-1)/2 + 2\delta_{m2}$ parameters, $m > 1$. (Note that comparison with our results is not straightforward due to the quommutators used in [65].)

Finally, we consider the case $m = n > 1$, when there are only $m + n - 2 = 2m - 2$ conditions. Here we fix $2m - 2$ parameters of the first row, that is, the parameters q_{1I} with $3 \le I \le 2m$. We have

$$q^0_{1I} \equiv q_{12}\left(\prod_{J=3}^{2m}\left(\frac{u}{q_{2J}}\right)^{d_J}\right)\left(\prod_{J=2}^{I-1}\left(\frac{u}{q_{JI}}\right)^{d_J}\right)\prod_{J=I+1}^{2m}\left(\frac{q_{IJ}}{u}\right)^{d_J}. \tag{4.72}$$

Then we obtain

$$(\tilde{q}_I)_{q_{1J}=q^0_{1J}} = 1, \quad 1 \le I \le 2m - 1. \tag{4.73}$$

To be more explicit, one should substitute (4.73) in the expressions of $\mathcal{P}_I$ to obtain the latter in terms of H_I; cf. [146].

We thus have the deformation of $U(\mathrm{sl}(m/m))$ $(m > 1)$, depending on $3 + m(2m - 3)$ parameters; cf. (4.53), (4.43b,c), (4.47), (4.50b,c), (4.60), and (4.61).

4.3 Induced representations of the multiparameter Hopf superalgebras $U_{u\mathbf{q}}(\mathrm{gl}(m/n))$ and $U_{u\mathbf{q}}(\mathrm{sl}(m/n))$

In the present section we construct the induced holomorphic representations of $\mathcal{U} \equiv \mathcal{U}_{u\mathbf{q}}(\mathrm{gl}(m/n))$ and $\mathcal{U}' \equiv \mathcal{U}_{u\mathbf{q}}(\mathrm{sl}(m/n))$ for generic deformation parameters. The representations are labeled by $m+n$ integer numbers, respectively, $m+n-1$ complex numbers and they act in the space of formal power series of $(m + n)(m + n - 1)/2$ noncommuting variables, of which mn are odd and the rest are even. These variables generate a q-deformation of a flag supermanifold of the supergroup $\mathrm{GL}(m/n)$, respectively, $\mathrm{SL}(m/n)$. The construction is achieved by using the Gauss decomposition of the generators of the multiparameter matrix quantum supergroup $\mathcal{A} \equiv \mathrm{GL}_{u\mathbf{q}}(m/n)$. We use it to give a new basis of $\mathcal{A}$ which we use as expansion basis for our functions and which has convenient properties w.r.t. the right action of $\mathcal{U}$. Namely, we impose the conditions of right covariance [114] in order to eliminate the dependence of our functions on the strictly upper diagonal generators in the Gauss decomposition, while the dependence on the diagonal generators in the Gauss decomposition is fixed for all functions. These fixed powers of the diagonal generators are the integer numbers which parametrize our representations. For $u = \mathbf{q} = 1$ our representations coincide with the holomorphic representations induced from the upper diagonal Borel subsupergroup B of $G \equiv \mathrm{GL}(m/n)$ and acting on the coset G/G^+, where G^+ is the strictly upper diagonal supergroup of G. That is why we call our representations induced. Further, we enforce the conditions under which $\mathcal{U}'$ is a Hopf superalgebra. Then we can set the superdeterminant to unity and consider the representations of $\mathcal{U}'$. Finally, we eliminate also the dependence on the diagonal generators of the Gauss decomposition. This is done invariantly, so that the representation parameters remain in the matrix elements. For $u = \mathbf{q} = 1$ the latter representations coincide with the standard holomorphic representations induced from B and acting on the coset G/B. These representations can be extended to arbitrary complex values of the $m + n - 1$ representations parameters. We finish by stressing that our representations are the most general holomorphic representations of $\mathcal{U}$ and $\mathcal{U}'$ (supposing that the deformation parameters are not nontrivial roots of unity). This is as in the classical case ($u = \mathbf{q} = 1$) where, in particular, all representations occurring in the applications may be obtained as subrepresentations of such representations.

4.3.1 Left and right actions of $\mathcal{U}$ and $\mathcal{U}'$

We begin by defining two actions of the dual algebra $\mathcal{U}$ on $\mathcal{A}$. First we introduce (as in [118]) the left regular representation of $\mathcal{U}$ by

$$\pi(y)T_{IL} = \sum_{N=1}^{m+n} \langle y_{\mathcal{U}}(y), T_{IN}\rangle T_{NL} \tag{4.74a}$$

$$= \langle y_{\mathcal{U}}(y), (T_{IL})_{(1)}\rangle (T_{IL})_{(2)}, \tag{4.74b}$$

where in the second line we have used (4.22a). From (4.74) we find the explicit action of the generators of $\mathcal{U}$:

$$\pi(K_I)T_{JL} = u^{d_I(\frac{d_I}{d_{I+1}}\delta_{I+1,J}-\delta_{IJ})/2}T_{JL} \tag{4.75a}$$

$$\pi(X_I^+)T_{JL} = -u^{(\tilde{I}+d_I+d_{I+1})/2}Q_{I,I+1}^{-1/2}\delta_{IJ}T_{J+1,L} \tag{4.75b}$$

$$\pi(X_I^-)T_{JL} = -(-1)^{\tilde{I}}u^{(\tilde{I}-3d_I-d_{I+1})/2}Q_{II}^{1/2}\delta_{I+1,J}T_{J-1,L}, \tag{4.75c}$$

from which follows

$$\pi(\mathcal{P}_I^{1/2})T_{JL} = Q_{IJ}^{-1/2}T_{JL}, \tag{4.75d}$$

$$\pi(\mathcal{Q}_I^{1/2})T_{JL} = u^{d_I(\frac{d_I}{d_{I+1}}\delta_{I+1,J}-\delta_{IJ})}Q_{IJ}^{1/2}T_{JL}. \tag{4.75e}$$

Finally,

$$\pi(\mathcal{K})T_{JL} = u^{-1/2}T_{JL} \quad m \neq n,$$
$$\pi(\mathcal{K})T_{JL} = u^{-d_J/2}T_{JL} \quad m = n. \tag{4.76}$$

The above is supplemented with the following action on the unit element of $\mathcal{A}$:

$$\pi(K_I)1_{\mathcal{A}} = 1_{\mathcal{A}}, \quad \pi(X_I^{\pm})1_{\mathcal{A}} = 0, \quad \pi(\mathcal{K})1_{\mathcal{A}} = 1_{\mathcal{A}}. \tag{4.77}$$

In order to derive the action of $\pi(y)$ on $\mathcal{A}$ we shall use the general form [119], which is the same as (4.74b) but for an arbitrary element ψ of $\mathcal{A}$:

$$\pi(y)\psi = \langle \gamma_{\mathcal{U}}(y), \psi_{(1)}\rangle \psi_{(2)}. \tag{4.78}$$

So the action on the product of two homogeneous elements may be calculated using the properties of pairing, the graded tensor product, coproduct and antipode, namely,

$$\begin{aligned} \langle y_1 \otimes y_2, \psi_1 \otimes \psi_2\rangle &= (-1)^{\hat{y}_2 \hat{\psi}_1} \langle y_1, \psi_1\rangle \langle y_2, \psi_2\rangle \\ (a\otimes b)(c\otimes d) &= (-1)^{\hat{b}\hat{c}} ac\otimes bd, \\ \delta(\phi\psi) &= (-1)^{\hat{\varphi}_{(2)}\hat{\mathcal{S}}_{(1)}} \phi_{(1)}\psi_{(1)} \otimes \phi_{(2)}\psi_{(2)}, \\ \gamma(ab) &= (-1)^{\hat{a}\hat{b}} \gamma(b)\gamma(a). \end{aligned} \tag{4.79}$$

We find using (4.78) and (4.79)

$$\pi(y)\phi\psi = (-1)^{\hat{Y}_{(1)}(\hat{\varphi}+\hat{Y}_{(2)})} (\pi(y_{(2)})\phi)(\pi(y_{(1)})\psi). \tag{4.80}$$

Thus, we have, for the generating elements $y = K_I, X_I^{\pm}$ (note that in all cases we have $\hat{Y}_{(1)}\hat{Y}_{(2)} = 0$):

$$\pi(K_I)\phi\psi = (\pi(K_I)\phi)\pi(K_I)\psi, \tag{4.81a}$$

$$\begin{aligned} \pi(X_I^+)\phi\psi &= (-1)^{\hat{\varphi}\tilde{I}} (\pi(\mathcal{P}_I^{1/2})\phi)\pi(X_I^+)\psi \\ &\quad + (\pi(X_I^+)\phi)\pi(\mathcal{P}_I^{-1/2})\psi, \end{aligned} \tag{4.81b}$$

$$\begin{aligned} \pi(X_I^-)\phi\psi &= (-1)^{\hat{\varphi}\tilde{I}} (\pi(\mathcal{Q}_I^{1/2})\phi)\pi(X_I^-)\psi \\ &\quad + (\pi(X_I^-)\phi)\pi(\mathcal{Q}_I^{-1/2})\psi. \end{aligned} \tag{4.81c}$$

From (4.81a) follows

$$\pi(\mathcal{P}_I^{1/2})\phi\psi = (\pi(\mathcal{P}_I^{1/2})\phi)\pi(\mathcal{P}_I^{1/2})\psi, \tag{4.81d}$$

$$\pi(\mathcal{Q}_I^{1/2})\phi\psi = (\pi(\mathcal{Q}_I^{1/2})\phi)\pi(\mathcal{Q}_I^{1/2})\psi. \tag{4.81e}$$

For $\mathcal{K}$ we have

$$\pi(\mathcal{K})\phi\psi = (\pi(\mathcal{K})\phi)\pi(\mathcal{K})\psi. \tag{4.82}$$

Applying the above rules one obtains

$$\pi(K_I)(T_{JL})^n = u^{nd_I(\frac{d_I}{d_{I+1}}\delta_{I+1,J}-\delta_{IJ})/2}(T_{JL})^n \tag{4.83}$$

$$\pi(\mathcal{P}_I^{1/2})(T_{JL})^n = Q_{IJ}^{-n/2}(T_{JL})^n,$$

$$\pi(\mathcal{Q}_I^{1/2})(T_{JL})^n = u^{nd_I(\frac{d_I}{d_{I+1}}\delta_{I+1,J}-\delta_{IJ})} Q_{IJ}^{n/2}(T_{JL})^n,$$

$$\pi(X_I^+)(T_{JL})^n = -\delta_{IJ} u^{(\tilde{I}+d_I+d_{I+1})/2} Q_{I,I+1}^{-1/2} c_I (T_{JL})^{n-1} T_{J+1,L}, \tag{4.84}$$

$$\pi(X_I^-)(T_{JL})^n = -\delta_{I+1,J}(-1)^{\tilde{I}} u^{(\tilde{I}-3d_I-d_{I+1})/2} Q_{II}^{1/2} u^{-(n-1)\tilde{I}} c_I T_{J-1,L}(T_{JL})^{n-1}, \tag{4.85}$$

$$c_I = \begin{cases} (q_{i,i+1})^{(n-1)/2}[n]_u, & \text{if } I = i \le m \\ (q'_{\alpha,\alpha+1})^{(n-1)/2}[n]_u, & \text{if } I = \alpha > m, \end{cases} \tag{4.86}$$

where $T_{JL} = a_{jl}$ or $d_{\alpha\beta}$ and $[n]_u = (u^n - u^{-n})/\lambda$.

The left action is consistent with $(b_{i\alpha})^2 = (c_{\alpha i})^2 = 0$. For $\mathcal{K}$ we have

$$\begin{aligned} \pi(\mathcal{K})(T_{JL})^n &= u^{-n/2}(T_{JL})^n, \quad m \neq n, \\ \pi(\mathcal{K})(T_{JL})^n &= u^{-nd_J/2}(T_{JL})^n, \quad m = n. \end{aligned} \tag{4.87}$$

Next we introduce the right action of $\mathcal{U}$ following [118] but taking into account the graded structure:

$$\pi_R(y)T_{IL} = \sum_{N=1}^{m+n} (-1)^{\hat{Y}\widehat{T_{IN}}} T_{IN}\langle y, T_{NL}\rangle \tag{4.88a}$$

$$= (-1)^{\hat{Y}\widehat{T_{IL(1)}}}(T_{IL})_{(1)}\langle y, (T_{IL})_{(2)}\rangle \tag{4.88b}$$

where $y \in \mathcal{U}$.

From (4.88) we find the explicit right action of the generators of $\mathcal{U}$:

$$\pi_R(K_I)T_{JL} = u^{d_I(\delta_{IL} - \frac{d_I}{d_{I+1}}\delta_{I+1,L})/2} T_{JL} \tag{4.89a}$$

$$\pi_R(X_I^+)T_{JL} = \delta_{I+1,L}(-1)^{\tilde{I}\widehat{T_{J,L-1}}} u^{\tilde{I}/2} Q_{I,I+1}^{-1/2} T_{J,L-1} \tag{4.89b}$$

$$\pi_R(X_I^-)T_{JL} = \delta_{IL}(-1)^{\tilde{I}(1+\widehat{T_{J,L+1}})} u^{(\tilde{I}-2d_I)/2} Q_{II}^{1/2} T_{J,L+1}. \tag{4.89c}$$

From (4.89a) follows

$$\pi_R(\mathcal{P}_I^{1/2})T_{JL} = Q_{IL}^{1/2} T_{JL}, \tag{4.89d}$$

$$\pi_R(\mathcal{Q}_I^{1/2})T_{JL} = u^{d_I(\delta_{IL} - \frac{d_I}{d_{I+1}}\delta_{I+1,L})} Q_{IL}^{-1/2} T_{JL}. \tag{4.89e}$$

Finally,

$$\begin{aligned} \pi_R(\mathcal{K})T_{JL} &= u^{1/2} T_{JL} \quad m \neq n, \\ \pi_R(\mathcal{K})T_{JL} &= u^{d_L/2} T_{JL} \quad m = n. \end{aligned} \tag{4.90}$$

The above are supplemented with the following action on the unit element of $\mathcal{A}$:

$$\pi_R(K_I)1_{\mathcal{A}} = 1_{\mathcal{A}}, \quad \pi_R(X_I^\pm)1_{\mathcal{A}} = 0, \quad \pi_R(\mathcal{K})1_{\mathcal{A}} = 1_{\mathcal{A}}. \tag{4.91}$$

In order to derive the action $\pi_R(y)$ on $\mathcal{A}$ we shall use the general form [119], which is the same as (4.88b) but for an arbitrary homogeneous element ψ of $\mathcal{A}$:

$$\pi_R(y)\psi = (-1)^{\hat{Y}\hat{\mathcal{S}}_{(1)}} \psi_{(1)}\langle y, \psi_{(2)}\rangle. \tag{4.92}$$

So the action of an arbitrary homogeneous element $y \in \mathcal{U}$ on the product of two homogeneous elements of $\mathcal{A}$ is given by

$$\begin{aligned} \pi_R(y)\phi\psi &= (-1)^{\hat{Y}\hat{\varphi}_{(1)}+\hat{Y}_{(2)}(\hat{\varphi}_{(2)}+\hat{\mathcal{S}}_{(1)})}\phi_{(1)}\langle y_{(1)},\phi_{(2)}\rangle\psi_{(1)}\langle y_{(2)},\psi_{(2)}\rangle \\ &= (-1)^{\hat{\varphi}\hat{Y}_{(2)}}(\pi_R(y_{(1)})\phi)\pi_R(y_{(2)})\psi. \end{aligned} \tag{4.93}$$

Thus we have for the generating elements $y = K_I, X_I^{\pm}$,

$$\pi_R(K_I)\phi\psi = (\pi_R(K_I)\phi)\pi_R(K_I)\psi, \tag{4.94a}$$

$$\begin{aligned} \pi_R(X_I^+)\phi\psi &= (\pi_R(X_I^+)\phi)\pi_R(\mathcal{P}_I^{1/2})\psi \\ &\quad + (-1)^{\hat{\varphi}\tilde{I}}(\pi_R(\mathcal{P}_I^{-1/2})\phi)\pi_R(X_I^+)\psi, \end{aligned} \tag{4.94b}$$

$$\begin{aligned} \pi_R(X_I^-)\phi\psi &= (\pi_R(X_I^-)\phi)\pi_R(\mathcal{Q}_I^{1/2})\psi \\ &\quad + (-1)^{\hat{\varphi}\tilde{I}}(\pi_R(\mathcal{Q}_I^{-1/2})\phi)\pi_R(X_I^-)\psi. \end{aligned} \tag{4.94c}$$

From (4.94a) follows

$$\pi_R(\mathcal{P}_I^{1/2})\phi\psi = (\pi_R(\mathcal{P}_I^{1/2})\phi)\pi_R(\mathcal{P}_I^{1/2})\psi, \tag{4.94d}$$

$$\pi_R(\mathcal{Q}_I^{1/2})\phi\psi = (\pi_R(\mathcal{Q}_I^{1/2})\phi)\pi_R(\mathcal{Q}_I^{1/2})\psi. \tag{4.94e}$$

For $\mathcal{K}$ we have

$$\pi_R(\mathcal{K})\phi\psi = (\pi_R(\mathcal{K})\phi)\pi_R(\mathcal{K})\psi. \tag{4.95}$$

Using this we find

$$\begin{aligned} \pi_R(K_I)(T_{JL})^n &= u^{nd_I(\delta_{IL}-\frac{d_I}{d_{I+1}}\delta_{I+1,L})/2}(T_{JL})^n, \\ \pi_R(\mathcal{P}_I^{1/2})(T_{JL})^n &= Q_{IL}^{n/2}(T_{JL})^n, \\ \pi_R(\mathcal{Q}_I^{1/2})(T_{JL})^n &= u^{nd_I(\delta_{IL}-\frac{d_I}{d_{I+1}}\delta_{I+1,L})}Q_{IL}^{-n/2}(T_{JL})^n, \end{aligned} \tag{4.96}$$

$$\pi_R(X_I^+)(T_{JL})^n = \delta_{I+1,L}(-1)^{\tilde{I}\widehat{T_{J,L-1}}}u^{\tilde{I}/2}Q_{I,I+1}^{-1/2}u^{(d_I+d_{I+1})(n-1)/2}\tilde{c}_I T_{J,L-1}(T_{JL})^{n-1}, \tag{4.97}$$

$$\pi_R(X_I^-)(T_{JL})^n = \delta_{IL}(-1)^{\tilde{I}(1+\widehat{T_{J,L+1}})}u^{(\tilde{I}-2d_I)/2}Q_{II}^{1/2}u^{d_I(n-1)}\tilde{c}_I(T_{JL})^{n-1}T_{J,L+1}, \tag{4.98}$$

$$\tilde{c}_I = \begin{cases} (q_{i,i+1})^{(1-n)/2}[n]_u, & \text{if } I = i \leq m \\ (q'_{\alpha,\alpha+1})^{(1-n)/2}[n]_u, & \text{if } I = \alpha > m, \end{cases} \tag{4.99}$$

where $T_{JL} = a_{jl}$ or $d_{\alpha\beta}$. The right action is consistent with $(b_{i\alpha})^2 = (c_{\alpha i})^2 = 0$. For $\mathcal{K}$ we have

$$\begin{aligned} \pi_R(\mathcal{K})(T_{JL})^n &= u^{n/2}(T_{JL})^n \quad m \neq n, \\ \pi_R(\mathcal{K})(T_{JL})^n &= u^{nd_L/2}(T_{JL})^n \quad m = n. \end{aligned} \tag{4.100}$$

4.3.2 Basis via Gauss decomposition

Until now we have used implicitly the basis for $\mathcal{A}$ given in [146] and in Section 4.2; however, it is not suitable for the construction of the induced representations following [114, 118]. From the latter references we know that the suitable basis is via the use of a Gauss decomposition. The point is that we shall use right covariance [114] to reduce the number of variables on which our functions depend. Right covariance with respect to the raising generators X_I^+ means that their right action will annihilate our functions. It so happens that this right action will annihilate automatically the 'lower triangular' and 'diagonal' entries of the Gauss decomposition. Thus, right covariance eliminates dependence on the 'upper triangular' entries of the Gauss decomposition. Right covariance with respect to the Cartan generators means that their right action will be scalar on our functions. For this purpose, it is sufficient that the right action of the Cartan generators will be scalar on the 'lower triangular' and 'diagonal' entries of the Gauss decomposition.

For the simple cases $m = n = 1$ and $m = 2$, $n = 1$ we refer to [147], Appendix A and Appendix B, respectively. Below we treat the general case.

The matrix T in (4.20) may be written as

$$T = \begin{pmatrix} A & B \\ C & D \end{pmatrix} = \begin{pmatrix} 1 & 0 \\ F & 1 \end{pmatrix} \begin{pmatrix} A & 0 \\ 0 & H \end{pmatrix} \begin{pmatrix} 1 & E \\ 0 & 1 \end{pmatrix} \tag{4.101}$$

where

$$H = D - CA^{-1}B, \quad E = A^{-1}B, \quad F = CA^{-1}, \tag{4.102}$$

and A^{-1} is the inverse of the quantum matrix A. Furthermore, the quantum matrices A and H may be decomposed as follows:

$$A = A_L A_D A_U, \quad H = H_L H_D H_U, \tag{4.103}$$

where the index L indicates the strictly lower triangular matrix (with units on the main diagonal), D for the diagonal matrix and U for the strictly upper triangular matrix (with units at the main diagonal). Then the quantum supermatrix T may be decomposed as follows:

$$T = \begin{pmatrix} A & B \\ C & D \end{pmatrix} = \begin{pmatrix} A_L & 0 \\ \Gamma & H_L \end{pmatrix} \begin{pmatrix} A_D & 0 \\ 0 & H_D \end{pmatrix} \begin{pmatrix} A_U & \Lambda \\ 0 & H_U \end{pmatrix} \tag{4.104}$$

where

$$\Lambda = A_U E = A_U A^{-1} B, \quad \Gamma = F A_L = C A^{-1} A_L. \tag{4.105}$$

In fact, the elements of the quantum matrix A are even and their commutation relations are that of $\mathrm{GL}_{u\mathbf{q}}(m)$, so we can get its Gauss decomposition directly from [118]. For this purpose, we have to suppose that the principal minor determinants of A,

$$D_r = \sum_{\rho\in S_r} \epsilon(\rho)a_{1\rho(1)}\dots a_{r\rho(r)} = \sum_{\rho\in S_r} \epsilon'(\rho)a_{\rho(1)1}\dots a_{\rho(r)r}, \quad r \le m,$$
$$\epsilon(\rho) = \prod_{\substack{j<k \\ \rho(j)>\rho(k)}} \left(\frac{-q_{\rho(k)\rho(j)}}{u^2}\right), \quad \epsilon'(\rho) = \prod_{\substack{j<k \\ \rho(j)>\rho(k)}} \left(\frac{-1}{q_{\rho(k)\rho(j)}}\right), \tag{4.106}$$

are invertible; note that D_m is just the quantum determinant of A (we will denote it by $\mathcal{D}_A$). Further, for the ordered set $I = \{i_1 < \dots < i_r\}$ and $J = \{j_1 < \dots < j_r\}$, let ξ^I_J be the r-minor determinant with respect to rows I and columns J such that

$$\xi^I_J = \sum_{\rho\in S_r} \epsilon'(\rho)a_{i_{\rho(1)}j_1}\dots a_{i_{\rho(r)}j_r}. \tag{4.107}$$

Note that $\xi^{1\dots i}_{1\dots i} = D_i$. Then one has as in [118] $(1 \le i, k, l \le m)$

$$a_{il} = Y_{ik}D_{kk}U_{kl} \tag{4.108}$$

where Y_{ik} are elements of A_L, D_{kk} are elements of A_D and U_{kl} are those of A_U. They are given explicitly by

$$Y_{ik} = \prod_{s=1}^{k-1} \frac{q_{si}}{q_{sk}} \xi^{1\dots k-1i}_{1\dots k} D_k^{-1}, \tag{4.109}$$
$$D_{kk} = D_k D_{k-1}^{-1} \quad (D_0 = 1), \quad U_{kl} = D_k^{-1}\xi^{1\dots k}_{1\dots k-1l}.$$

Now let us calculate the right action of X_I^+ on Y_{il} and D_{ll}. From (4.94b) we deduce that

$$\pi_R(X_I^+)\phi\psi = (\pi_R(X_I^+)\phi)\pi_R(\mathcal{P}_I^{1/2})\psi + (\pi_R(\mathcal{P}_I^{-1/2})\phi)\pi_R(X_I^+)\psi \tag{4.110}$$

where ϕ is an arbitrary product of a_{jl} with $1 \le j, l \le m$. Then, using (4.89b,d), one can prove by a direct calculation that

$$\pi_R(X_I^+)\xi^N_L = 0, \quad \text{for } L = \{1,\dots l\},\ \forall N, \tag{4.111}$$

and in the particular case we have

$$\pi_R(X_I^+)D_j = 0. \tag{4.112}$$

Then using (4.110) we get

$$\pi_R(X_I^+)Y_{jl} = 0, \quad \pi_R(X_I^+)D_{ll} = 0. \tag{4.113}$$

To calculate the right action of X_I^+ on Γ, we first introduce the left and right quantum cofactor matrices A_{ij} and A'_{ij} associated to A:

$$A_{ij} = \sum_{\rho(i)=j} \frac{\epsilon(\rho\sigma_i)}{\epsilon(\sigma_i)} a_{1\rho(1)} \dots \check{a}_{ij} \dots a_{m\rho(m)},$$

$$A'_{ij} = \sum_{\rho(j)=i} \frac{\epsilon'(\rho\sigma'_j)}{\epsilon'(\sigma'_j)} a_{\rho(1)1} \dots \check{a}_{ij} \dots a_{\rho(m)m}, \tag{4.114}$$

where σ_i and σ'_j denote the cyclic permutations

$$\sigma_i = \{i, \dots 1\}, \quad \sigma'_j = \{j, \dots m\}, \tag{4.115}$$

and the notation $\check{x}$ in (4.114) indicates that x is to be omitted. Then one can show that

$$\sum a_{ij} A_{kj} = \sum A'_{ji} a_{jk} = \delta_{ik} \mathcal{D}_A \tag{4.116}$$

and obtain the left and right inverse of A as

$$M_{ij} = \mathcal{D}_A^{-1} A'_{ji} = A_{ji} \mathcal{D}_A^{-1}. \tag{4.117}$$

One can calculate the following:

$$\pi_R(\mathcal{P}_I^{1/2}) \mathcal{D}_A = \prod_{s=1}^{m} Q_{Is}^{1/2} \mathcal{D}_A, \tag{4.118a}$$

$$\pi_R(\mathcal{P}_I^{1/2}) \mathcal{D}_A^{-1} = \prod_{s=1}^{m} Q_{Is}^{-1/2} \mathcal{D}_A^{-1}, \tag{4.118b}$$

$$\pi_R(\mathcal{P}_I^{1/2}) M_{ij} = Q_{Ii}^{-1/2} M_{ij}, \tag{4.118c}$$

and using (4.110), (4.117) and (4.118b), we note

$$\pi_R(X_I^+) M_{jl} = -Q_{II}^{1/2} Q_{I,I+1}^{-1} \delta_{Ij} M_{j+1,l} \tag{4.119}$$

and then we get

$$\begin{aligned} \pi_R(X_I^+) F_{\alpha l} &= \pi_R(X_I^+) C_{\alpha j} M_{jl} \\ &= (\pi_R(X_I^+) C_{\alpha j}) \pi_R(\mathcal{P}_I^{1/2}) M_{jl} \\ &\quad + (-1)^{\tilde{I}\widehat{C_{\alpha j}}} (\pi_R(\mathcal{P}_I^{-1/2}) C_{\alpha j}) \pi_R(X_I^+) M_{jl} = 0, \end{aligned} \tag{4.120}$$

from which it follows that

$$\pi_R(X_I^+) \Gamma_{\alpha l} = 0. \tag{4.121}$$

It remains now to calculate the right action of X_I^+ on the lower triangular matrix H_L and the diagonal one H_D. Note that the defining commutation relations of $\mathrm{GL}_{u\mathbf{q}}(m/n)$ in (4.21) are in fact the explicit form of the following super-RTT equation:

$$(-1)^{\widehat{N}(\widehat{N}+\hat{\ell})}R^{IJ}{}_{MN}T_{MN}T_{NL} = (-1)^{\hat{M}(\hat{J}+\widehat{N})}T_{IM}T_{JN}R^{MN}{}_{NL} \tag{4.122}$$

where the finite-dimensional R-matrix is given by

$$R^{IJ}{}_{NL} = \delta_L^I\delta_N^J\left\{(-u^2)^{\hat{I}}\delta^{IJ} + \tau^{IJ}(-1)^{\hat{I}\hat{J}}q_{JI} + \tau^{JI}(-1)^{\hat{I}\hat{J}}\frac{u^2}{q_{IJ}}\right\} + \delta_N^I\delta_L^J\tau^{JI}(1-u^2) \tag{4.123}$$

where $\tau^{IJ} = 1$ if $I > J$ and 0 otherwise. (For $n = 0$ and $q_i = u$, $\forall i$, the above relations will reduce to the RTT relations for $\mathrm{GL}_u(m)$, [162].) On the other hand, starting from (4.122) one can prove that the matrix H satisfies the same super-RTT equation with all indices being odd. This is proved in [219]. So the elements of H satisfy the defining commutation relations of $\mathrm{GL}_{u'\mathbf{q}'}(n)$. Furthermore, one can prove that the right action on H is as follows:

$$\begin{aligned}
\pi_R(K_I)h_{\alpha\beta} &= u^{d_I(\delta_{I\beta}-\frac{d_I}{d_{I+1}}\delta_{I+1,\beta})/2}h_{\alpha\beta}, \\
\pi_R(X_I^+)h_{\alpha\beta} &= \begin{cases} u^{\tilde{I}/2}Q_{I,I+1}^{-1/2}\delta_{I+1,\beta}h_{\alpha,\beta-1}, & \beta > m+1, \\ 0, & \beta = m+1, \end{cases} \\
\pi_R(X_I^-)h_{\alpha\beta} &= \delta_{I\beta}uQ_{\beta\beta}^{1/2}h_{\alpha\beta+1} - \delta_{Im}u^{-1/2}h_{\alpha,m+1}E_{m\beta}, \\
\pi_R(\mathcal{K})h_{\alpha\beta} &= \begin{cases} u^{1/2}h_{\alpha\beta}, & \text{for } m \neq n \\ u'^{1/2}h_{\alpha\beta}, & \text{for } m = n, \end{cases} \\
\pi_R(\mathcal{P}_I^{1/2})h_{\alpha\beta} &= Q_{I\beta}^{1/2}h_{\alpha\beta}, \\
\pi_R(\mathcal{Q}_I^{1/2})h_{\alpha\beta} &= u^{d_I(\delta_{I\beta}-\frac{d_I}{d_{I+1}}\delta_{I+1,\beta})}Q_{I\beta}^{-1/2}h_{\alpha\beta}.
\end{aligned} \tag{4.124}$$

Now, one can get the Gauss decomposition of H in the same way as it was done for the quantum matrix A. For this purpose, we have to suppose that the principal minor determinant of H,

$$\begin{aligned}
G_\alpha &= \sum_{\rho\in S_{\alpha-m}} \tilde{\epsilon}(\rho)h_{m+1\rho(m+1)}\cdots h_{\alpha\rho(\alpha)} \\
&= \sum_{\rho\in S_{\alpha-m}} \tilde{\epsilon}'(\rho)h_{\rho(m+1)m+1}\cdots h_{\rho(\alpha)\alpha}, \quad m+1 \le \alpha \le m+n,
\end{aligned} \tag{4.125}$$

$$\tilde{\epsilon}(\rho) = \prod_{\substack{\alpha<\beta \\ \rho(\alpha)>\rho(\beta)}}\left(\frac{-q'_{\rho(\beta)\rho(\alpha)}}{u'^2}\right), \quad \tilde{\epsilon}'(\rho) = \prod_{\substack{\alpha<\beta \\ \rho(\alpha)>\rho(\beta)}}\left(\frac{-1}{q'_{\rho(\beta)\rho(\alpha)}}\right), \tag{4.126}$$

is invertible; note that G_{m+n} is just the quantum determinant of H (we will denote it by $\mathcal{D}_H$). Furthermore, for the ordered set $I = \{\alpha_1 < \cdots < \alpha_r\}$ and $J = \{\beta_1 < \cdots < \beta_r\}$, let ξ'^I_J

be the r-minor determinant with respect to rows I and columns J such that

$$\xi'^{I}_{J} = \sum_{\rho \in S_r} \tilde{\epsilon}'(\rho) h_{\alpha_{\rho(1)}\beta_1} \dots h_{\alpha_{\rho(r)}\beta_r}. \tag{4.127}$$

Note that $\xi'^{m+1\dots\alpha}_{m+1\dots\alpha} = G_\alpha$. Then one has $(m+1 \le \alpha, \beta, \gamma \le m+n)$:

$$h_{\alpha\gamma} = Z_{\alpha\beta} G_{\beta\beta} V_{\beta\gamma} \tag{4.128}$$

where $Z_{\alpha\beta}$ are elements of H_L, $G_{\beta\beta}$ are elements of H_D, and $V_{\beta\gamma}$ are elements of H_U. They are given explicitly by

$$\begin{aligned} Z_{\alpha\beta} &= \prod_{\gamma=m+1}^{\beta-1} \frac{q_{\gamma\alpha}}{q_{\gamma\beta}} \xi'^{m+1\dots\beta-1\alpha}_{m+1\dots\beta} G_\beta^{-1}, \\ G_{\beta\beta} &= G_\beta G_{\beta-1}^{-1} \quad (G_m = 1), \\ V_{\beta\gamma} &= G_\beta^{-1} \xi'^{m+1\dots\beta}_{m+1\dots\beta-1\gamma}. \end{aligned} \tag{4.129}$$

Now let us calculate the right action of X_I^+ on $Z_{\alpha\beta}$ and $G_{\alpha\alpha}$. Using (4.94b) and $\widehat{h_{\alpha\beta}} = 0 \pmod 2$ we get

$$\pi_R(X_I^+) h_{\alpha\beta}\psi = (\pi_R(X_I^+) h_{\alpha\beta}) \pi_R(\mathcal{P}_I^{1/2})\psi + (\pi_R(\mathcal{P}_I^{-1/2}) h_{\alpha\beta}) \pi_R(X_I^+)\psi, \tag{4.130}$$

from which we deduce

$$\pi_R(X_I^+) \phi\psi = (\pi_R(X_I^+)\phi) \pi_R(\mathcal{P}_I^{1/2})\psi + (\pi_R(\mathcal{P}_I^{-1/2})\phi) \pi_R(X_I^+)\psi, \tag{4.131}$$

where ϕ is an arbitrary product of $h_{\alpha\beta}$ with $m+1 \le \alpha, \beta \le m+n$. Then one can prove in the same way as for A that

$$\pi_R(X_I^+) \xi'^{N}_{L} = 0, \quad \text{for } L = \{m+1, \dots \alpha\},\ \forall N \tag{4.132}$$

and in this particular case we have

$$\pi_R(X_I^+) G_\alpha = 0. \tag{4.133}$$

Then using (4.131) we get

$$\pi_R(X_I^+) Z_{\alpha\beta} = 0, \quad \pi_R(X_I^+) G_{\beta\beta} = 0. \tag{4.134}$$

Finally, we write down the superdeterminant:

$$\mathcal{F} = \prod_{s=1}^{m} D_{ss} \prod_{\alpha=m+1}^{m+n} G_{\alpha\alpha}^{-1} = D_m G_{m+n}^{-1}, \tag{4.135}$$

for which we also obtain

$$\pi_R(X_I^+)\mathcal{F} = 0. \tag{4.136}$$

Thus, we have proved that the right action of X_I^+ on the strictly lower and diagonal matrices in the Gauss decomposition of T is zero. On the other hand the right action of X_I^+ on the strictly upper diagonal matrices in the Gauss decomposition of T is nontrivial.

We have now for the right action of the Cartan generators:

$$\begin{aligned}\pi_R(K_I)\xi_L^N &= u^{\delta_{Il}/2}\xi_L^N, \quad L = \{1,\dots l\}\ \forall N \\ \pi_R(K_I)\xi'^N_L &= u^{\delta_{Im}/2}u'^{\delta_{I\alpha}/2}\xi'^N_L, \quad L = \{m+1,\dots \alpha\}\ \forall N,\end{aligned} \tag{4.137}$$

from which follows

$$\begin{aligned}&\pi_R(K_I)D_j = u^{\delta_{Ij}/2}D_j, \quad \pi_R(K_I)G_\beta = u^{\delta_{Im}/2}u'^{\delta_{I\beta}/2}G_\beta, \\ &\pi_R(K_I)Y_{jl} = Y_{jl}, \quad \pi_R(K_I)Z_{\alpha\beta} = Z_{\alpha\beta},\end{aligned} \tag{4.138}$$

we have also

$$\pi_R(K_I)\Gamma_{\alpha l} = \Gamma_{\alpha l}. \tag{4.139}$$

Now we give the action of $\mathcal{K}$ in the two cases $m \neq n$ and $m = n$.
For $m \neq n$ we have

$$\begin{aligned}\pi_R(\mathcal{K})\mathcal{D}_{\mathcal{A}} &= u^{m/2}\mathcal{D}_{\mathcal{A}}, \quad \pi_R(\mathcal{K})M_{jl} = u^{-1/2}M_{jl}, \\ \pi_R(\mathcal{K})\xi_L^N &= u^{l/2}\xi_L^N, \quad L = \{1,\dots l\}\ \forall N, \\ \pi_R(\mathcal{K})\xi'^N_L &= u^{(\beta-m)/2}\xi'^N_L, \quad L = \{m+1,\dots \beta\}\ \forall N,\end{aligned} \tag{4.140}$$

from which it follows that

$$\begin{aligned}&\pi_R(\mathcal{K})D_j = u^{j/2}D_j, \quad \pi_R(\mathcal{K})G_\beta = u^{(\beta-m)/2}G_\beta \\ &\pi_R(\mathcal{K})Y_{jl} = Y_{jl}, \quad \pi_R(\mathcal{K})\Gamma_{\alpha l} = \Gamma_{\alpha l}, \quad \pi_R(\mathcal{K})Z_{\alpha\beta} = Z_{\alpha\beta}.\end{aligned} \tag{4.141}$$

For $m = n$ we have

$$\begin{aligned}\pi_R(\mathcal{K})\mathcal{D}_{\mathcal{A}} &= u^{m/2}\mathcal{D}_{\mathcal{A}}, \quad \pi_R(\mathcal{K})M_{jl} = u^{-1/2}M_{jl}, \\ \pi_R(\mathcal{K})\xi_L^N &= u^{l/2}\xi_L^N, \quad L = \{1,\dots l\}\ \forall N, \\ \pi_R(\mathcal{K})\xi'^N_L &= u'^{(\beta-m)/2}\xi'^N_L, \quad L = \{m+1,\dots \beta\}\ \forall N,\end{aligned} \tag{4.142}$$

from which it follows that

$$\pi_R(\mathcal{K})D_j = u^{j/2}D_j, \quad \pi_R(\mathcal{K})G_\beta = u'^{(\beta-m)/2}G_\beta \tag{4.143}$$

$$\pi_R(\mathcal{K})Y_{jl} = Y_{jl}, \quad \pi_R(\mathcal{K})\Gamma_{al} = \Gamma_{al}, \quad \pi_R(\mathcal{K})Z_{\alpha\beta} = Z_{\alpha\beta}.$$

Thus, we have shown that the right action of the Cartan generators is scalar on all entries of the Gauss decomposition.

The generators Y_{lj}, $\Gamma_{\alpha l}$, $Z_{\beta\alpha}$ are the q-analogs of the strictly lower triangular supermatrices of GL(m/n), while the generators U_{jl}, $\Lambda_{i\alpha}$, $V_{\alpha\beta}$ are the q-analogs of the strictly upper triangular supermatrices of GL(m/n). The generators D_{jj}, $G_{\alpha\alpha}$, $\mathcal{F}$ are the q-analogs of the diagonal supermatrices of GL(m/n). In the following we shall use their commutation relations. Since these are rather lengthy we do not reproduce them here—they are given in Appendix C of [147].

Clearly one can replace the basis of $\mathcal{A}$ in terms of T_{JL} with a basis in terms of $X_{LJ} = (Y_{lj}, \Gamma_{\alpha j}, Z_{\beta\alpha})$ with $(L > J)$, D_i, G_α, $(\alpha \le m+n-1)$, $\mathcal{F}$, and $W_{JL} = (U_{jl}, \Lambda_{j\alpha}, V_{\alpha\beta})$. More precisely, the basis will be given as follows:

$$\begin{aligned}
f_{\bar{v},\bar{k},\bar{w}} &\doteq (Y_{21})^{v_{21}} \dots (Y_{m,m-1})^{v_{m,m-1}} (\Gamma_{m+1,1})^{v_{m+1,1}} \dots (\Gamma_{m+n,m})^{v_{m+n,m}} \\
&\quad \times (Z_{m+2,m+1})^{v_{m+2,m+1}} \dots (Z_{m+n,m+n-1})^{v_{m+n,m+n-1}} \\
&\quad \times (D_1)^{k_1} \dots (D_m)^{k_m} (G_{m+1})^{k_{m+1}} \dots (G_{m+n-1})^{k_{m+n-1}} (\mathcal{F})^{k_{m+n}} \\
&\quad \times (V_{m+n-1,m+n})^{w_{m+n-1,m+n}} \dots (V_{m+1,m+2})^{w_{m+1,m+2}} \\
&\quad \times (\Lambda_{m,m+n})^{w_{m,m+n}} \dots (\Lambda_{1,m+1})^{w_{1,m+1}} (U_{m-1,m})^{w_{m-1,m}} \dots (U_{12})^{w_{12}}, \\
\bar{v} &\doteq \{v_{IJ} \mid 1 \le J < I \le m+n\}, \ v_{IJ} \in \mathbb{Z}_+, \ v_{\alpha i} \le 1 \\
\bar{k} &\doteq \{k_I \mid 1 \le I \le m+n\}, \ k_I \in \mathbb{Z}, \\
\bar{w} &\doteq \{w_{IJ} \mid 1 \le I < J \le m+n\}, \ w_{IJ} \in \mathbb{Z}_+, \ w_{i\alpha} \le 1,
\end{aligned} \tag{4.144}$$

and we are using the normal ordering similar to [118], namely, we first put the elements Y_{ij} in lexicographic order, (i. e., if $i < k$ then Y_{ij} is before $Y_{k\ell}$ and Y_{ti} is before Y_{tk}), then the elements $\Gamma_{\alpha i}$ in lexicographic order, then the elements $Z_{\alpha\beta}$ in lexicographic order, then the elements D_I and $\mathcal{F}$, then the elements $V_{\alpha\beta}$ in antilexicographic order (i. e., if $\alpha > \gamma$ then $V_{\alpha\beta}$ is before $V_{\gamma\delta}$ and $V_{\tau\alpha}$ is before $V_{\tau\gamma}$), then the elements $\Lambda_{i\alpha}$ in antilexicographic order; finally, the elements U_{ij} in antilexicographic order. Note that the basis includes the unit element of $\mathcal{A}$:

$$f_{0,0,0} = 1_{\mathcal{A}}. \tag{4.145}$$

4.3.3 Representations of $\mathcal{U}$ and $\mathcal{U}'$

We have already seen that the basis introduced in (4.144) has the necessary right covariance properties we mentioned earlier. Thus, we consider as candidates for our representation spaces the formal power series

$$\varphi = \sum_{\substack{k_I \in \mathbb{Z}, v_{\alpha i}, w_{i\alpha} \in \{0,1\} \\ v_{ji}, v_{\beta\alpha}, w_{ij}, w_{\alpha\beta} \in \mathbb{Z}_+}} \bar{m}_{\bar{v},\bar{k},\bar{w}} f_{\bar{v},\bar{k},\bar{w}}, \quad \bar{m}_{\bar{v},\bar{k},\bar{w}} \in \mathbb{C}. \tag{4.146}$$

We impose now right covariance with respect to X_I^+, i. e., we require

$$\pi_R(X_I^+)\varphi = 0. \tag{4.147}$$

This means that our functions φ do not depend on W_{IJ}, since (4.147) is fulfilled automatically for the other elements of the basis, as we saw in Section 4.3. Thus, the functions obeying (4.147) are

$$\varphi = \sum_{\substack{k_I\in\mathbb{Z},\nu_{\alpha i}\in\{0,1\}\\ \nu_{ji},\nu_{\beta\alpha}\in\mathbb{Z}_+}} \bar{m}_{\bar{\nu},\bar{k}} f_{\bar{\nu},\bar{k}}, \quad \bar{m}_{\bar{\nu},\bar{k}} \doteq \bar{m}_{\bar{\nu},\bar{k},0}, \; f_{\bar{\nu},\bar{k}} \doteq f_{\bar{\nu},\bar{k},0}. \tag{4.148}$$

Next we impose right covariance with respect to K_I and $\mathcal{K}$:

$$\begin{aligned}\pi_R(K_I)\varphi &= u^{d_I r_I/2}\varphi,\\ \pi_R(\mathcal{K})\varphi &= u^{\hat{r}/2}\varphi \quad \text{if } m \neq n,\\ \pi_R(\mathcal{K})\varphi &= u^{\tilde{r}/2}\varphi \quad \text{if } m = n,\end{aligned} \tag{4.149}$$

where r_I and $\hat{r}$, $\tilde{r}$ are parameters to be specified below. Using the following:

$$\begin{aligned}\pi_R(K_I)\mathcal{F} &= \mathcal{F},\\ \pi_R(\mathcal{K})\mathcal{F} &= u^{(m-n)/2}\mathcal{F} \quad \text{if } m \neq n,\\ \pi_R(\mathcal{K})\mathcal{F} &= u^{m}\mathcal{F}, \quad \text{if } m = n,\end{aligned} \tag{4.150}$$

and for the action of K_I and $\mathcal{K}$ on the new generators and their products we find

$$\begin{aligned}\pi_R(K_I)\varphi &= u^{d_I k_I/2}\varphi, \quad \text{for } I \leq m+n-1,\ I \neq m,\\ \pi_R(K_m)\varphi &= u^{\frac{1}{2}(k_m+\sum_{\beta=m+1}^{m+n-1}k_\beta)}\varphi,\\ \pi_R(\mathcal{K})\varphi &= u^{\frac{1}{2}(\sum_{j=1}^{m} jk_j+\sum_{\beta=m+1}^{m+n-1}(\beta-m)k_\beta+(m-n)k_{m+n})}\varphi \quad \text{if } m \neq n,\\ \pi_R(\mathcal{K})\varphi &= u^{\frac{1}{2}(\sum_{j=1}^{m} jk_j-\sum_{\beta=m+1}^{m+n-1}(\beta-m)k_\beta+2mk_{m+n})}\varphi \quad \text{if } m = n.\end{aligned} \tag{4.151}$$

Comparing the right covariance (4.149) with the direct calculations (4.151) we obtain

$$\begin{aligned}k_I &= r_I, \quad \text{for } I \leq m+n-1,\ I \neq m,\\ k_m &= r_m - \sum_{\beta=m+1}^{m+n-1} r_\beta,\\ \hat{r} &= \sum_{j=1}^{m} jk_j + \sum_{\beta=m+1}^{m+n-1}(\beta-m)k_\beta + (m-n)k_{m+n}\\ &= \sum_{j=1}^{m} jr_j + \sum_{\beta=m+1}^{m+n-1}(\beta-2m)r_\beta + (m-n)k_{m+n}, \quad \text{if } m \neq n,\end{aligned}$$

$$\tilde{r} = \sum_{j=1}^{m} jk_j - \sum_{\beta=m+1}^{2m-1} (\beta - m)k_\beta + 2mk_{2m}$$
$$= \sum_{J=1}^{2m-1} Jd_J r_J + 2mk_{2m}, \quad \text{if } m = n. \tag{4.152}$$

This means that $r_I, \hat{r}, \tilde{r} \in \mathbb{Z}$ and there is no summation in k_I; also we have

$$k_{m+n} = \frac{1}{m-n}\left(\hat{r} - \sum_{j=1}^{m} jr_j - \sum_{\beta=m+1}^{m+n-1} (\beta - 2m)r_\beta\right) \quad \text{if } m \neq n,$$
$$k_{2m} = \frac{1}{2m}\left(\tilde{r} - \sum_{J=1}^{2m-1} Jd_J r_J\right) \quad \text{if } m = n. \tag{4.153}$$

Thus, the reduced functions obeying (4.147) and (4.149) are

$$\varphi = \sum_{\substack{\nu_{\alpha i} \in \{0,1\} \\ \nu_{ji}, \nu_{\beta\alpha} \in \mathbb{Z}_+}} \bar{m}_{\bar{\nu}} f_{\bar{\nu}} \Xi_{\bar{r}}, \quad \bar{m}_{\bar{\nu}} \doteq \bar{m}_{\bar{\nu},0}, \; f_{\bar{\nu}} \doteq f_{\bar{\nu},0}, \tag{4.154}$$
$$\Xi_{\bar{r}} \doteq (D_1)^{r_1} \dots (D_{m-1})^{r_{m-1}} (D_m)^{\widehat{S}} (G_{m+1})^{r_{m+1}} \dots (G_{m+n-1})^{r_{m+n-1}} (\mathcal{F})^{\hat{t}},$$
$$\bar{r} = \{r_1, \dots, r_{m+n-1}, \hat{r} \text{ (or } \tilde{r})\},$$

where

$$\widehat{S} = r_m - \sum_{\beta=m+1}^{m+n-1} r_\beta \tag{4.155}$$
$$\hat{t} = \begin{cases} \frac{1}{m-n}(\hat{r} - \sum_{j=1}^{m} jr_j - \sum_{\beta=m+1}^{m+n-1} (\beta - 2m)r_\beta) & \text{if } m \neq n \\ \frac{1}{2m}(\tilde{r} - \sum_{J=1}^{2m-1} Jd_J r_J) & \text{if } m = n. \end{cases}$$

Next we shall give the $\mathcal{U}$ representation (left) action π on φ. Besides the action of the 'Chevalley' generators K_I, $X_I^\pm$, $\mathcal{K}$ we shall give for the reader's convenience also the action of $\mathcal{P}_I$, $\mathcal{Q}_I$, though it follows from that of K_I. We have

$$\pi(K_I)Y_{lj} = u^{(\delta_{I+1,l} - \delta_{I+1,j} - \delta_{Il} + \delta_{Ij})/2} Y_{lj},$$
$$\pi(X_I^+)Y_{lj} = -u^{(\bar{I} + d_I + d_{I+1})/2} Q_{I,I+1}^{-1/2} Q_{Ij}^{-1/2} \delta_{Il} (\delta_{lm} \Gamma_{m+1,m}$$
$$+ (1 - \delta_{lm}) Y_{l+1,j}) +$$
$$+ u Q_{I,I+1}^{-1/2} Q_{Il}^{-1/2} \left(\frac{q_{j,j+1} q_{j+1,l}}{q_{jl}}\right)^{(1-\delta_{l,j+1})} \delta_{Ij} Y_{j+1,j} Y_{lj}$$
$$+ u Q_{I,I+1}^{-1/2} Q_{Il}^{1/2} Q_{I,j-1}^{-1/2} Q_{Ij}^{-1/2} \delta_{I+1,j}$$
$$\times \left\{ \frac{q_{j-1,l}}{q_{j-1,j} q_{jl}} Y_{l,j-1} - Y_{j,j-1} Y_{lj} \right\},$$

$$
\begin{aligned}
\pi(X_I^-)Y_{lj} &= -u^{-2}Q_{II}^{1/2}Q_{Ij}^{1/2}u^{-\delta_{Ij}}\delta_{I+1,l}Y_{l-1,j},\\
\pi(\mathcal{K})Y_{lj} &= Y_{lj},\\
\pi(\mathcal{P}_I^{1/2})Y_{lj} &= Q_{Il}^{-1/2}Q_{Ij}^{1/2}Y_{lj},\\
\pi(\mathcal{Q}_I^{1/2})Y_{lj} &= u^{(\delta_{I+1,l}-\delta_{I+1,j}-\delta_{Il}+\delta_{Ij})}Q_{Il}^{1/2}Q_{Ij}^{-1/2}Y_{lj},
\end{aligned}
\tag{4.156}
$$

$$
\begin{aligned}
\pi(K_I)\Gamma_{\alpha j} &= u^{d_I(\frac{d_I}{d_{I+1}}(\delta_{I+1,\alpha}-\delta_{I+1,j})-\delta_{I\alpha}+\delta_{Ij})/2}\Gamma_{\alpha j},\\
\pi(X_I^+)\Gamma_{\alpha j} &= -u^{-1}Q_{I,I+1}^{-1/2}Q_{Ij}^{-1/2}\delta_{I\alpha}\Gamma_{\alpha+1,j}\\
&\quad + u^{(\tilde{I}+d_I+d_{I+1})/2}Q_{I,I+1}^{-1/2}Q_{I\alpha}^{-1/2}\left(\frac{q_{j,j+1}q_{j+1,\alpha}}{q_{j\alpha}}\right)^{(1-\delta_{\alpha,j+1})}\delta_{Ij}\\
&\quad \times((1-\delta_{jm})Y_{j+1,j}-\delta_{jm}\Gamma_{m+1,m})\Gamma_{\alpha j}\\
&\quad + uQ_{I,I+1}^{-1/2}Q_{I\alpha}^{1/2}Q_{I,j-1}^{-1/2}Q_{Ij}^{-1/2}\delta_{I+1,j}\\
&\quad \times\left\{\frac{q_{j-1,\alpha}}{q_{j-1,j}q_{j\alpha}}\Gamma_{\alpha,j-1}-Y_{j,j-1}\Gamma_{\alpha j}\right\},\\
\pi(X_I^-)\Gamma_{\alpha j} &= -(-1)^{\tilde{I}}u^{(\tilde{I}-3d_I-d_{I+1})1/2}Q_{Ij}^{1/2}u^{-\delta_{Ij}}\delta_{I+1,\alpha}\\
&\quad \times\{\delta_{\alpha,m+1}Y_{mj}+(1-\delta_{\alpha,m+1})\Gamma_{\alpha-1,j}\},\\
\pi(\mathcal{K})\Gamma_{\alpha j} &= \begin{cases}\Gamma_{\alpha j} & \text{if } m\neq n,\\ u\Gamma_{\alpha j} & \text{if } m=n,\end{cases}\\
\pi(\mathcal{P}_I^{1/2})\Gamma_{\alpha j} &= Q_{I\alpha}^{-1/2}Q_{Ij}^{1/2}\Gamma_{\alpha j},\\
\pi(\mathcal{Q}_I^{1/2})\Gamma_{\alpha j} &= u^{d_I(\frac{d_I}{d_{I+1}}(\delta_{I+1,\alpha}-\delta_{I+1,j})-\delta_{I\alpha}+\delta_{Ij})}Q_{I\alpha}^{1/2}Q_{Ij}^{-1/2}\Gamma_{\alpha j},
\end{aligned}
\tag{4.157}
$$

$$
\begin{aligned}
\pi(K_I)Z_{\beta\alpha} &= u^{d_I(\frac{d_I}{d_{I+1}}(\delta_{I+1,\beta}-\delta_{I+1,\alpha})-\delta_{I\beta}+\delta_{I\alpha})/2}Z_{\beta\alpha},\\
\pi(X_I^+)Z_{\beta\alpha} &= -u'Q_{I,I+1}^{-1/2}Q_{I\alpha}^{-1/2}\delta_{I\beta}Z_{\beta+1,\alpha}\\
&\quad + u'Q_{I,I+1}^{-1/2}Q_{I\beta}^{-1/2}\left(\frac{q'_{\alpha,\alpha+1}q'_{\alpha+1,\beta}}{q'_{\alpha\beta}}\right)^{(1-\delta_{\beta,\alpha+1})}\delta_{I\alpha}Z_{\alpha+1,\alpha}Z_{\beta\alpha}\\
&\quad + (-1)^{\hat{I}+\widehat{I+1}}u^{2\tilde{I}}u^{(\tilde{I}+d_I+d_{I+1})/2}Q_{I,I+1}^{-1/2}Q_{I\beta}^{1/2}Q_{I,\alpha-1}^{-1/2}Q_{I\alpha}^{-1/2}\delta_{I+1,\alpha}\\
&\quad \times\left\{\frac{q'_{\alpha-1,\beta}}{q'_{\alpha-1,\alpha}q'_{\alpha\beta}}(\delta_{Im}\Gamma_{\beta m}+(1-\delta_{Im})Z_{\beta,\alpha-1})\right.\\
&\quad \left. -\delta_{Im}\Gamma_{m+1,m}Z_{\beta\alpha}-(1-\delta_{Im})Z_{\alpha,\alpha-1}Z_{\beta\alpha}\right\},\\
\pi(X_I^-)Z_{\beta\alpha} &= -u'^{-2}Q_{II}^{1/2}Q_{I\alpha}^{1/2}u'^{-\delta_{I\alpha}}\delta_{I+1,\beta}Z_{\beta-1,\alpha},\\
\pi(\mathcal{K})Z_{\beta\alpha} &= Z_{\beta\alpha},\\
\pi(\mathcal{P}_I^{1/2})Z_{\beta\alpha} &= Q_{I\beta}^{-1/2}Q_{I\alpha}^{1/2}Z_{\beta\alpha},\\
\pi(\mathcal{Q}_I^{1/2})Z_{\beta\alpha} &= u^{d_I(\frac{d_I}{d_{I+1}}(\delta_{I+1,\beta}-\delta_{I+1,\alpha})-\delta_{I\beta}+\delta_{I\alpha})}Q_{I\beta}^{1/2}Q_{I\alpha}^{-1/2}Z_{\beta\alpha},
\end{aligned}
\tag{4.158}
$$

$$\begin{aligned}
\pi(K_I)D_j &= u^{-\delta_{Ij}/2}D_j, \\
\pi(X_I^+)D_j &= -u^{(\tilde{I}+d_I+d_{I+1})/2}Q_{I,I+1}^{-1/2}\prod_{s=1}^{j-1}Q_{Is}^{1/2}\delta_{Ij} \\
&\quad \times (\delta_{jm}\Gamma_{m+1,m} + (1-\delta_{jm})Y_{j+1,j})D_j, \\
\pi(X_I^-)D_j &= 0, \\
\pi(\mathcal{K})D_j &= u^{-j/2}D_j, \\
\pi(\mathcal{P}_I^{1/2})D_j &= \prod_{s=1}^{j}Q_{Is}^{-1/2}D_j, \\
\pi(\mathcal{Q}_I^{1/2})D_j &= u^{-\delta_{Ij}}\prod_{s=1}^{j}Q_{Is}^{1/2}D_j,
\end{aligned} \tag{4.159}$$

$$\begin{aligned}
\pi(K_I)G_\beta &= u^{(-\delta_{Im}+\delta_{I\beta})/2}G_\beta, \\
\pi(X_I^+)G_\beta &= -u^{(\tilde{I}+d_I+d_{I+1})/2}Q_{I,I+1}^{-1/2}\left\{\prod_{\alpha=m+1}^{\beta-1}Q_{I\alpha}^{1/2}\delta_{I\beta}Z_{\beta+1,\beta}\right. \\
&\quad \left. + \prod_{\alpha=m+2}^{\beta}Q_{m\alpha}^{1/2}\delta_{Im}\Gamma_{m+1,m}\right\}G_\beta, \\
\pi(X_I^-)G_\beta &= 0, \\
\pi(\mathcal{K})G_\beta &= \begin{cases} u^{-(\beta-m)/2}G_\beta & \text{if } m \neq n, \\ u'^{-(\beta-m)/2}G_\beta & \text{if } m = n, \end{cases} \\
\pi(\mathcal{P}_I^{1/2})G_\beta &= \prod_{\alpha=m+1}^{\beta}Q_{I\alpha}^{-1/2}G_\beta, \\
\pi(\mathcal{Q}_I^{1/2})G_\beta &= u^{(-\delta_{Im}+\delta_{I\beta})}\prod_{\alpha=m+1}^{\beta}Q_{I\alpha}^{1/2}G_\beta,
\end{aligned} \tag{4.160}$$

$$\begin{aligned}
\pi(K_I)\mathcal{F} &= \mathcal{F}, \\
\pi(X_I^+)\mathcal{F} &= 0, \\
\pi(X_I^-)\mathcal{F} &= 0, \\
\pi(\mathcal{K})\mathcal{F} &= \begin{cases} u^{(n-m)/2}\mathcal{F} & \text{if } m \neq n, \\ u^{-m}\mathcal{F} & \text{if } m = n, \end{cases} \\
\pi(\mathcal{P}_I^{1/2})\mathcal{F} &= \mathcal{F}, \\
\pi(\mathcal{Q}_I^{1/2})\mathcal{F} &= \mathcal{F}.
\end{aligned} \tag{4.161}$$

Now we note that from (4.159), (4.160), and (4.161) we have the important consequence that the degrees of variables D_j, G_β, $\mathcal{F}$ are not changed by the action of $\mathcal{U}$. Thus, the parameters r_I and $\hat{r}$ (or $\check{r}$) indeed characterize the action of $\mathcal{U}$, i. e., we have obtained representations of $\mathcal{U}$.

- Thus, by equations (4.156), (4.157), (4.158), (4.159), (4.160), and (4.161), we have given the induced representations of $\mathcal{U}$ labeled by the $m+n$ integer numbers r_I and $\hat{r}$ (or $\tilde{r}$) and acting in the space of formal power series of $(m+n)(m+n-1)/2$ non-commuting variables, of which the mn variables $\Gamma_{\alpha i}$ are odd and the variables Y_{ij} and $Z_{\alpha\beta}$ are even.

Remark. For $u = \mathbf{q} = 1$ our representations coincide with the holomorphic representations induced from the upper diagonal Borel subsupergroup B of $G \equiv \mathrm{GL}(m/n)$ and acting on the coset G/G^+, where G^+ is the strictly upper diagonal supergroup of G. That is why we call our representations induced. ◇

To obtain our representation more explicitly, one is using these formulas together with the rules (4.81) and (4.82). In particular, we see that

$$\pi(\mathcal{K})\varphi = \begin{cases} u^{-\hat{r}/2}\varphi, & \text{if } m \neq n \\ u^{-\tilde{r}/2}\varphi', & \text{if } m - n \end{cases}$$

$$\varphi' = \sum_{\substack{\nu_{\alpha i} \in \{0,1\} \\ \nu_{ji}, \nu_{\beta\alpha} \in \mathbb{Z}_+}} \bar{m}_{\bar{\nu}} u^{\sum_{\alpha,i} \nu_{\alpha i}} f_{\bar{\nu}} \Xi_{\bar{r}}. \tag{4.162}$$

We notice from (4.161) that $\mathcal{U}'$ acts trivially on $\mathcal{F}$. Thus, the action of $\mathcal{U}'$ involves only the parameters r_I, $I \leq m + n - 1$. On the other hand by (4.162) we see that the action of $\mathcal{K}$ involves only the parameter $\tilde{r}'$ ($\tilde{r}' = \hat{r}$ if $m \neq n$, $\tilde{r}' = \tilde{r}$ if $m = n$). Thus we can consistently also from the representation theory point of view restrict ourselves to $\mathrm{SL}_{u\mathbf{q}}(m/n)$, i. e., we set

$$\mathcal{F} = \mathcal{F}^{-1} = 1_{\mathcal{A}}. \tag{4.163}$$

Note that in order to enforce this condition it is also necessary that $\mathcal{F}$ commutes with all generators, which follows from the decoupling conditions (4.71).

With (4.163) enforced the dual algebra is $\mathcal{U}' \equiv \mathcal{U}_{u\mathbf{q}}(\mathrm{sl}(m/n))$. Thus, the reduced functions for the $\mathcal{U}'$ action are

$$\varphi = \sum_{\substack{\nu_{\alpha i} \in \{0,1\} \\ \nu_{ji}, \nu_{\beta\alpha} \in \mathbb{Z}_+}} \bar{m}_{\bar{\nu}} f_{\bar{\nu}} \Xi^0_{\bar{r}},$$

$$\Xi^0_{\bar{r}} \doteq D_1^{r_1} \dots D_{m-1}^{r_{m-1}} D_m^{\widehat{S}} G_{m+1}^{r_{m+1}} \dots G_{m+n-1}^{r_{m+n-1}}. \tag{4.164}$$

- Thus, by equations (4.156), (4.157), (4.158), (4.159), and (4.160), we have given the induced representations of $\mathcal{U}'$ labeled by the $m + n - 1$ integer numbers r_I. For $u = \mathbf{q} = 1$ our representations coincide with the standard holomorphic representations induced from B and acting on the coset G/B.

To obtain the representations more explicitly, one uses these formulas together with the rules (4.81). In particular, we have

$$\pi(K_I)(Y_{lj})^k = u^{k(\delta_{I+1,l} - \delta_{I+1,j} - \delta_{Il} + \delta_{Ij})/2}(Y_{lj})^k,$$

$$
\begin{aligned}
\pi(X_I^+)(Y_{lj})^k &= -u^{(\tilde{I}+d_I+d_{I+1})/2} Q_{I,I+1}^{-1/2} Q_{Ij}^{(k-2)/2} c_l \delta_{Il} (Y_{lj})^{k-1} \\
&\quad \times (\delta_{lm} \Gamma_{m+1,m} + (1-\delta_{lm}) Y_{l+1,j}) \\
&\quad + u Q_{I,I+1}^{-1/2} Q_{Il}^{(k-2)/2} c_j \left(\frac{q_{j,j+1} q_{j+1,l}}{q_{jl}} \right)^{(1-\delta_{l,j+1})} \delta_{Ij} Y_{j+1,j} (Y_{lj})^k \\
&\quad + u Q_{I,I+1}^{-1/2} Q_{Il}^{k/2} \left(\frac{q_{j-1,j}}{u} \right)^k \tilde{c}_{j-1} \delta_{I+1,j} \\
&\quad \times \left\{ \frac{q_{j-1,l}}{q_{j-1,j} q_{jl}} Y_{l,j-1} (Y_{lj})^{k-1} - Y_{j,j-1} (Y_{lj})^k \right\}, \\
\pi(X_I^-)(Y_{lj})^k &= -u^{-2} Q_{II}^{1/2} Q_{Ij}^{k/2} u^{-k\delta_{Ij}} c_{l-1} \delta_{I+1,l} Y_{l-1,j} (Y_{lj})^{k-1}, \\
\pi(\mathcal{P}_I^{1/2})(Y_{lj})^k &= Q_{Il}^{-k/2} Q_{Ij}^{k/2} (Y_{lj})^k, \\
\pi(\mathcal{Q}_I^{1/2})(Y_{lj})^k &= u^{k(\delta_{I+1,l}-\delta_{I+1,j}-\delta_{Il}+\delta_{Ij})} Q_{Il}^{k/2} Q_{Ij}^{-k/2} (Y_{lj})^k,
\end{aligned}
\tag{4.165}
$$

$$
\begin{aligned}
\pi(K_I)(Z_{\beta\alpha})^k &= u^{kd_I(\frac{d_I}{d_{I+1}}(\delta_{I+1,\beta}-\delta_{I+1,\alpha})-\delta_{I\beta}+\delta_{I\alpha})/2} (Z_{\beta\alpha})^k, \\
\pi(X_I^+)(Z_{\beta\alpha})^k &= -u' Q_{I,I+1}^{-1/2} Q_{I\alpha}^{(k-2)/2} c_\beta \delta_{I\beta} (Z_{\beta\alpha})^{k-1} Z_{\beta+1,\alpha} \\
&\quad + u' Q_{I,I+1}^{-1/2} Q_{I\beta}^{(k-2)/2} c_\alpha \left(\frac{q'_{\alpha,\alpha+1} q'_{\alpha+1,\beta}}{q'_{\alpha\beta}} \right)^{(1-\delta_{\beta,\alpha+1})} \\
&\quad \times \delta_{I\alpha} Z_{\alpha+1,\alpha} (Z_{\beta\alpha})^k \\
&\quad + (-1)^{\hat{I}+\widehat{I+1}} u^{(\tilde{I}+d_I+d_{I+1})/2} Q_{I,I+1}^{-1/2} Q_{I\beta}^{k/2} \tilde{c}_I \delta_{I+1,\alpha} \\
&\quad \times \left\{ \frac{q'_{\alpha-1,\beta}}{q'_{\alpha-1,\alpha} q'_{\alpha\beta}} \left(\delta_{Im} (q_{m,m+1})^k \Gamma_{\beta m} \right. \right. \\
&\quad \left. + (1-\delta_{Im}) \left(\frac{q'_{\alpha-1,\alpha}}{u'} \right)^k Z_{\beta,\alpha-1} \right) (Z_{\beta\alpha})^{k-1} \\
&\quad - \delta_{Im} (q_{m,m+1})^k \Gamma_{m+1,m} (Z_{\beta\alpha})^k \\
&\quad \left. - (1-\delta_{Im}) \left(\frac{q'_{\alpha-1,\alpha}}{u'} \right)^k Z_{\alpha,\alpha-1} (Z_{\beta\alpha})^k \right\}, \\
\pi(X_I^-)(Z_{\beta\alpha})^k &= -u'^{-2} Q_{II}^{1/2} Q_{I\alpha}^{k/2} u'^{-k\delta_{I\alpha}} c_{\beta-1} \delta_{I+1,\beta} Z_{\beta-1,\alpha} (Z_{\beta\alpha})^{k-1}, \\
\pi(\mathcal{P}_I^{1/2})(Z_{\beta\alpha})^k &= Q_{I\beta}^{-k/2} Q_{I\alpha}^{k/2} (Z_{\beta\alpha})^k, \\
\pi(\mathcal{Q}_I^{1/2})(Z_{\beta\alpha})^k &= u^{kd_I(\frac{d_I}{d_{I+1}}(\delta_{I+1,\beta}-\delta_{I+1,\alpha})-\delta_{I\beta}+\delta_{I\alpha})} Q_{I\beta}^{k/2} Q_{I\alpha}^{-k/2} (Z_{\beta\alpha})^k,
\end{aligned}
\tag{4.166}
$$

$$
\begin{aligned}
\pi(K_I)(D_j)^k &= u^{-k\delta_{Ij}/2} (D_j)^k, \\
\pi(X_I^+)(D_j)^k &= -u^{(\tilde{I}+d_I+d_{I+1})/2} Q_{I,I+1}^{-1/2} \prod_{s=1}^{j-1} Q_{Is}^{k/2} \tilde{c}_j \delta_{Ij} \\
&\quad \times (\delta_{jm} \Gamma_{m+1,m} + (1-\delta_{jm}) Y_{j+1,j}) (D_j)^k, \\
\pi(X_I^-)(D_j)^k &= 0,
\end{aligned}
$$

$$\pi(\mathcal{P}_I^{1/2})(D_j)^k = \prod_{s=1}^{j} Q_{Is}^{-k/2}(D_j)^k,$$
$$\pi(\mathcal{Q}_I^{1/2})(D_j)^k = u^{-k\delta_{Ij}} \prod_{s=1}^{j} Q_{Is}^{k/2}(D_j)^k, \tag{4.167}$$
$$\pi(K_I)(G_\beta)^k = u^{k(-\delta_{Im}+\delta_{I\beta})/2}(G_\beta)^k,$$
$$\pi(X_I^+)(G_\beta)^k = -u^{(\tilde{I}+d_I+d_{I+1})/2} Q_{I,I+1}^{-1/2} \tilde{c}_I \Bigg\{ \prod_{\alpha=m+1}^{\beta-1} Q_{I\alpha}^{k/2} \delta_{I\beta} Z_{\beta\,1,\beta} + \prod_{\alpha=m+2}^{\beta} Q_{m\alpha}^{k/2} \delta_{Im} \Gamma_{m+1,m} \Bigg\} (G_\beta)^k,$$
$$\pi(X_I^-)(G_\beta)^k = 0,$$
$$\pi(\mathcal{P}_I^{1/2})(G_\beta)^k = \prod_{\alpha=m+1}^{\beta} Q_{I\alpha}^{-k/?}(G_\beta)^k,$$
$$\pi(\mathcal{Q}_I^{1/2})(G_\beta)^k = u^{k(-\delta_{Im}+\delta_{I\beta})} \prod_{\alpha=m+1}^{\beta} Q_{I\alpha}^{k/2}(G_\beta)^k. \tag{4.168}$$

As a consequence we have, e. g.,

$$\begin{aligned}\pi(K_I)\varphi = u^{-\frac{1}{2}d_I r_I} &\sum_{\substack{\nu_{\gamma k}\in\{0,1\}\\ \nu_{ji},\nu_{\beta\alpha}\in\mathbb{Z}_+}} u^{\frac{1}{2}\nu_{ji}(\delta_{I+1,j}-\delta_{I+1,i}-\delta_{Ij}+\delta_{Ii})} \\ &\times u^{\frac{1}{2}\nu_{\gamma k} d_I(\frac{d_I}{d_{I+1}}(\delta_{I+1,\beta}-\delta_{I+1,\alpha})-\delta_{I\beta}+\delta_{I\alpha})} \\ &\times u^{\frac{1}{2}\nu_{\beta\alpha} d_I(\frac{d_I}{d_{I+1}}(\delta_{I+1,\beta}-\delta_{I+1,\alpha})-\delta_{I\beta}+\delta_{I\alpha})} \bar{m}_{\bar{\nu}} f_{\bar{\nu}} \Xi_{\bar{r}}^0.\end{aligned} \tag{4.169}$$

Finally, since the action of $\mathcal{U}'$ is not affecting the degrees of D_j and G_β, we may introduce (as in [114, 118]) the restricted functions

$$\tilde{\varphi} = \sum_{\substack{\nu_{\alpha i}\in\{0,1\}\\ \nu_{ji},\nu_{\beta\alpha}\in\mathbb{Z}_+}} \bar{m}_{\bar{\nu}} f_{\bar{\nu}} \tag{4.170}$$

using the intertwining operator:

$$\tilde{\varphi} \equiv \mathcal{I}\varphi \doteq \varphi|_{D_i=G_\alpha=1_{\mathcal{A}}}. \tag{4.171}$$

We denote the representation space of φ by $C_{\bar{r}}$, the representation space of $\tilde{\varphi}$ by $\widetilde{C}_{\bar{r}}$, and the representation acting on $\tilde{\varphi}$ by $\tilde{\pi}$. Thus, the operator $\mathcal{I}$ acts from $C_{\bar{r}}$ to $\widetilde{C}_{\bar{r}}$. The properties of $\widetilde{C}_{\bar{r}}$ follow from the intertwining requirement for $\mathcal{I}$ [114]:

$$\tilde{\pi}\mathcal{I} = \mathcal{I}\pi. \tag{4.172}$$

In particular, we have

$$\begin{aligned}\tilde{\pi}(K_I)\tilde{\varphi} = u^{-\frac{1}{2}d_I r_I} & \sum_{\substack{\nu_{\gamma k}\in\{0,1\}\\ \nu_{ji},\nu_{\beta\alpha}\in\mathbb{Z}_+}} u^{\frac{1}{2}\nu_{ji}(\delta_{I+1,j}-\delta_{I+1,i}-\delta_{Ij}+\delta_{Ii})} \\ & \times u^{\frac{1}{2}\nu_{\gamma k}d_I(\frac{d_I}{d_{I+1}}(\delta_{I+1,\beta}-\delta_{I+1,\alpha})-\delta_{I\beta}+\delta_{I\alpha})} \\ & \times u^{\frac{1}{2}\nu_{\beta\alpha}d_I(\frac{d_I}{d_{I+1}}(\delta_{I+1,\beta}-\delta_{I+1,\alpha})-\delta_{I\beta}+\delta_{I\alpha})}\bar{m}_{\bar{\nu}}f_{\bar{\nu}}.\end{aligned} \tag{4.173}$$

- We finish by noting that the functions $\tilde{\varphi}$ have the important advantage that the representation action $\tilde{\pi}$ can be extended to arbitrary complex r_I. This is seen, e. g., from (4.173).

Outlook

The representations constructed in this section have many applications. The most interesting ones seem to be connected with the case of the multiparameter quantum conformal supergroup which is a real form of $\mathcal{U}'$ for $m = 4$, i. e., of $U_{u\mathbf{q}}(\mathrm{sl}(4/N))$. In this case the non-commuting variables Y_{ij} contain a deformation of Minkowski space (see Volume 2) which together with the variables $\Gamma_{\alpha i}$ will give a noncommutative N-extended Minkowski superspace. Furthermore, one may analyze the reducibility of our representations and construct invariant super-q-difference equations (e. g., deformed super-Yang–Mills equations), generalizing to the supersymetric case the construction of invariant q-difference equations given in [118, 120]. A separate line of investigation may be the construction of the positive energy unitary irreducible representations of deformed extended conformal supersymmetry, e. g., the massless superconformal representations, generalizing, e. g., the massless conformal representations; cf. Section 2.2.4. Naturally, we expect as usually the generalizations to supersymmetry to exhibit much richer structures than the nonsupersymmetric ones. We finish by stressing that these are only some examples of future developments and applications.

Bibliography

[1] E. Abe, *Hopf Algebras*, Cambridge Tracts in Math., Vol. 74, (Cambridge Univ. Press, 1980).

[2] O. Aharony, L. Berdichevsky, M. Berkooz, Y. Hochberg, D. Robles-Llana, "Inverted sparticle hierarchies from natural particle hierarchies", Phys. Rev. D **81** (2010) 085006. O. Aharony, M. Berkooz, S. J. Rey, "Rigid holography and six-dimensional N = (2,0) theories on $AdS_5 \times S^1$", J. High Energy Phys. **1503** (2015) 121. O. Aharony, M. Evtikhiev, "On four dimensional N = 3 superconformal theories", J. High Energy Phys. **1604** (2016) 040.

[3] O. Aharony, O. Bergman, D. L. Jafferis, J. Maldacena, "N = 6 superconformal Chern-Simons-matter theories, M2-branes and their gravity duals", J. High Energy Phys. **0810** (2008) 091.

[4] O. Aharony, S. S. Gubser, J. Maldacena, H. Ooguri, Y. Oz, "Large N field theories, string theory and gravity", Phys. Rep. **323** (2000) 184–386.

[5] M. Ali-Akbari, M. M. Sheikh-Jabbari, M. Torabian, "Tiny graviton matrix theory/SYM correspondence: analysis of BPS states", Phys. Rev. D **74** (2006) 066005.

[6] A. Alldridge, "Fréchet globalisations of Harish-Chandra supermodules", Int. Math. Res. Not. **17** (2017), 5182–5232.

[7] S. Ananth, S. Kovacs, S. Parikh, "Gauge-invariant correlation functions in light-cone superspace", J. High Energy Phys. **1205** (2012) 96.

[8] T. Andrade, C. F. Uhlemann, "Beyond the unitarity bound in AdS/CFT(A)dS", J. High Energy Phys. **1201** (2012) 123.

[9] L. Andrianopoli, S. Ferrara, E. Sokatchev, B. Zupnik, "Shortening of primary operators in N-extended $SCFT_4$ and harmonic-superspace analyticity", Adv. Theor. Math. Phys. **4** (2000) 1149–1197.

[10] D. Anselmi, M. Grisaru, A. Johansen, "A critical behaviour of anomalous currents, electric-magnetic universality and CFT in four dimensions", Nucl. Phys. B **491** (1997) 221–248.

[11] O. Antipin, K. Tuominen, "Resizing the conformal window: a beta function ansatz", Phys. Rev. D **81** (2010) 076011; "Constraints on conformal windows from holographic duals", Mod. Phys. Lett. A **26** (2011) 2227–2246.

[12] Y. Aoki, T. Aoyama, M. Kurachi, T. Maskawa, K.-i. Nagai, H. Ohki, A. Shibata, K. Yamawaki (LatKMI Collaboration), "Lattice study of conformality in twelve-flavor QCD", Phys. Rev. D **86** (2012) 054506.

[13] F. Aprile, J. M. Drummond, P. Heslop, H. Paul, "Unmixing supergravity", J. High Energy Phys. **1802** (2018) 133.

[14] A. A. Ardehali, J. T. Liu, P. Szepietowski, "The spectrum of IIB supergravity on $AdS_5 \times S^5/Z_3$ and a $1/N^2$ test of AdS/CFT", J. High Energy Phys. **1306** (2013) 024; "The shortened KK spectrum of IIB supergravity on $Y^{p,q}$", J. High Energy Phys. **1402** (2014) 064.

[15] P. Argyres, M. Lotito, Y. Lü, M. Martone, "Geometric constraints on the space of N = 2 SCFTs. I: physical constraints on relevant deformations", J. High Energy Phys. **1802** (2018) 001; "II: Construction of special Kähler geometries and RG flows", J. High Energy Phys. **1802** (2018) 002; "III: Enhanced Coulomb branches and central charges", J. High Energy Phys. **1802** (2018) 003.

[16] P. C. Argyres, M. Presser, N. Seiberg, E. Witten, "New $N = 2$ superconformal field theories in four dimensions", Nucl. Phys. B **461** (1996) 71–84.

[17] N. Arkani-Hamed, F. Cachazo, C. Cheung, "The Grassmannian origin of dual superconformal invariance", J. High Energy Phys. **1003** (2010) 036.

[18] D. Arnaudon, J. Avan, N. Crampe, A. Doikou, L. Frappat, E. Ragoucy, "General boundary conditions for the sl(N) and sl(M/N) open spin chains", J. Stat. Mech. **0408** (2004) P08005.

https://doi.org/10.1515/9783110527490-005

[19] G. Arutyunov, B. Eden, A. C. Petkou, E. Sokatchev, “Exceptional nonrenormalization properties and OPE analysis of chiral four point functions in N = 4 SYM(4)”, Nucl. Phys. B **620** (2002) 380–404.

[20] G. Arutyunov, B. Eden, E. Sokatchev, “On nonrenormalization and OPE in superconformal field theories”, Nucl. Phys. B **619** (2001) 359–372.

[21] G. Arutyunov, S. Penati, A. C. Petkou, A. Santambrogio, E. Sokatchev, “Non-protected operators in N = 4 SYM and multiparticle states of AdS_5 SUGRA”, Nucl. Phys. B **643** (2002) 49–78.

[22] G. Arutyunov, E. Sokatchev, “Implications of superconformal symmetry for interacting (2,0) tensor multiplets”, Nucl. Phys. B **635** (2002) 3–32; “A note on the perturbative properties of BPS operators”, Class. Quantum Gravity **20** (2003) L123–L131.

[23] V. Asnin, “On metric geometry of conformal moduli spaces of four-dimensional superconformal theories”, J. High Energy Phys. **1009** (2010) 012.

[24] A. Babichenko, B. Stefanski, K. Zarembo, “Integrability and the AdS(3)/CFT(2) correspondence”, J. High Energy Phys. **1003** (2010) 058.

[25] V. Balasubramanian, P. Kraus, “A Stress tensor for Anti-de Sitter gravity”, Commun. Math. Phys. **208** (1999) 413–428.

[26] I. Bandos, J. Lukierski, “Tensorial central charges and new superparticle models with fundamental spinor coordinates”, Mod. Phys. Lett. A **14** (1999) 1257–1272.

[27] P. Banerjee, P. K. Dhani, M. Mahakhud, V. Ravindran, S. Seth, “Finite remainders of Konishi at two loops in $\mathcal{N} = 4$ SYM”, J. High Energy Phys. **1705** (2017) 085.

[28] I. Bars and M. Gunaydin, “Unitary representations of noncompact supergroups, ” Commun. Math. Phys. **91** (1983) 31–51.

[29] M. Batchelor, “The structure of supermanifolds”, Trans. Am. Math. Soc. **253** (1979) 329–338; “Two approaches to supermanifolds”, Trans. Amer. Math. Soc. **268** (1980) 257–270.

[30] L. Baulieu, G. Bossard, “Superconformal invariance from N = 2 supersymmetry Ward identities”, J. High Energy Phys. **0802** (2008) 075.

[31] A. Barabanschikov, L. Grant, L. L. Huang, S. Raju, “The spectrum of Yang Mills on a sphere”, J. High Energy Phys. **0601** (2006) 160.

[32] M. Beccaria, A. A. Tseytlin, “Higher spins in AdS_5 at one loop: vacuum energy, boundary conformal anomalies and AdS/CFT”, J. High Energy Phys. **1411** (2014) 114; “Conformal anomaly c-coefficients of superconformal 6d theories”, J. High Energy Phys. **1601** (2016) 001; “C_T for higher derivative conformal fields and anomalies of (1,0) superconformal 6d theories”, J. High Energy Phys. **1706** (2017) 002.

[33] M. Bednar, C. Burdik, M. Couture, L. Hlavaty, “On the quantum symmetries associated with the two parameter free fermion model”, J. Phys. A **25** (1992) L346–L431.

[34] C. Beem, M. Lemos, P. Liendo, et al., “Infinite chiral symmetry in four dimensions”, Commun. Math. Phys. **336** (2015) 1359–1443. C. Beem, M. Lemos, P. Liendo, L. Rastelli, B. C. van Rees, “The $\mathcal{N} = 2$ superconformal bootstrap”, J. High Energy Phys. **1603** (2016) 183. C. Beem, M. Lemos, L. Rastelli, B. C. van Rees, “The (2,0) superconformal bootstrap”, Phys. Rev. D **93** (2016) 025016; C. Beem, L. Rastelli, B. C. van Rees, “The $\mathcal{N} = 4$ superconformal bootstrap”, Phys. Rev. Lett. **111** (2013) 071601; “W Symmetry in six dimensions”, J. High Energy Phys. **1505** (2015) 017.

[35] N. Beisert, “BMN operators and superconformal symmetry”, Nucl. Phys. B **659** (2003) 79–118; “The complete one-loop dilatation operator of N = 4 Super Yang-Mills theory”, Nucl. Phys. B **676** (2004) 3–42; “The dilatation operator of N = 4 Super Yang-Mills theory and integrability”, Phys. Rept. **405** (2005) 1–202; “Review of AdS/CFT integrability, chapter VI.1: superconformal symmetry”, Lett. Math. Phys. **99** (2012) 529–545.

[36] N. Beisert, "The SU(2/2) dynamic S-matrix", Adv. Theor. Math. Phys. **12** (2008) 945–979; "The analytic Bethe Ansatz for a chain with centrally extended su(2/2) symmetry", J. Stat. Mech. **0701** (2007) P01017.

[37] N. Beisert, M. Bianchi, J. F. Morales, H. Samtleben, "On the spectrum of AdS/CFT beyond supergravity", J. High Energy Phys. **0402** (2004) 001; "Higher spin symmetry and N = 4 SYM", J. High Energy Phys. **0407** (2004) 058.

[38] N. Beisert, H. Elvang, D. Z. Freedman, M. Kiermaier, A.Morales, S. Stieberger, "E7(7) constraints on counterterms in N = 8 supergravity", Phys. Lett. B **694** (2010) 265–271.

[39] N. Beisert, A. Garus, "Yangian algebra and correlation functions in planar gauge theories", arXiv:1804.09110 [hep-th].

[40] N. Beisert, C. Kristjansen, M. Staudacher, "The dilatation operator of N = 4 Super Yang-Mills theory", Nucl. Phys. B **664** (2003) 131–184. N. Beisert, M. Staudacher, "The N = 4 SYM integrable super spin chain", Nucl. Phys. B **670** (2003) 439–463.

[41] A. A. Belavin, A. M. Polyakov, A. B. Zamolodchikov, "Infinite conformal symmetry in two-dimensional quantum field theory", Nucl. Phys. B **241** (1984) 333–380.

[42] M. Benna, I. Klebanov, T. Klose, M. Smedback, "Superconformal Chern Simons theories and AdS(4)/CFT(3) correspondence", J. High Energy Phys. **0809** (2008) 072.

[43] S. Benvenuti, G. Bonelli, M. Ronzani, A. Tanzini, "Symmetry enhancements via 5d instantons, qW-algebrae and (1,0) superconformal index", J. High Energy Phys. **1609** (2016) 053.

[44] S. Benvenuti, D. Rodriguez-Gomez, E. Tonni, H. Verlinde, "N = 8 superconformal gauge theories and M2 branes", J. High Energy Phys. **0901** (2009) 078.

[45] F. A. Berezin, "Mathematical base of supersymmetrical field theories. (in Russian), " Yad. Fiz. **29** (1979) 1670–1687; "Differential forms on supermanifolds. (in Russian)", Yad. Fiz. **30** (1979) 1168–1174.

[46] F. A. Berezin, G. Kac, "Lie groups with commuting and anticommuting parameters", Math. USSR Sb. **11** (1970) 311–326.

[47] F. A. Berezin, D. A. Leites, "Supervarieties", Sov. Math. Dokl. **16** (1975) 1218–1222.

[48] F. A. Berezin, V. N. Tolstoy, "The group with Grassmann structure Uosp(1,2)", Commun. Math. Phys. **78** (1981) 409–428.

[49] E. Bergshoeff, E. Sezgin, A. Van Proeyen, "Superconformal tensor calculus and matter couplings in six-dimensions", Nucl. Phys. B **264** (1986) 653–686; Erratum: [Nucl. Phys. B **598** (2001) 667].

[50] E. Bergshoeff, A. Van Proeyen, "The Many faces of Osp(1|32)", Class. Quantum Gravity **17** (2000) 3277–3304.

[51] M. Berkooz, "A comment on nonchiral operators in SQCD and its dual", Nucl. Phys. B **466** (1996) 75–84.

[52] N. Berkovits, J. Maldacena, "Fermionic T-duality, dual superconformal symmetry, and the amplitude/Wilson loop connection", J. High Energy Phys. **0809** (2008) 062.

[53] I. N. Bernstein, I. M. Gel'fand, S. I. Gel'fand, "Structure of representations generated by highest weight vectors", Funkc. Anal. Prilozh. **5**(1) 1–9(1971); English translation: Funct. Anal. Appl. **5** (1971) 1–8.

[54] I. N. Bernstein, D. A. Leites, "A formula for the characters of the irreducible finite-dimensional representations of Lie superalgebras of series gl and sl", C. R. Acad. Bulg. Sci. **33** (1980) 1049–1051.

[55] J. Bhattacharya, S. Bhattacharyya, S. Minwalla, S. Raju, "Indices for superconformal field theories in 3, 5 and 6 dimensions", J. High Energy Phys. **0802** (2008) 064.

[56] M. Bianchi, "Higher spin symmetry (breaking) in N = 4 SYM theory and holography", C. R. Phys. **5** (2004) 1091–1099; "Higher spins and stringy $AdS_5 \times S^5$", Fortsch. Phys. **53** (2005) 665–691. M. Bianchi, V. Didenko, "Massive higher spin multiplets and holography",

in: Presented at 1st Solvay Workshop on Higher Spin Gauge Theories Brussels, 12–14 May 2004, (Brussels: Intl. Solvay Inst. 2004) pp. 1–20. M. Bianchi, B. Eden, G. Rossi, Ya. S. Stanev, "On Operator Mixing in N = 4 SYM", Nucl. Phys. B **646** (2002) 69–101. M. Bianchi, P. J. Heslop, F. Riccioni, "More on la Grande Bouffe", J. High Energy Phys. **0508** (2005) 088. M. Bianchi, S. Kovacs, "Nonrenormalization of extremal correlators in N = 4 SYM Theory", Phys. Lett. B **468** (1999) 102–110. M. Bianchi, S. Kovacs, G. Rossi, "Instantons and supersymmetry", Lect. Notes in Physics **737** (2008) 303–470.

[57] M. Bianchi, F. A. Dolan, P. J. Heslop, H. Osborn, "N = 4 superconformal characters and partition functions", Nucl. Phys. B **767** (2007) 163–226.

[58] M. Bianchi, S. Kovacs, G. Rossi, Y. S. Stanev, "On the logarithmic behaviour in N = 4 SYM theory", J. High Energy Phys. **9908** (1999) 020; "Anomalous dimensions in $N = 4$ SYM theory at order G^4", Nucl. Phys. B **584** (2000) 216–232; "Properties of the Konishi multiplet in N = 4 SYM theory", J. High Energy Phys. **0105** (2001) 042.

[59] A. Bissi, T. Lukowski, "Revisiting N = 4 superconformal blocks", J. High Energy Phys. **1602** (2016) 115.

[60] N. Bobev, S. El-Showk, D. Mazac, M. F. Paulos, "Bootstrapping SCFTs with four supercharges", J. High Energy Phys. **1508** (2015) 142. N. Bobev, E. Lauria, D. Mazac, "Superconformal blocks for SCFTs with eight supercharges", J. High Energy Phys. **1707** (2017) 061.

[61] A. Bohm, "Spectrum supersymmetry of Regge trajectories", Phys. Rev. Lett. **57** (1986) 1203–1206; Proceedings XV Intern. Conf. on Diff. Geom. Methods in Theor. Physics, Clausthal, Germany, July 1986, eds. H. D. Doebner, J. Hennig (World Sci, Singapore, 1987) pp. 317–342. A. Bohm, M. Kmiecik, L. J. Boya, "Representation theory of superconformal quantum mechanics", J. Math. Phys. **29** (1988) 1163–1170.

[62] F. Bonetti, T. W. Grimm, S. Hohenegger, "Non-Abelian tensor towers and (2,0) superconformal theories", J. High Energy Phys. **1305** (2013) 129.

[63] G. Bossard, P. S. Howe, K. S. Stelle, P. Vanhove, "The vanishing volume of D = 4 superspace", Class. Quantum Gravity **28** (2011) 215005. G. Bossard, P. S. Howe, K. S. Stelle, "Invariants and divergences in half-maximal supergravity theories", J. High Energy Phys. **1307** (2013) 117.

[64] A. J. Bracken, M. D. Gould, R. B. Zhang, "Quantum supergroups and solutions of the Yang-Baxter equation", Mod. Phys. Lett. A **5** (1990) 831–840.

[65] Y. Brihaye, "Quommutator deformations of spl(N,1)", J. Math. Phys. **39** (1997) 4923–4927.

[66] L. Brink, S. Deser, B. Zumino, P. Di Vecchia, P. S. Howe, "Local supersymmetry for spinning particles", Phys. Lett. B **64** (1976) 435–438; Erratum: [Phys. Lett. B **68** (1977) 488]. L. Brink, P. Di Vecchia, P. S. Howe, "A locally supersymmetric and reparametrization invariant action for the spinning string", Phys. Lett. B **65** (1976) 471–474. L. Brink, J. H. Schwarz, J. Scherk, "Supersymmetric Yang-Mills theories", Nucl. Phys. B **121** (1977) 77–92. L. Brink, J. H. Schwarz, "Local complex supersymmetry in two-dimensions", Nucl. Phys. B **121** (1977) 285–295.

[67] T. W. Brown, "Cut-and-join operators and N = 4 super Yang-Mills", J. High Energy Phys. **1005** (2010) 058.

[68] J. Brundan, "Kazhdan-Lusztig polynomials and character formulae for the Lie superalgebra gl(m/n)", J. Am. Math. Soc. **16** (2002) 185–231.

[69] J. Brundan, "Kazhdan-Lusztig polynomials and character formulae for the Lie superalgebra q(n)", Adv. Math. **182** (2004) 28–77.

[70] M. Buican, "Minimal distances between SCFTs", J. High Energy Phys. **1401** (2014) 155. M. Buican, J. Hayling, C. Papageorgakis, "Aspects of superconformal multiplets in D > 4", J. High Energy Phys. **1611** (2016) 091. M. Buican, Z. Laczko, T. Nishinaka, "N = 2 S-duality revisited", J. High Energy Phys. **1709** (2017) 087. M. Buican, T. Nishinaka, "Compact conformal manifolds", J. High Energy Phys. **1501** (2015) 112; "Argyres-Douglas theories, S^1 reductions, and topological symmetries", J. Phys. A **49** (2016) 045401; "Argyres-Douglas theories, the

Macdonald index, and an RG inequality", J. High Energy Phys. **1602** (2016) 159; "Conformal manifolds in four dimensions and chiral algebras", J. Phys. A **49** (2016) 465401; "On irregular singularity wave functions and superconformal indices", J. High Energy Phys. **1709** (2017) 066. M. Buican, T. Nishinaka, C. Papageorgakis, "Constraints on chiral operators in N = 2 SCFTs", J. High Energy Phys. **1412** (2014) 095.

[71] C. Burdik, R. Tomasek, "The two parameter deformation of the supergroup GL(1/1), its differential calculus and its Lie algebra", Lett. Math. Phys. **26** (1992) 97–103.

[72] P. Candelas, X. C. De La Ossa, P. S. Green, L. Parkes, "A pair of Calabi-Yau manifolds as an exactly soluble superconformal theory", Nucl. Phys. B **359** (1991) 21–74.

[73] L. Carbone, M. Cederwall, J. Palmkvist, "Generators and relations for Lie superalgebras of Cartan type", arXiv:1802.05767 [math.RT].

[74] C. Carmeli, G. Cassinelli, A. Toigo, V. S. Varadarajan, "Unitary representations of super Lie groups and applications to the classification and multiplet structure of super particles", Commun. Math. Phys. **263** (2006) 217–258. C. Carmeli, G. Cassinelli, A. Toigo, "Unitary representations of super groups and Mackey theory", in: Lie Theory and Its Applications in Physics VI, eds. H.-D. Doebner, V. K. Dobrev (Heron Press, Sofia, 2006) pp. 269–279. V. S. Varadarajan, "Unitary representations of super Lie groups", Lectures given in Oporto, Portugal, July 20–23, (2006), UCLA preprint, http://www.math.ucla.edu/~vsv/urtr.pdf.

[75] M. Cederwall, J. Palmkvist, "L_∞ algebras for extended geometry from Borcherds superalgebras", arXiv:1804.04377 [hep-th].

[76] A. Ceresole, G. Dall'Agata, R. D'Auria, S. Ferrara, "Superconformal field theories from IIB spectroscopy on $AdS_5 \times T^{11}$", Class. Quantum Gravity **17** (2000) 1017–1025; "Spectrum of type IIB supergravity on $AdS_5 \times T^1 1$", Phys. Rev. D **61** (2000) 066001.

[77] M. Chaichian, P. P. Kulish, "On the method of inverse scattering problem and backlund transformations for supersymmetric equations", Phys. Lett. B **78** (1978) 413–416.

[78] M. Chaichian, P. P. Kulish, "Quantum Lie superalgebras and q oscillators", Phys. Lett. B **234** (1990) 72–80.

[79] A. Chakrabarti, "q analogs of Iu(n) and U(n,1)", J. Math. Phys. **32** (1991) 1227–1234.

[80] D. Chang, I. Philips, L. Rozansky, "R-matrix approach to quantum superalgebras $su_q(m/n)$", J. Math. Phys. **33** (1992) 3710–3715.

[81] S.-J. Cheng, N. Lam, R. B. Zhang, "Character formula for infinite dimensional unitarizable modules of the general linear superalgebra", J. Algebra **273** (2004) 780–805. S.-J. Cheng, W. Wang, "Dualities for Lie superalgebras", Lecture notes for ECNU summer school 2009 in Shanghai, Lie Theory and Representation Theory, Surveys of Modern Mathematics #2, (International Press, Boston, 2012) pp. 1–46.

[82] D. Chicherin, J. Drummond, P. Heslop, E. Sokatchev, "All three-loop four-point correlators of half-BPS operators in planar N = 4 SYM", J. High Energy Phys. **1608** (2016) 053.

[83] C. Closset, T. T. Dumitrescu, G. Festuccia, Z. Komargodski, N. Seiberg, "Contact terms, unitarity, and F-maximization in three-dimensional superconformal theories", J. High Energy Phys. **1210** (2012) 053.

[84] C. Cordova, T. T. Dumitrescu, K. Intriligator, "Deformations of superconformal theories", J. High Energy Phys. **1611** (2016) 135; "Multiplets of Superconformal Symmetry in Diverse Dimensions", arXiv:1612.00809 [hep-th].

[85] M. Cornagliotto, M. Lemos, V. Schomerus, "Long multiplet bootstrap", J. High Energy Phys. **1710** (2017) 119.

[86] J. F. Cornwell, *Group Theory in Physics, Vol. III*, (Academic Press, London & San Diego, 1989).

[87] E. Corrigan, D. B. Fairlie, P. Fletcher, R. Sasaki, "Some aspects of quantum groups and supergroups", J. Math. Phys. **31** (1990) 776–780.

[88] K. Coulembier, "Bernstein-Gelfand-Gelfand resolutions for basic classical Lie superalgebras", J. Algebra, **399** (2014) 131–169.

[89] E. Cremmer, S. Ferrara, L. Girardello, A. Van Proeyen, "Yang-Mills theories with local supersymmetry: Lagrangian, transformation laws and SuperHiggs effect", Nucl. Phys. B **212** (1983) 413–442.

[90] T. Creutzig, P. Gao, A. R. Linshaw, "Fermionic coset, critical level $W_4^{(2)}$-algebra and higher spins", J. High Energy Phys. **1204** (2012) 031.

[91] C. Csaki, P. Meade, J. Terning, "A mixed phase of SUSY gauge theories from A-maximization", J. High Energy Phys. **0404** (2004) 040.

[92] W. Cui, "On crystal bases of two-parameter (v,t)-quantum groups", arXiv:1411.4727 [math.QA] (2014).

[93] T. Curtright, "Massless field supermultiplets with arbitrary spin", Phys. Lett. B **85** (1979) 219–224.

[94] L. Dabrowski, P. Parashar, "h deformation of GL(1/1)", Lett. Math. Phys. **38** (1996) 331–336.

[95] L. Dabrowski, L. Wang, "Two parameter quantum deformation of GL(1/1)", Phys. Lett. B **266** (1991) 51–54.

[96] M. D'Alessandro, L. Genovese, "A wide class of four point functions of BPS operators in N = 4 SYM at order g^4", Nucl. Phys. B **732** (2006) 64–88.

[97] R. D'Auria, S. Ferrara, M. A. Lledo, V. S. Varadarajan, "Spinor algebras", J. Geom. Phys. **40** (2001) 101–128. R. D'Auria, S. Ferrara, M. A. Lledo, "On the embedding of space-time symmetries into simple superalgebras", Lett. Math. Phys. **57** (2001) 123–133. S. Ferrara, M. A. Lledo, "Considerations on superPoincare algebras and their extensions to simple superalgebras", Rev. Math. Phys. **14** (2002) 519–530.

[98] N. Debergh, J. Van der Jeugt, "Realizations of the Lie superalgebra q(2) and applications", J. Phys. A **34** (2001) 8119–8134.

[99] F. Delduc, M. Magro, B. Vicedo, "Derivation of the action and symmetries of the q-deformed $AdS_5 \times S^5$ superstring", J. High Energy Phys. **1410** (2014) 132.

[100] B. de Wit, D. Z. Freedman, "On combined supersymmetric and gauge invariant field theories", Phys. Rev. D **12** (1975) 2286–2297.

[101] E. D'Hoker, J. Erdmenger, D. Z. Freedman, M. Perez-Victoria, "Near extremal correlators and vanishing supergravity couplings in AdS/CFT", Nucl. Phys. B **589** (2000) 3–37.

[102] E. D'Hoker, D. Z. Freedman, "Supersymmetric gauge theories and the AdS/CFT correspondence", Lectures given at Theoretical Advanced Study Institute in Elementary Particle Physics (TASI 2001): Strings, Branes and Extra Dimensions, Boulder, Colorado, 3–29 Jun 2001, in "Boulder 2001, Strings, branes and extra dimensions" pp. 3–158; "General scalar exchange in AdS(d+1)", Nucl. Phys. B **550** (1999) 261–288.

[103] E. D'Hoker, D. Z. Freedman, S. D. Mathur, A. Matusis, L. Rastelli, "Extremal correlators in the AdS/CFT correspondence", in: The Many Faces of the Superworld, Yuri Golfand Memorial Volume, ed. M. Shifman (World Scient., 2000) pp. 332–360.

[104] E. D'Hoker, P. Heslop, P. Howe, A. V. Ryzhov, "Systematics of quarter BPS operators in N = 4 SYM", J. High Energy Phys. **0304** (2003) 038.

[105] E. D'Hoker, B. Piolin, "Near extremal correlators and generalised consistent truncation in $AdS(4/7) \times S^{(7/4)}$", J. High Energy Phys. **0007** (2000) 021.

[106] E. D'Hoker, A. V. Ryzhov, "Three-point functions of quarter BPS operators in N = 4 SYM", J. High Energy Phys. **0202** (2002) 047.

[107] D. D. Dietrich, "A mass-dependent beta-function", Phys. Rev. D **80** (2009) 065032; "Quasiconformality and mass", Phys. Rev. D **82** (2010) 065007.

[108] R. Ding, L. Wang, B. Zhu, "Neutralino dark matter in gauge mediation after run I of LHC and LUX", Phys. Lett. B **733** (2014) 373–379.

[109] P. Di Vecchia, S. Ferrara, "Classical solutions in two-dimensional supersymmetric field theories", Nucl. Phys. B **130** (1977) 93–104. A. D'Adda, P. Di Vecchia, "Supersymmetry and Instantons", Phys. Lett. B **73** (1978) 162–166. A. D'Adda, R. Horsley, P. Di Vecchia, "Supersymmetric magnetic monopoles and dyons", Phys. Lett. B **76** (1978) 298–302.

[110] J. Dixmier, *Enveloping Algebras*, (North Holland, New York, 1977).

[111] V. K. Dobrev, "Elementary representations and intertwining operators for SU(2,2): I", J. Math. Phys. **26** (1985) 235–251; first as ICTP Trieste preprint IC/83/36 (1983).

[112] V. K. Dobrev, "Multiplet classification of the reducible elementary representations of real semisimple Lie groups : the $SO_e(p,q)$ example", Lett. Math. Phys. **9** (1985) 205–211.

[113] V. K. Dobrev, "Multiplet classification of the indecomposable highest weight modules over affine Lie algebras and invariant differential operators: the $A_\ell^{(1)}$ example", Talk at the Conference on Algebraic Geometry and Integrable Systems, Oberwolfach, July 1984 and ICTP, Trieste, preprint, IC/85/9 (1985).

[114] V. K. Dobrev, "Canonical construction of intertwining differential operators associated with representations of real semisimple Lie groups", Rept. Math. Phys. **25** (1988) 159–181; first as ICTP Trieste preprint IC/86/393 (1986).

[115] V. K. Dobrev, "Singular vectors of quantum groups representations for straight Lie algebra roots", Lett. Math. Phys. **22** (1991) 251–266.

[116] V. K. Dobrev, "Duality for the matrix quantum group $GL_{p,q}(2,\mathbb{C})$", J. Math. Phys. **33** (1992) 3419–3430.

[117] V. K. Dobrev, "Canonical q-deformations of noncompact Lie (super-) algebras", J. Phys. A **26** (1993) 1317–1334; first as Göttingen University preprint, (July 1991).

[118] V. K. Dobrev, "q-difference intertwining operators for $U_q(sl(n))$: general setting and the case n = 3", J. Phys. A **27** (1994) 4841–4857 & 6633–6634.

[119] V. K. Dobrev, *Representations of Quantum Groups and q-Deformed Invariant Wave Equations*, Dr. Habil. Thesis, Tech. Univ. Clausthal 1994, (Papierflieger Verlag, Clausthal-Zellerfeld, 1995) ISBN 3-930697-59-9.

[120] V. K. Dobrev, "Subsingular vectors and conditionally invariant (q-deformed) equations", J. Phys. A **28** (1995) 7135–7155.

[121] V. K. Dobrev, "Kazhdan-Lusztig polynomials, subsingular vectors, and conditionally invariant (q-deformed) equations" (invited talk), in: Symposium "Symmetries in Science IX", Bregenz, Austria, (August 1996), Proceedings, eds. B. Gruber and M. Ramek (Plenum Press, New York and London, 1997) pp. 47–80.

[122] V. K. Dobrev, "Positive energy unitary irreducible representations of D = 6 conformal supersymmetry", J. Phys. A **35** (2002) 7079–7100.

[123] V. K. Dobrev, "Characters of the positive energy UIRs of D = 4 conformal supersymmetry", Phys. Part. Nucl. **38** (2007) 564–609.

[124] V. K. Dobrev, "Decompositions of superfields in D = 4 conformal SUSY", Czechoslov. J. Phys. **56** (2006) 1131–1136.

[125] V. K. Dobrev, "Positive energy representations, holomorphic discrete series and finite-dimensional irreps", J. Phys. A **41** (2008) 425206.

[126] V. K. Dobrev, "Note on centrally extended su(2/2) and Serre relations", Fortschr. Phys. **57** (2009) 542–545.

[127] V. K. Dobrev, "Group-theoretical classification of BPS and possibly protected states in D = 4 conformal supersymmetry", Nucl. Phys. B **854**(3) (2012) 878–893.

[128] V. K. Dobrev, "Group-theoretical classification of BPS states in D = 4 conformal supersymmetry: the case of (1/N)-BPS", Phys. Part. Nucl. **43** (2012) 616–620.

[129] V. K. Dobrev, "Explicit character formulae for positive energy UIRs of D = 4 conformal supersymmetry", J. Phys. A **46** (2013) 405202.

[130] V. K. Dobrev, *Invariant Differential Operators, Volume 1: Noncompact Semisimple Lie Algebras and Groups*, De Gruyter Studies in Mathematical Physics, Vol. 35 (De Gruyter, Berlin, Boston, 2016, ISBN 978-3-11-042764-6), 409 pages.

[131] V. K. Dobrev, *Invariant Differential Operators, Volume 2: Quantum Groups*, De Gruyter Studies in Mathematical Physics, Vol. 39 (De Gruyter, Berlin, Boston, 2017, ISBN 978-3-11-043543-6 (h. c.), 978-3-11-042770-7), 395 pages.

[132] V. K. Dobrev, *Invariant Differential Operators, Volume 4: AdS/CFT, Virasoro and Affine (Super-)Algebras*, De Gruyter Studies in Mathematical Physics (De Gruyter, Berlin, Boston, 2018), in preparation.

[133] V. K. Dobrev, J. Lukierski, J. Sobczyk, V. N. Tolstoy, "q-deformed conformal superalgebra and its Hopf subalgebras", ICTP Trieste IC/92/188 (July 1992).

[134] V. K. Dobrev, G. Mack, V. B. Petkova, S. G. Petrova, I. T. Todorov, *Harmonic Analysis on the n-Dimensional Lorentz Group and Its Applications to Conformal Quantum Field Theory*, Lecture Notes in Physics, Vol. 63 (Springer-Verlag, Berlin-Heidelberg-New York, 1977).

[135] V. K. Dobrev, A. M. Miteva, R. B. Zhang, B. S. Zlatev, "On the unitarity of D = 9, 10, 11 conformal supersymmetry", Czechoslov. J. Phys. **54** (2004) 1249–1256.

[136] V. K. Dobrev, P. Parashar, "Duality for multiparametric quantum GL(n)", J. Phys. A **26** (1993) 6991–7002.

[137] V. K. Dobrev, V. B. Petkova, "Elementary representations and intertwining operators for the group $SU^*(4)$", Rep. Math. Phys. **13** (1978) 233–277.

[138] V. K. Dobrev, V. B. Petkova, "On the group-theoretical approach to extended conformal supersymmetry: classification of multiplets", Lett. Math. Phys. **9** (1985) 287–298.

[139] V. K. Dobrev, V. B. Petkova, "On the group-theoretical approach to extended conformal supersymmetry: function space realizations and invariant differential operators", Fortschr. Phys. **35** (1987) 537–572, first as ICTP Trieste preprint IC/85/29 (March 1985).

[140] V. K. Dobrev, V. B. Petkova, "All positive energy unitary irreducible representations of extended conformal supersymmetry", Phys. Lett. B **162** (1985) 127–132.

[141] V. K. Dobrev, V. B. Petkova, "Reducible representations of the extended conformal superalgebra and invariant differential operators" (talk by Petkova V. B.), in: Symposium on Conformal Groups and Structures (Clausthal, August 12–14 1985), Proceedings, eds. A. O. Barut and H. D. Doebner, Lecture Notes in Physics, Vol. 261 (Springer-Verlag, Berlin, 1986) pp. 291–299.

[142] V. K. Dobrev, V. B. Petkova, "All positive energy unitary irreducible representations of the extended conformal superalgebra" (talk by Dobrev V. K.), in: Symposium on Conformal Groups and Structures (Clausthal, August 12–14 1985), Proceedings, eds. A. O. Barut and H. D. Doebner, Lecture Notes in Physics, Vol. 261 (Springer-Verlag, Berlin, 1986) pp. 300–308.

[143] V. K. Dobrev, I. Salom, "Positive energy unitary irreducible representations of the superalgebras osp(1|2n,R) and character formulae" (plenary talk by V. K. D.), in: VIII Mathematical Physics Meeting (Belgrade, 24–31 August 2014), Proceedings, SFIN XXVIII (A1), eds. B. Dragovich et al. (Belgrade Inst. Phys., 2015, ISBN 978-86-82441-43-4), pp.59–81.

[144] V. K. Dobrev, I. Salom, "Positive energy unitary irreducible representations of the superalgebra osp(1|8,R)", Publ. Inst. Math. (Belgr.) **102**(116) (2017) 49–60.

[145] V. K. Dobrev, I. Salom, "Positive energy unitary irreducible representations of the superalgebras osp(1|2n,R) and character formulae for n = 3" (plenary talk by V. K. D.), in: 24-th International Conference on Integrable Systems and Quantum Symmetries (Prague, June 2016), J. Phys.: Conf. Ser., Vol. 804 (2017) 012015.

[146] V. K. Dobrev, E. H. Tahri, "Duality for multiparametric quantum deformation of the supergroup GL(m/n)", Int. J. Mod. Phys. A **13** (1998) 4339–4366.

[147] V. K. Dobrev, E. H. Tahri, "Induced representations of the multiparameter Hopf superalgebras $U_{uq}(gl(m/n))$ and $U_{uq}(sl(m/n))$", J. Phys. A **32** (1999) 4209–4237.
[148] V. K. Dobrev, R. B. Zhang, "Positive energy unitary irreducible representations of the superalgebras osp(1|2n,R)", Phys. At. Nucl. **68** (2005) 1660–1669.
[149] F. A. Dolan, "Character formulae and partition functions in higher dimensional conformal field theory", J. Math. Phys. **47** (2006) 062303; "Counting BPS operators in N = 4 SYM", Nucl. Phys. B **790** (2008) 432–464.
[150] F. A. Dolan, H. Osborn, "On short and semi-short representations for four dimensional superconformal symmetry", Ann. Phys. **307** (2003) 41–89.
[151] L. Dolan, C. R. Nappi, E. Witten, "A relation between approaches to integrability in superconformal Yang-Mills theory", J. High Energy Phys. **0310** (2003) 017; "Yangian symmetry in D = 4 superconformal Yang-Mills theory", in: Proc. 3rd International Symposium "Quantum Theory and Symmetries", Cincinnati, 10–14.9.2003, eds. R. Wijewardhana et al. (World Scientific, Singapore, 2004) pp. 300–315.
[152] N. Dorey, T. J. Hollowood, V. V. Khoze, M. P. Mattis, S. Vandoren, "Multi-instanton calculus and the AdS/CFT correspondence in N = 4 superconformal field theory", Nucl. Phys. B **552** (1999) 88–168.
[153] B. Drabant, M. Schlieker, W. Weich, B. Zumino, "Complex quantum groups and their quantum enveloping algebras", Commun. Math. Phys. **147** (1992) 625–634.
[154] V. G. Drinfeld, "Hopf algebras and the quantum Yang-Baxter equation", Dokl. Akad. Nauk SSSR **283** (1985) 1060–1064 (in Russian); English translation: Soviet. Math., Dokl. **32** (1985) 254–258.
[155] J. M. Drummond, J. Henn, G. P. Korchemsky, E. Sokatchev, "Dual superconformal symmetry of scattering amplitudes in N = 4 super-Yang-Mills theory", Nucl. Phys. B **828** (2010) 317–374.
[156] J. M. Drummond, P. J. Heslop, P. S. Howe, S. F. Kerstan, "Integral invariants in N = 4 SYM and the effective action for coincident D-branes", J. High Energy Phys. **0308** (2003) 016.
[157] B. Eden, S. Ferrara, E. Sokatchev, "(2,0) superconformal OPEs in D = 6, selection rules and non-renormalization theorems", J. High Energy Phys. **0111** (2001) 020.
[158] B. Eden, C. Jarczak, E. Sokatchev, "A three-loop test of the dilatation operator in N = 4 SYM", Nucl. Phys. B **712** (2005) 157–195.
[159] B. Eden, E. Sokatchev, "On the OPE of 1/2 BPS short operators in $N = 4$ $SCFT_4$", Nucl. Phys. B **618** (2001) 259–276.
[160] T. Eguchi, K. Hori, K. Ito, S. K. Yang, "Study of N = 2 superconformal field theories in four-dimensions", Nucl. Phys. B **471** (1996) 430–444.
[161] J. Erdmenger, Z. Guralnik, I. Kirsch, "Four-dimensional superconformal theories with interacting boundaries or defects", Phys. Rev. D **66** (2002) 025020.
[162] L. D. Faddeev, N. Yu. Reshetikhin, L. A. Takhtajan, "Quantization of Lie groups and Lie algebras", Leningr. Math. J. **1** (1990) 193–225; Algebra Anal. **1**(1) (1989) 178–206.
[163] P. Fayet, "Fermi-Bose hypersymmetry", Nucl. Phys. B **113** (1976) 135–155; "Supersymmetry and weak, electromagnetic and strong interactions", Phys. Lett. B **64** (1976) 159–162; "Spontaneous generation of massive multiplets and central charges in extended supersymmetric theories", Nucl. Phys. B **149** (1979) 137–169. G. R. Farrar, P. Fayet, "Phenomenology of the production, decay, and detection of new hadronic states associated with supersymmetry", Phys. Lett. B **76** (1978) 575–579.
[164] A. Ferber, "Supertwistors and conformal supersymmetry", Nucl. Phys. B **132** (1978) 55–64.
[165] S. Fernando, M. Gunaydin, "Minimal unitary representation of SU(2,2) and its deformations as massless conformal fields and their supersymmetric extensions", J. Math. Phys. **51** (2010) 082301; "Minimal unitary representation of SO*(8) = SO(6,2) and its SU(2) deformations as massless 6D conformal fields and their supersymmetric extensions", Nucl. Phys. B **841**

(2010) 339–387; "SU(2) deformations of the minimal unitary representation of OSp(8*/2N) as massless 6D conformal supermultiplets", Nucl. Phys. B **843** (2011) 784–815. S. Fernando, M. Gunaydin, O. Pavlyk, "Spectra of PP-wave limits of M-superstring theory on $AdS_p \times S^q$ spaces", J. High Energy Phys. **0210** (2002) 007.
[166] S. Ferrara, "Superspace representations of SU(2,2/N) superalgebras and multiplet shortening", Talk given at the TMR Conference on Quantum Aspects of Gauge Theories, Supersymmetry and Unification, Paris, 1–7 September 1999, hep-th/0002141.
[167] S. Ferrara, C. Fronsdal, "Conformal fields in higher dimensions", in: Moscow 2000, Quantization, Gauge Theory, and Strings, Vol. 1, eds. A. Semikhatov et al. (Scientific World, Moscow, 2001) pp. 405–426.
[168] S. Ferrara, C. Fronsdal, A. Zaffaroni, "On N = 8 supergravity on AdS(5) and N = 4 superconformal Yang-Mills theory", Nucl. Phys. B **532** (1998) 153–162.
[169] S. Ferrara, A. F. Grillo, G. Parisi, R. Gatto, "The shadow operator formalism for conformal algebra. Vacuum expectation values and operator products", Lett. Nuovo Cimento **4** (1972) 115–120.
[170] S. Ferrara, M. Kaku, P. K. Townsend, P. van Nieuwenhuizen, "Gauging the graded conformal group with unitary internal symmetries", Nucl. Phys. B **129** (1977) 125–134.
[171] S. Ferrara, E. Sokatchev, "Short representations of SU(2,2/N) and harmonic superspace analyticity", Lett. Math. Phys. **52** (2000) 247–262.
[172] S. Ferrara, E. Sokatchev, "Conformal primaries of $osp(8/4, R)$ and BPS states in AdS(4)", J. High Energy Phys. **0005** (2000) 038.
[173] S. Ferrara, E. Sokatchev, "Conformal superfields and BPS states in $AdS_{4/7}$ geometries", Int. J. Mod. Phys. B **14** (2000) 2315–2333.
[174] S. Ferrara, E. Sokatchev, "Superconformal interpretation of BPS states in AdS geometries", Int. J. Theor. Phys. **40** (2001) 935–984.
[175] S. Ferrara, E. Sokatchev, "Representations of superconformal algebras in the $AdS_{7/4}/CFT_{6/3}$ correspondence", J. Math. Phys. **42** (2001) 3015–3026.
[176] S. Ferrara, E. Sokatchev, "Universal properties of superconformal OPEs for 1/2 BPS operators in $3 \leq D \leq 6$", New J. Phys. **4** (2002) #2.
[177] S. Ferrara, A. Zaffaroni, "Superconformal field theories, multiplet shortening, and the $AdS_5/SCFT_4$ correspondence", in: Proceedings of the Moshe Flato Conference, Dijon, 5–8 September 1999, in Dijon 1999, Quantization, Deformations and Symmetries, Vol. 1, pp. 177–188.
[178] S. Ferrara, B. Zumino, "Supergauge invariant Yang-Mills theories", Nucl. Phys. B **79** (1974) 413–421; "Transformation properties of the supercurrent", Nucl. Phys. B **87** (1975) 207–220. S. Ferrara, J. Wess, B. Zumino, "Supergauge multiplets and superfields", Phys. Lett. B **51** (1974) 239–241. P. Fayet, S. Ferrara, "Supersymmetry", Phys. Rept. **32** (1977) 249–334.
[179] R. Fioresi, M. A. Lledo, V. S. Varadarajan, "The Minkowski and conformal superspaces", J. Math. Phys. **48** (2007) 105017.
[180] A. L. Fitzpatrick, J. Kaplan, Z. U. Khandker, D. L. Li, D. Poland, D. Simmons-Duffin, "Covariant approaches to superconformal blocks", J. High Energy Phys. **1408** (2014) 129.
[181] M. Flato, C. Fronsdal, "Quantum field theory of singletons: the rac", J. Math. Phys. **22** (1981) 1100–1105.
[182] M. Flato, C. Fronsdal, "Representations of conformal supersymmetry", Lett. Math. Phys. **8** (1984) 159–162.
[183] R. Floreanini, D. A. Leites, L. Vinet, "On the defining relations of quantum superalgebras", Lett. Math. Phys. **23** (1991) 127–131.
[184] R. Floreanini, V. P. Spiridonov, L. Vinet, "Bosonic realization of the quantum superalgebra $osp_q(1,2n)$", Phys. Lett. B **242** (1990) 383–386; "q-Oscillator realizations of the

quantum superalgebras $sl_q(m,n)$ and $osp_q(m,2n)$", Commun. Math. Phys. **137** (1991) 149–160.

[185] J. Fokken, C. Sieg, M. Wilhelm, "The complete one-loop dilatation operator of planar real beta-deformed N = 4 SYM theory", J. High Energy Phys. **1407** (2014) 150.

[186] E. S. Fradkin, A. A. Tseytlin, "Conformal anomaly in Weyl theory and anomaly free superconformal theories", Phys. Lett. B **134** (1984) 187–193.

[187] L. Frappat, V. Hussin, G. Rideau, "Classification of the quantum deformation of the superalgebra gl(1/1)", J. Phys. A **31** (1998) 4049–4072.

[188] L. Frappat, A. Sciarrino, P. Sorba, "Dynkin-like diagrams and representation of the strange superalgebra P(n)", J. Math. Phys. **32** (1991), 3268–3277.

[189] D. Z. Freedman, S. S. Gubser, K. Pilch, N. P. Warner, "Renormalization group flows from holography supersymmetry and a *c* theorem", Adv. Theor. Math. Phys. **3** (1999) 363–417.

[190] D. Z. Freedman, S. D. Mathur, A. Matusis, L. Rastelli, "Correlation functions in the CFT(d)/AdS(d+1) correspondence", Nucl. Phys. B **546** (1999) 96–118.

[191] D. Z. Freedman, H. Nicolai, "Multiplet shortening in Osp(N,4), " Nucl. Phys. B **237** (1984) 342–366.

[192] P. G. O. Freund, I. Kaplansky, "Simple supersymmetries", J. Math. Phys. **17** (1976) 228–231.

[193] D. Friedan, E. J. Martinec, S. H. Shenker, "Conformal invariance, supersymmetry and string theory", Nucl. Phys. B **271** (1986) 93–165.

[194] D. Gaiotto, S. Giombi, X. Yin, "Spin chains in N = 6 superconformal Chern-Simons-Matter theory", J. High Energy Phys. **0904** (2009) 066.

[195] D. Gaiotto, J. Maldacena, "The gravity duals of N = 2 superconformal field theories", J. High Energy Phys. **1210** (2012) 189.

[196] J. Galloway, J. McRaven, J. Terning, "Anomalies, unparticles, and Seiberg duality", Phys. Rev. D **80** (2009) 105017.

[197] A. Galperin, E. Ivanov, S. Kalitsyn, V. Ogievetsky, E. Sokatchev, "Unconstrained N = 2 matter, Yang-Mills and supergravity theories in harmonic superspace", Class. Quantum Gravity **1** (1984) 469–498; Erratum: [Class. Quantum Gravity **2** (1985) 127]. "Unconstrained off-shell N = 3 supersymmetric Yang-Mills theory", Class. Quantum Gravity **2** (1985) 155–166. A. Galperin, E. Ivanov, V. Ogievetsky, E. Sokatchev, "Harmonic supergraphs. Green functions", Class. Quantum Gravity **2** (1985) 601–616; "Harmonic supergraphs. Feynman rules and examples", Class. Quantum Gravity **2** (1985) 617–630.

[198] E. Gava, G. Milanesi, K. S. Narain, M. O'Loughlin, "1/8 BPS states in AdS/CFT", J. High Energy Phys. **0705** (2007) 030.

[199] L. Genovese, Ya. S. Stanev, "Rationality of the anomalous dimensions in N = 4 SYM theory", Nucl. Phys. B **721** (2005) 212–228.

[200] D. Gepner, E. Witten, "String theory on group manifolds", Nucl. Phys. B **278** (1986) 493–549.

[201] A. Ghodsi, B. Khavari, A. Naseh, "Holographic two-point functions in conformal gravity", J. High Energy Phys. **1501** (2015) 137.

[202] P. H. Ginsparg, "Curiosities at c = 1", Nucl. Phys. B **295** (1988) 153–170.

[203] L. Girardello, A. Giveon, M. Porrati, A. Zaffaroni, "S-duality in N = 4 Yang-Mills theories with general gauge groups", Nucl. Phys. B **448** (1995) 127–165.

[204] W. D. Goldberger, W. Skiba, M. Son, "Superembedding methods for 4d N = 1 SCFTs", Phys. Rev. **D86** (2012) 025019.

[205] Y. A. Golfand, E. P. Likhtman, "Extension of the algebra of Poincare group generators and violation of p invariance", JETP Lett. **13** (1971) 323–326; [Pisma Zh. Eksp. Teor. Fiz. **13** (1971) 452–455].

[206] M. D. Gould, "Classification of infinite dimensional irreducible modules for type I Lie superalgebras", Rep. Math. Phys. **30** (1991) 363–380.

[207] D. Green, Z. Komargodski, N. Seiberg, Yu. Tachikawa, B. Wecht, "Exactly marginal deformations and global symmetries", J. High Energy Phys. **1006** (2010) 106.

[208] D. Green, D. Shih, "Bounds on SCFTs from conformal perturbation theory", J. High Energy Phys. **1209** (2012) 026.

[209] N. Gromov, V. Kazakov, Z. Tsuboi, "PSU(2,2/4) character of quasiclassical AdS/CFT", J. High Energy Phys. **1007** (2010) 097.

[210] P. Grozman, D. A. Leites, "Defining relations for Lie superalgebras with Cartan matrix", Czechoslov. J. Phys. **51** (2001) 1–22.

[211] S. S. Gubser, C. P. Herzog, S. S. Pufu, T. Tesileanu, "Superconductors from superstrings", Phys. Rev. Lett. **103** (2009) 141601.

[212] S. S. Gubser, I. R. Klebanov, A. M. Polyakov, "Gauge theory correlators from noncritical string theory", Phys. Lett. B **428** (1998) 105–114.

[213] D. R. Gulotta, "Properly ordered dimers, *R*-charges, and an efficient inverse algorithm", J. High Energy Phys. **0810** (2008) 014.

[214] M. Gunaydin, "Unitary supermultiplets of OSp(1/32,R) and M-theory", Nucl. Phys. B **528** (1998) 432–450.

[215] M. Gunaydin, N. Marcus, "The spectrum of the S^5 compactification of the chiral N = 2, D = 10 supergravity and the unitary supermultiplets of U(2,2/4)", Class. Quantum Gravity **2** (1985) L11–L18.

[216] M. Gunaydin, P. van Nieuwenhuizen, N. P. Warner, "General construction of the unitary representations of anti-de Sitter superalgebras and the spectrum of the S^4 compactification of eleven-dimensional supergravity", Nucl. Phys. B **255** (1985) 63–92.

[217] M. Gunaydin, L. J. Romans, N. P. Warner, "Compact and noncompact gauged supergravity theories in five-dimensions", Nucl. Phys. B **272** (1986) 598–646.

[218] R. Haag, J. T. Lopuszanski, M. Sohnius, "All possible generators of supersymmetries of the S-matrix", Nucl. Phys. B **88** (1975) 257–274.

[219] P. H. Hai, "On structure of quantum supergroup $GL_q(m/n)$", J. Algebra **211** (1999) 363–383.

[220] A. Hanany, V. Jejjala, S. Ramgoolam, R.-K. Seong, "Consistency and derangements in brane tilings", J. Phys. A **49** (2016) 355401.

[221] T. Harmark, "Interacting giant gravitons from spin matrix theory", Phys. Rev. D **94** (2016) 066001.

[222] I. Heckenberger, F. Spill, A. Torrielli, H. Yamane, "Drinfeld second realization of the quantum affine superalgebras of $D^{(1)}(2,1:x)$ via the Weyl groupoid", Publ. Res. Inst. Math. Sci. Kyoto B **8** (2008) 171–216.

[223] I. Heckenberger, H. Yamane, "A generalization of Coxeter groups, root systems, and Matsumoto's theorem", Math. Z. **259** (2008) 255–276.

[224] W. Heidenreich, "All linear unitary irreducible representations of de Sitter supersymmetry with positive energy", Phys. Lett. B **110** (1982) 461–464.

[225] S. Hellerman, Sh. Maeda, M. Watanabe, "Operator dimensions from moduli", J. High Energy Phys. **1710** (2017) 089.

[226] J. Henn, C. Jarczak, E. Sokatchev, "On twist-two operators in N = 4 SYM", Nucl. Phys. B **730** (2005) 191–209.

[227] P. J. Heslop, "Superfield representations of superconformal groups", Class. Quantum Gravity **19** (2002) 303–346; "Aspects of superconformal field theories in six dimensions", J. High Energy Phys. **0407** (2004) 056.

[228] P. J. Heslop, P. S. Howe, "On harmonic superspaces and superconformal fields in four dimensions", Class. Quantum Gravity **17** (2000) 3743–3768; "A note on composite operators in N = 4 SYM", Phys. Lett. B **516** (2001) 367–375; "OPEs and 3-point correlators of protected

operators in N = 4 SYM", Nucl. Phys. B **626** (2002) 265–286; "Aspects of N = 4 SYM", J. High Energy Phys. **0401** (2004) 058.

[229] H. Hinrichsen, V. Rittenberg, "A two parameter deformation of the SU(1/1) superalgebra and the XY quantum chain in a magnetic field", Phys. Lett. B **275** (1992) 350–354.

[230] L. Hlavaty, "A remark on quantum supergroups", Mod. Phys. Lett. A **7** (1992) 3365–3372.

[231] P. Horava, E. Witten, "Heterotic and type I string dynamics from eleven dimensions", Nucl. Phys. B **460** (1996) 506–524.

[232] P. S. Howe, U. Lindström, "Super-Laplacians and their symmetries", J. High Energy Phys. **1705** (2017) 119.

[233] P. S. Howe, K. S. Stelle, P. K. Townsend, "Supercurrents", Nucl. Phys. B **192** (1981) 332–352.

[234] S. Hu, T. Li, "Radial quantization of the 3d CFT and the higher spin/vector model duality", Int. J. Mod. Phys. A **29** (2014) 1450147.

[235] K. Intriligator, N. Seiberg, "Lectures on supersymmetric gauge theory of electric-magnetic duality", Nucl. Phys. B, Proc. Suppl. **45** (1996) 1–28.

[236] K. Iohara, Y. Koga, "Central extension of Lie superalgebras", Comment. Math. Helv. **76** (2001) 110–154.

[237] E. A. Ivanov, A. A. Kapustnikov, "General relationship between linear and nonlinear realizations of supersymmetry", J. Phys. A **11** (1978) 2375–2384. E. A. Ivanov, A. S. Sorin, "The structure of representations of conformal supergroup in the osp(1,4) basis", Teor. Mat. Fiz. **45** (1980) 30–45; [Theor. Math. Phys. **45** (1980) 862–873]; "Superfield formulation of osp(1,4) supersymmetry", J. Phys. A **13** (1980) 1159–1188.

[238] M. Jarvinen, F. Sannino, "Holographic conformal window – a bottom up approach", J. High Energy Phys. **1005** (2010) 041.

[239] C.-Y. Ju, W. Siegel, "Systematizing semi-shortening", Phys. Rev. D **90** (2014) 125004.

[240] V. G. Kac, "Simple irreducible graded Lie algebras of finite growth", Math. USSR, Izv. **2** (1968) 1271–1311.

[241] V. G. Kac, "Lie superalgebras", Adv. Math. **26** (1977) 8–96. V. G. Kac, "A sketch of Lie superalgebra theory", Commun. Math. Phys. **53** (1977) 31–64; the second paper is an adaptation for physicists of the first paper.

[242] V. G. Kac, "Characters of typical representations of classical Lie superalgebras", Commun. Algebra **5** (1977) 889–897.

[243] V. G. Kac, *Representations of Classical Lie Superalgebras*, Lect. Notes in Math., Vol. **676** (Springer-Verlag, Berlin, 1978) pp. 597–626.

[244] V. G. Kac, *Infinite-Dimensional Lie Algebras. An Introduction*, Progr. Math., Vol. 44 (Birkhäuser, Boston, 1983).

[245] V. G. Kac, private communication (May 1991).

[246] V. G. Kac, D. Kazhdan, "Structure of representations with highest weight of infinite-dimensional Lie algebras", Adv. Math. **34** (1979) 97–108.

[247] V. G. Kac, I. T. Todorov, "Superconformal current algebras and their unitary representations", Commun. Math. Phys. **102** (1985) 337–347.

[248] M. Kaku, P. K. Townsend, P. van Nieuwenhuizen, "Gauge theory of the conformal and superconformal group", Phys. Lett. B **69** (1977) 304–308.

[249] V. Karimipour, A. Aghamohammadi, "Multiparametric quantization of the special linear superalgebra", J. Math. Phys. **34** (1993) 2561–2571.

[250] A. Kato, "Classification of modular invariant partition functions in two-dimensions", Mod. Phys. Lett. A **2** (1987) 585–600; "Zonotopes and four-dimensional superconformal field theories", J. High Energy Phys. **0706** (2007) 037.

[251] L. H. Kauffman, H. Saleur, "Free fermions and the Alexander-Conway polynomial", Commun. Math. Phys. **141** (1991) 293–327.

[252] T. Kawano, F. Yagi, "a-maximization in N = 1 supersymmetric Spin(10) gauge theories", Int. J. Mod. Phys. A **25**(31) (2010) 5595–5645.

[253] D. Kazhdan, G. Lusztig, "Representations of Coxeter groups and Hecke algebras", Invent. Math. **53** (1979) 165–184.

[254] S. M. Khoroshkin, V. N. Tolstoy, "Universal R-matrix for quantized (super) algebras", Commun. Math. Phys. **141** (1991) 599–617.

[255] S. M. Khoroshkin, V. N. Tolstoy, "The Cartan-Weyl basis and the universal R-matrix for quantum Kac-Moody algebras and superalgebras", in: "Quantum Symmetries", Proceedings, Workshop on Quantum Groups at II Wigner Symposium (Goslar, 1991), eds. H.-D. Doebner, V. K. Dobrev (World Sci., Singapore, 1993) pp. 336–351.

[256] H. J. Kim, L. J. Romans, P. van Nieuwenhuizen, "The mass spectrum of chiral N = 2 D = 10 supergravity on S^5", Phys. Rev. D **32** (1985) 389–399.

[257] J. Kinney, J. Maldacena, Sh. Minwalla, S. Raju, "An index for 4 dimensional superconformal theories", Commun. Math. Phys. **275** (2007) 209–254.

[258] I. R. Klebanov, E. Witten, "Superconformal field theory on three-branes at a Calabi-Yau singularity", Nucl. Phys. B **536** (1998) 199–218.

[259] A. W. Knapp, G. J. Zuckerman, *Classification Theorems for Representations of Semisimple Groups*, Lecture Notes in Math., Vol. 587 (Springer, Berlin, 1977) pp. 138–159; "Classification of irreducible tempered representations of semisimple groups", Ann. Math. **116** (1982) 389–501.

[260] H. Knuth, "On invariants and scalar chiral correlation functions in N = 1 superconformal field theories", Int. J. Mod. Phys. A **26** (2011) 2007–2025.

[261] T. Kobayashi, "Algebraic analysis of minimal representations", Publ. RIMS **47** (2011) 585–611.

[262] T. Kobayashi, T. Uematsu, "Differential calculus on the quantum superspace and deformation of phase space", Z. Phys. C **56** (1992) 193–200.

[263] S. Komata, K. Mohri, H. Nohara, "Classical and quantum extended superconformal algebra", Nucl. Phys. B **359** (1991) 168–200.

[264] K. Konishi, "Anomalous supersymmetry transformation of some composite operators in SQCD", Phys. Lett. B **135** (1984) 439–444.

[265] B. Kostant, *Graded Manifolds, Graded Lie Theory and Prequantization*, Springer Lect. Notes Math., Vol. 570 (1977) pp. 177–306; "Harmonic analysis on graded or super Lie groups", Springer Lect. Notes Phys. Vol. 79 (1978) pp. 47–50.

[266] S. Kovacs, "On instanton contributions to anomalous dimensions in N = 4 superymetric Yang-Mills theory", Nucl. Phys. B **684** (2004) 3–74.

[267] J. Kubo, D. Suematsu, "Suppressing the μ and neutrino masses by a superconformal force", Phys. Rev. D **64** (2001) 115014.

[268] T. Kugo, S. Uehara, "N = 1 superconformal tensor calculus: multiplets with external Lorentz indices and spinor derivative operators", Prog. Theor. Phys. **73** (1985) 235–264.

[269] J. Kujawa, "Crystal structures arising from representations of GL(m/n)", Represent. Theory **10** (2006) 49–85.

[270] R. P. Langlands, "On the classification of irreducible representations of real algebraic groups", in: Representation Theory and Harmonic Analysis on Semi-Simple Lie Groups, eds. P. Sally, D. A. Vogan, Jr., Math. Surveys and Monographs, Vol. 31 (AMS, 1989) pp. 101–170; (first as IAS Princeton preprint, 1973).

[271] J. Lee, M. Yamazaki, "Gauging and decoupling in 3d $N = 2$ dualities", J. High Energy Phys. **1606** (2016) 077.

[272] S. Lee, S. Minwalla, "Three point functions of chiral operators in d = 4, N = 4 SYM at large N", Adv. Theor. Math. Phys. **2** (1998) 697–718. S. Lee, J. H. Park, "Noncentral extension of the

$AdS_5 \times S^5$ superalgebra: supermultiplet of brane charges", J. High Energy Phys. **0406** (2004) 038.
[273] S. Lee, S. Minwalla, M. Rangamani, N. Seiberg, "Three point functions of chiral operators in D = 4, N = 4 SYM at large N", Adv. Theor. Math. Phys. **2** (1998) 697–718.
[274] M. de Leeuw, T. Matsumoto, S. Moriyama, V. Regelskis, A. Torrielli, "Secret symmetries in AdS/CFT", Phys. Scr. **86** (2012) , 028502.
[275] R. G. Leigh, A. C. Petkou, "Holography of the N = 1 higher-spin theory on AdS4", J. High Energy Phys. **0306** (2003) 011; "SL(2,Z) action on three-dimensional CFTs and holography", J. High Energy Phys. **0312** (2003) 020.
[276] R. G. Leigh, M. J. Strassler, "Exactly marginal operators and duality in four-dimensional n = 1 supersymmetric gauge theory", Nucl. Phys. B **447** (1995) 95–136.
[277] D. A. Leites, "Representations of Lie superalgebras", Theor. Math. Phys. **52** (1982) 764–766; [Teor. Mat. Fiz. **52** (1982) 225–228].
[278] D. A. Leites, V. V. Serganova, "Defining relations for classical Lie superalgebras. I. Superalgebras with Cartan matrix or Dynkin-type diagram", in: Proc. Topological and Geometrical Methods in Field Theory, ed. J. Mickelsson et al. (World Sci., Singapore, 1992) pp. 194–201.
[279] M. Lemos, P. Liendo, "Bootstrapping N = 2 chiral correlators", J. High Energy Phys. **1601** (2016) 025.
[280] M. Lemos, P. Liendo, C. Meneghelli, V. Mitev, "Bootstrapping $\mathcal{N} = 3$ superconformal theories", J. High Energy Phys. **1704** (2017) 032.
[281] D. Li, A. Stergiou, "Two-point functions of conformal primary operators in $\mathcal{N} = 1$ superconformal theories", J. High Energy Phys. **1410** (2014) 037. D. Li, D. Meltzer, A. Stergiou, "Bootstrapping mixed correlators in 4D $\mathcal{N} = 1$ SCFTs", J. High Energy Phys. **1707** (2017) 029. A. Manenti, A. Stergiou, A. Vichi, "R-current three-point functions in 4d $\mathcal{N} = 1$ superconformal theories", arXiv:1804.09717 [hep-th].
[282] P. Liendo, E. Pomoni, L. Rastelli, "The complete one-loop dilation operator of N = 2 SuperConformal QCD", J. High Energy Phys. **1207** (2012) 3.
[283] S. Lievens, N. I. Stoilova, J. Van der Jeugt, "The paraboson Fock space and unitary irreducible representations of the Lie superalgebra osp(1|2n)", Commun. Math. Phys. **281** (2008) 805–826; "A class unitary irreducible representations of the Lie superalgebra osp(1|2n)", J. Gen. Lie Theory Appl. **2** (2008) 206–210.
[284] U. Lindstrom, M. Rocek, "Constrained local superfields", Phys. Rev. D **19** (1979) 2300–2303.
[285] H. Liu, A. A. Tseytlin, "D = 4 super Yang Mills, D = 5 gauged supergravity and D = 4 conformal supergravity", Nucl. Phys. B **533** (1998) 88–108.
[286] J. T. Liu, B. McPeak, "One-loop holographic Weyl anomaly in six dimensions", J. High Energy Phys. **1801** (2018) 149.
[287] J. Louis, S. Lust, "Supersymmetric AdS_7 backgrounds in half-maximal supergravity and marginal operators of (1,0) SCFTs", J. High Energy Phys. **1510** (2015) 120.
[288] O. Lunin, "Gravitational description of field theories", Nucl. Phys. B, Proc. Suppl. **171** (2007) 99–118.
[289] C. L. Luo, "On polynomial representations of strange Lie superalgebras of Q-type", Acta Math. Sin. Engl. Ser. **32** (2016) 559–570.
[290] M. Luty, M. Schmaltz, "A sequence of duals for Sp(2N) supersymmetric gauge theories with adjoint matter", Phys. Rev. D **54** (1996) 7815–7824.
[291] G. Mack, "All unitary ray representations of the conformal group SU(2,2) with positive energy", Commun. Math. Phys. **55** (1977) 1–28.
[292] J. M. Maldacena, "The large N limit of superconformal field theories and supergravity", Adv. Theor. Math. Phys. **2** (1998) 231–252.

[293] Y. I. Manin, "Multiparametric quantum deformation of the general linear supergroup", Commun. Math. Phys. **123** (1989) 163–175.

[294] M. Mansour, "Super star products and quantum superalgebras", Acta Phys. Pol. B **31** (2000) 1639–1654.

[295] C. Marboe, D. Volin, "The full spectrum of AdS5/CFT4 I: representation theory and one-loop Q-system", J. Phys. A **51** (2018) 165401.

[296] T. Matsumoto, A. Molev, "Representations of centrally extended Lie superalgebra psl(2/2)", J. Math. Phys. **55** (2014) 091704.

[297] R. R. Metsaev, "Massive totally symmetric fields in AdS(d)", Phys. Lett. B **590** (2004) 95–104; "Eleven dimensional supergravity in light cone gauge", Phys. Rev. D **71** (2005) 085017; "Mixed symmetry massive fields in AdS(5)", Class. Quantum Gravity **22** (2005) 2777–2796; "Light-cone formulation of conformal field theory adapted to AdS/CFT correspondence", Phys. Lett. B **636** (2006) 227–233; "Ordinary-derivative formulation of conformal totally symmetric arbitrary spin bosonic fields", J. High Energy Phys. **1206** (2012) 062; "Ordinary-derivative formulation of conformal low spin fields", J. High Energy Phys. **1201** (2012) 064; "Light-cone AdS/CFT-adapted approach to AdS fields/currents, shadows, and conformal fields", J. High Energy Phys. **1510** (2015) 110.

[298] S. G. Mikhov, D. T. Stoyanov, B. L. Aneva, "On some representations of the conformal superalgebra," Theor. Math. Phys. **27** (1977) 502–508; [Teor. Mat. Fiz. **27** (1976) 307–316]; "On some properties of the conformal superalgebra representations", Theor. Math. Phys. **31** (1977) 394–402; [Teor. Mat. Fiz. **31** (1977) 177–189].

[299] G. Milanesi, M. O'Loughlin, "Singularities and closed time-like curves in type IIB 1/2 BPS geometries", J. High Energy Phys. **0509** (2005) 008; "Holography and chronology protection", in: Proc. 2nd Time and Matter conference, August 2007, Bled, Slovenia, (University of Nova Gorica Press, 2008) pp. 235–250, http://tam.ung.si/2007/papers/TAM2007-Proceedings.pdf.

[300] S. P. Milian, "Supermultiplet of β-deformations from twistors", Int. J. Mod. Phys. A **32** (2017) 1750157.

[301] J. A. Minahan, K. Zarembo, "The Bethe ansatz for superconformal Chern-Simons", J. High Energy Phys. **0809** (2008) 040.

[302] S. Minwalla, "Restrictions imposed by superconformal invariance on quantum field theories", Adv. Theor. Math. Phys. **2** (1998) 781–846.

[303] V. V. Molotkov, S. G. Petrova, D. T. Stoyanov, "Representation of superconformal algebra. Two point and three point green functions of scalar superfields, " Theor. Math. Phys. **26** (1976) 125–131; [Teor. Mat. Fiz. **26** (1976) 188–197].

[304] P. Moylan, "Localization and the connection between and $U_q(so(3))$ and $U_q(osp(1/2))$", J. Phys. Conf. Ser. **597** (2015) 012061.

[305] H. Murayama, Ya. Nomura, D. Poland, "More visible effects of the hidden sector", Phys. Rev. D **77** (2008) 015005.

[306] W. Nahm, "Supersymmetries and their representations", Nucl. Phys. B **135** (1978) 149–166.

[307] Y. Nakayama, "Index for orbifold quiver gauge theories", Phys. Lett. B **636** (2006) 132–136; "Index for supergravity on $AdS_5 \times T^{1,1}$ and conifold gauge theory", Nucl. Phys. B **755** (2006) 295–312; "SUSY unparticle and conformal sequestering", Phys. Rev. D **76** (2007) 105009; "Index for non-relativistic superconformal field theories", J. High Energy Phys. **0810** (2008) 083.

[308] S. Nawata, "Localization of N = 4 superconformal field theory on $S^1 \times S^3$ and index", J. High Energy Phys. **1111** (2011) 144.

[309] N. Nekrasov, A. S. Schwarz, "Instantons on noncommutative R^4 and (2,0) superconformal six-dimensional theory", Commun. Math. Phys. **198** (1998) 689–703.

[310] K. H. Neeb, H. Salmasian, “Positive definite superfunctions and unitary representations of Lie supergroups”, Transform. Groups **18** (2013) 803–844.

[311] A. E. Nelson, M. J. Strassler, “A one scale model of dynamical supersymmetry breaking”, Phys. Rev. D **60** (1999) 015004; “Suppressing flavor anarchy”, J. High Energy Phys. **0009** (2000) 030.

[312] T. H. Newton, M. Spradlin, “Quite a character: the spectrum of Yang-Mills on S^3”, Phys. Lett. B **672** (2009) 382–385.

[313] P. van Nieuwenhuizen, “Supergravity”, Phys. Rep. **68** (1981) 189–398; “The actions of the $N = 1$ and $N = 2$ spinning strings as conformal supergravities”, Int. J. Mod. Phys. A **1** (1986) 155–192.

[314] H. P. Nilles, “Supersymmetry, supergravity and particle physics”, Phys. Rep. **110** (1984) 1–162.

[315] V. Ogievetsky, E. Sokatchev, “On vector superfield generated by supercurrent”, Nucl. Phys. B **124** (1977) 309–316; “Structure of supergravity group”, Phys. Lett. B **79** (1978) 222–224; [Czech. J. Phys. B **29** (1979) 68].

[316] T. Ohl, Ch. F. Uhlemann, “Saturating the unitarity bound in AdS/CFT$_{(AdS)}$”, J. High Energy Phys. **1205** (2012) 161.

[317] H. Osborn, “N = 1 superconformal symmetry in four-dimensional quantum field theory”, Ann. Phys. **272** (1999) 243–294.

[318] T. D. Palev, N. I. Stoilova, “Finite dimensional reprsentations of the Lie superalgebra gl(2/2) in a gl(2) × gl(2) basis. 2. Nontypical representations”, J. Math. Phys. **31** (1990) 953–988; “A description of the quantum superalgebra U_q[sl(n+1/m)] via creation and annihilation generators”, J. Phys. A **32** (1999) 1053–1064. T. D. Palev, N. I. Stoilova, J. Van der Jeugt, “Finite-dimensional representations of the quantum superalgebra U_q[gl(n/m)] and related q-identities”, Commun. Math. Phys. **166** (1994) 367–378.

[319] T. D. Palev, V. N. Tolstoy, “Finite-dimensional irreducible representations of the quantum superalgebra U_q[gl (n/1)]”, Commun. Math. Phys. **141** (1991) 549–558.

[320] T. D. Palev, J. Van der Jeugt, “Fock representations of the Lie superalgebra q(n+1)”, J. Phys. A **33** (2000) 2527–2544.

[321] G. Parisi, N. Sourlas, “Random magnetic fields, supersymmetry and negative dimensions”, Phys. Rev. Lett. **43** (1979) 744–745.

[322] J.-H. Park, “Superconformal symmetry and correlation functions”, Nucl. Phys. B **559** (1999) 455–501.

[323] A. Passias, A. Tomasiello, “Spin-2 spectrum of six-dimensional field theories”, J. High Energy Phys. **1612** (2016) 050.

[324] S. Penati, A. Santambrogio, “Superspace approach to anomalous dimensions in $\mathcal{N} = 4$ SYM”, Nucl. Phys. B **614** (2001) 367–387.

[325] I. Penkov, I. Skornyakov, “Cohomologie des D-modules tordus typiques sur les supervarietes de drapeaux”, C. R. Acad. Sci., Sér. 1 Math. **299** (1984) 1005–1008. I. Penkov, “Borel-Weil-Bott theory for classical Lie supergroups” (Russian), Translated in J. Sov. Math. **51** (1990) 2108–2140.

[326] M. Pernici and P. van Nieuwenhuizen, “A covariant action for the SU(2) spinning string as a hyperkahler or quaternionic nonlinear σ model”, Phys. Lett. B **169** (1986) 381–385.

[327] V. B. Petkova, G. M. Sotkov, “The six-point families of exceptional representations of the conformal group”, Lett. Math. Phys. **8** (1984) 217–226.

[328] D. Poland, “The phase structure of supersymmetric $Sp(2N_c)$ gauge theories with an adjoint”, J. High Energy Phys. **0911** (2009) 049; “Superconformal flavor simplified”, J. High Energy Phys. **1005** (2010) 079. D. Poland, D. Simmons-Duffin, “Bounds on 4D conformal and superconformal field theories”, J. High Energy Phys. **1105** (2011) 017. D. Poland, A. Stergiou, “Exploring the minimal 4D N = 1 SCFT”, J. High Energy Phys. **1512** (2015) 121.

[329] G. Policastro, D. T. Son, A. O. Starinets, “The Shear viscosity of strongly coupled N = 4 supersymmetric Yang-Mills plasma”, Phys. Rev. Lett. **87** (2001) 081601.
[330] I. A. Ramírez, “Mixed OPEs in N = 2 superconformal theories”, J. High Energy Phys. **1605** (2016) 043.
[331] L. Rastelli, S. S. Razamat, “The superconformal index of theories of class S”, in: New Dualities of Supersymmetric Gauge Theories, ed. J. Teschner (Springer, 2015) pp. 261–305.
[332] A. Rej, “Integrability and the AdS/CFT correspondence”, J. Phys. A **42** (2009) 254002.
[333] V. Rittenberg, *A Guide to Lie Superalgebras*, Springer Lecture Notes Phys., Vol. 79 (1978) pp. 3–21.
[334] T. A. Ryttov, “The conformal window and walking technicolor”, Nucl. Phys. B, Proc. Suppl. **192–193** (2009) 176–178; “Conformal behavior at four loops and scheme (in)dependence”, Phys. Rev. D **90** (2014) 056007; “Consistent perturbative fixed point calculations in QCD and SQCD”, Phys. Rev. Lett. **117** (2016) 071601.
[335] T. A. Ryttov, F. Sannino, “Conformal windows of SU(N) gauge theories, higher dimensional representations and the size of the unparticle world”, Phys. Rev. D **76** (2007) 105004; “Supersymmetry inspired QCD beta function”, Phys. Rev. D **78** (2008) 065001; “Ultra minimal technicolor and its dark matter TIMP”, Phys. Rev. D **78** (2008) 115010; “Conformal house”, Int. J. of Mod. Phys. A **25** (2010) 4603–4621.
[336] T. A. Ryttov, R. Shrock, “Higher-loop corrections to the infrared evolution of a gauge theory with fermions”, Phys. Rev. D **83** (2011) 056011; “Comparison of some exact and perturbative results for a supersymmetric SU(Nc) gauge theory”, Phys. Rev. D **85** (2012) 076009; “Scheme-independent calculations of physical quantities in an N = 1 supersymmetric gauge theory”, Phys. Rev. D **96** (2017) 105018.
[337] S. Ryu, T. Takayanagi, “Holographic derivation of entanglement entropy from AdS/CFT”, Phys. Rev. Lett. **96** (2006) 181602.
[338] A. V. Ryzhov, “Quarter BPS operators in N = 4 SYM”, J. High Energy Phys. **0111** (2001) 046; “Operators in the d = 4, N = 4 SYM and the AdS/CFT correspondence”, hep-th/0307169, (PhD thesis).
[339] D. Sadri, M. M. Sheikh-Jabbari, “The plane-wave/super Yang-Mills duality”, Rev. Mod. Phys. **76** (2004) 853–907.
[340] A. Salam, J. A. Strathdee, “Supergauge transformations”, Nucl. Phys. B **76** (1974) 477–482; “Unitary representations of supergauge symmetries”, Nucl. Phys. B **80** (1974) 499–505; “Supersymmetry and nonabelian gauges”, Phys. Lett. B **51** (1974) 353–355; “On superfields and fermi-bose symmetry”, Phys. Rev. D **11** (1975) 1521–1535; “Feynman rules for superfields”, Nucl. Phys. B **86** (1975) 142–152; “Supersymmetry and superfields”, Fortsch. Phys. **26** (1978) 57–142.
[341] I. Salom, “Representations and particles of orthosymplectic supersymmetry generalization”, Phys. Part. Nucl. Lett. **11** (2014) 968–970.
[342] F. Sannino, *Dynamical Stabilization of the Fermi Scale: Towards a Composite Universe*, Springer Briefs in Physics, (Springer, 2013); “Conformal windows of SP(2N) and SO(N) gauge theories”, Phys. Rev. D **79** (2009) 096007; “Conformal dynamics for TeV physics and cosmology”, Acta Phys. Polon. B **40** (2009) 3533–3743; “Phase diagrams of strongly interacting theories”, Int. J. Mod. Phys. A **25** (2010) 5145–5161.
[343] J. Scherk, J. H. Schwarz, “Spontaneous breaking of supersymmetry through dimensional reduction”, Phys. Lett. B **82** (1979) 60–64. F. Gliozzi, J. Scherk, D. I. Olive, “Supersymmetry, supergravity theories and the dual spinor model”, Nucl. Phys. B **122** (1977) 253–290.
[344] M. Scheunert, “Serre-type relations for special linear Lie superalgebras”, Lett. Math. Phys. **24** (1992) 173–181; “The quantum supergroup $SPO_q(2n/2m)$ and an $SPO_q(2n/2m)$-covariant quantum Weyl superalgebra”, Bonn-TH-2000-04; math.QA/0004033, 34 pages.

[345] W. B. Schmidke, S. P. Vokos, B. Zumino, "Differential geometry of the quantum supergroup $GL_q(1/1)$", Z. Phys. C **48** (1990) 249–256.
[346] A. S. Schwarz, "On the definition of superspace", Teor. Mat. Fiz. **60** (1984) 37–42; [Theor. Math. Phys. **60** (1985) 657–660].
[347] N. Seiberg, "Observations on the moduli space of superconformal field theories", Nucl. Phys. B **303** (1988) 286–304; "Electric-magnetic duality in supersymmetric non-abelian gauge theories", Nucl. Phys. B **435** (1995) 129–146.
[348] D. Serban, "Integrability and AdS/CFT correspondence", Memoire D'Habilitation, J. Phys. A **44** (2011) 124001.
[349] V. V. Serganova, "Classification of bases of simple Lie superalgebras", Appendix II to the paper: D. A. Leites, M. V. Saveliev, V. V. Serganova, in: Group Theoretical Methods in Physics, Proc. of 3rd Yurmala seminar, 1985 (Nauka, Moscow, 1985, in Russian) pp. 377–394 (English translation: VNU Sci. Press, Utrecht, 1986; pp. 255–298).
[350] V. V. Serganova, "Kazhdan-Lusztig polynomials and character formula for the Lie superalgebra gl(m/n)", Sel. Math. Sov. **2** (1996) 607–654.
[351] E. Sezgin, P. Sundell, "Massless higher spins and holography", Nucl. Phys. B **644** (2002) 303–370; Erratum: [Nucl. Phys. B **660** (2003) 403].
[352] A. D. Shapere, Y. Tachikawa, "Central charges of N = 2 superconformal field theories in four dimensions", J. High Energy Phys. **0809** (2008) 109.
[353] N. N. Shapovalov, "On a bilinear form on the universal enveloping algebra of a complex semisimple Lie algebra", Funkc. Anal. Prilozh. **6**(4) (1972) 65–70; English translation: Funkt. Anal. Appl. **6** (1972) 307–312.
[354] R. Shrock, "Analysis of a zero of a beta function using all-orders summation of diagrams", Phys. Rev. D **91** (2015) 125039.
[355] W. Siegel, "On-shell O(N) supergravity in superspace", Nucl. Phys. B **177** (1981) 325–332.
[356] E. Sokatchev, "Superconformal kinematics and dynamics in the AdS/CFT correspondence", in: Proc. V International Workshop, Varna, 2003, Lie Theory and its Applications in Physics V, eds. H.-D. Doebner, V. K. Dobrev (2004) pp. 135–153.
[357] E. Sokatchev, D. T. Stoyanov, "Nonrelativistic supersymmetry and Lorentz invariance, " Mod. Phys. Lett. A **1** (1986) 577–584. doi:10.1142/S0217732386000737.
[358] A. Solovyov, "Bethe ansatz equations for general orbifolds of N = 4 SYM", J. High Energy Phys. **0804** (2008) 013.
[359] M. F. Sohnius, "Bianchi identities for supersymmetric gauge theories", Nucl. Phys. B **136** (1978) 461–474; "Supersymmetry and central charges", Nucl. Phys. B **138** (1978) 109–121. R. Grimm, M. F. Sohnius, J. Wess, "Extended supersymmetry and gauge theories", Nucl. Phys. B **133** (1978) 275–284.
[360] F. Spill, A. Torrielli, "On Drinfeld's second realization of the AdS/CFT su(2/2) Yangian", J. Geom. Phys. **59** (2009) 489–502.
[361] V. P. Spiridonov, G. S. Vartanov, "Elliptic hypergeometry of supersymmetric dualities II. Orthogonal groups, knots, and vortices", Commun. Math. Phys. **325** (2014) 421–486.
[362] Y. S. Stanev, "Perturbative corrections to anomalous dimensions in N = 4 SYM theory", Published in Rome 2000, Recent developments in theoretical and experimental general relativity, gravitation and relativistic field theories, Pt. B, pp. 1146–1150.
[363] N. I. Stoilova, J. Van der Jeugt, "Gelfand-Zetlin basis and Clebsch-Gordan coefficients for covariant representations of the Lie superalgebra gl(m/n)", J. Math. Phys. **51** (2010) 093523. N. A. Ky, N. I. Stoilova, "Finite dimensional representations of the quantum superalgebra $U_q(gl(2/2))$. 2. Nontypical representations at generic q", J. Math. Phys. **36** (1995) 5979–6003.
[364] Yucai Su, R. B. Zhang, "Character and dimension formulae for general linear superalgebra", Adv. Math. **211** (2007) 1–33.

[365] A. Sudbery, "Canonical differential calculus on quantum general linear groups and supergroups", Phys. Lett. B **284** (1992) 61–65; Erratum: [Phys. Lett. B **291** (1992) 519].

[366] E. H. Tahri, A. El Hassouni, "On the two-parameter quantum supergroups and quantum superplanes", J. Phys. A **31** (1998) 2065–2074.

[367] J. Terning, *Modern Supersymmetry: Dynamics and Duality*, International Series of Monographs on Physics, Vol. 132 (Oxford University Press, 2005) 336 pages; "Non-perturbative supersymmetry", TASI-2002 Lectures, in: Particle Physics and Cosmology, eds. H. Haber, A. Nelson (World Scientific, River Edge, USA, 2004) pp. 343–443.

[368] V. N. Tolstoy, "Extremal projectors for quantized Kac-Moody superalgebras and some of their applications", in: Proceedings of Quantum Groups Workshop (Clausthal, 1989), eds. H. D. Doebner and J. D. Hennig, Lect. Notes in Physics, Vol. 370 (Springer-Verlag, 1990) pp. 118–125.

[369] A. Torrielli, "The Hopf superalgebra of AdS/CFT", J. Geom. Phys. **61** (2011) 230–236; "Yangians, S-matrices and AdS/CFT", J. Phys. A **44** (2011) 263001.

[370] P. K. Townsend, "P-brane democracy", in: "Particles, Strings and Cosmology", Proceedings of PASCOS/Hopkins 1995, eds. J. Bagger et al. (World Scientific, Singapore, 1996) pp. 271–285.

[371] Z. Tsuboi, "Analytic Bethe ansatz and functional equations associated with any simple root systems of the Lie superalgebra sl(r+1/s+1)", Physica A **252** (1998) 565–585.

[372] D. V. Uvarov, "Ambitwistors, oscillators and massless fields on AdS_5", Phys. Lett. B **762** (2016) 415–420.

[373] S. Valatka, "Exact results in supersymmetric gauge theories", arXiv:1501.00111 (2015), (PhD thesis).

[374] J. Van der Jeugt, "Character formulae for the Lie superalgebra C(n)", Commun. Algebra **19** (1991) 199–222.

[375] J. Van der Jeugt, J. W. B. Hughes, R. C. King, J. Thierry-Mieg, "A character formula for singly atypical modules of the Lie superalgebra sl(m/n)", Commun. Algebra **18** (1990) 3453–3480; "Character formulae for irreducible modules of the Lie superalgebras sl(m/n)", J. Math. Phys. **31** (1990) 2278–2304. J. Van der Jeugt, R. B. Zhang, "Characters and composition factor multiplicities for the Lie superalgebra gl(m/n)", Lett. Math. Phys. **47** (1999) 49–61.

[376] J. W. Van De Leur, "A classification of contragredient Lie superalgebras of finite growth", Commun. Algebra **17** (1989) 1815-1841.

[377] A. Vichi, "Improved bounds for CFT's with global symmetries", J. High Energy Phys. **1201** (2012) 162.

[378] M. A. Virasoro, "Subsidiary conditions and ghosts in dual-resonance models", Phys. Rev. D **1** (1970) 2933–2936.

[379] D. Volin, "String hypothesis for gl(n/m) spin chains: a particle/hole democracy", Lett. Math. Phys. **102** (2012) 1–29.

[380] D. V. Volkov, V. P. Akulov, "Is the neutrino a goldstone particle?", Phys. Lett. B **46** (1973) 109–110.

[381] A. A. Voronov, "Mappings of supermanifolds", Theor. Math. Phys. **60** (1984) 660–664; [Teor. Mat. Fiz. **60** (1984) 43–48].

[382] J. Wess, B. Zumino, "Supergauge transformations in four-dimensions", Nucl. Phys. B **70** (1974) 39–50; "A Lagrangian model invariant under supergauge transformations", Phys. Lett. B **49** (1974) 52–54. "Supergauge invariant extension of quantum electrodynamics", Nucl. Phys. B **78** (1974) 1–13.

[383] P. West, "The supersymmetric effective potential", Nucl. Phys. B **106** (1976) 219–227; "A review of non-renormalisation theorems in supersymmetric theories", Nucl. Phys. B. Proc. Suppl. **101** (2001) 112–128. A. H. Chamseddine, P. C. West, "Supergravity as a gauge theory of supersymmetry", Nucl. Phys. B **129** (1977) 39–44.

[384] E. Witten, “A supersymmetric form of the nonlinear sigma model in two-dimensions”, Phys. Rev. D **16** (1977) 2991–2994. E. Witten, D. I. Olive, “Supersymmetry algebras that include topological charges”, Phys. Lett. B **78** (1978) 97–101. R. Shankar, E. Witten, “The S matrix of the supersymmetric nonlinear sigma model”, Phys. Rev. D **17** (1978) 2134–2143.

[385] E. Witten, “Nonabelian bosonization in two-dimensions”, Commun. Math. Phys. **92** (1984) 455–472.

[386] E. Witten, “Anti-de Sitter space, thermal phase transition, and confinement in gauge theories”, Adv. Theor. Math. Phys. **2** (1998) 505–532.

[387] E. Witten, “Gauge theories, vertex models and quantum groups”, Nucl. Phys. B **330** (1990) 285–346.

[388] S. Woronowicz, “Compact matrix pseudogroups”, Commun. Math. Phys. **111** (1987) 613–665.

[389] H. Yamane, “Universal R-matrices for quantum groups associated to simple Lie superalgebras”, Proc. Jpn. Acad., Ser. A, Math. Sci. **67** (1991) 108–112; “Quantized enveloping algebras associated with simple Lie superalgebras and their universal R-matrices”, Publ. Res. Inst. Math. Sci. **30** (1994) 15–87.

[390] H. Yamane, “On defining relations of affine Lie superalgebras and affine quantized universal enveloping superalgebras”, Publ. Res. Inst. Math. Sci. **35** (1999) 321–390 and Errata, ibid. **37** (2001) 615–619.

[391] M. Yamazaki, “Comments on determinant formulas for general CFTs”, J. High Energy Phys. **1610** (2016) 035.

[392] K. Yonekura, “Notes on operator equations of supercurrent multiplets and anomaly puzzle in supersymmetric field theories”, J. High Energy Phys. **1009** (2010) 049; “Supersymmetric gauge theory, (2,0) theory and twisted 5d Super-Yang-Mills”, J. High Energy Phys. **1401** (2014) 142.

[393] K. Zarembo, “Algebraic curves for integrable string backgrounds”, talk at “Gauge Fields. Yesterday, today, tomorrow”, Moscow, 19–24.01.2010, arXiv:1005.1342 [hep-th].

[394] R. B. Zhang, “The quantum super-Yangian and Casimir operators of $U_q(gl(M/N))$”, Lett. Math. Phys. **33** (1995) 263–272; “Quantum enveloping superalgebras and link invariants”, J. Math. Phys. **43** (2002) 2029–2048.

[395] S. Zheng, “A note on bounds of scalar operators in perturbative SCFTs”, Nucl. Phys. B **870** (2013) 78–91.

[396] B. Zumino, “Supersymmetry and Kahler manifolds”, Phys. Lett. B **87** (1979) 203–206.

Author Index

Subject Index

De Gruyter Studies in Mathematics

Volume 48
Jared Maruskin
Dynamical Systems and Geometric Mechanics: An Introduction, 2018
ISBN 978-3-11-059729-5, e-ISBN (PDF) 978-3-11-059780-6,
e-ISBN (EPUB) 978-3-11-059803-2

Volume 47
Eugene Stefanovich
Elementary Particle Theory: Volume 3: Relativistic Quantum Dynamics, 2018
ISBN 978-3-11-049090-9, e-ISBN (PDF) 978-3-11-049322-1,
e-ISBN (EPUB) 978-3-11-049139-5

Volumc 46
Eugene Stefanovich
Elementary Particle Theory: Volume 2: Quantum Electrodynamics, 2018
ISBN 978-3-11-049089-3, e-ISBN (PDF) 978-3-11-049320-7,
e-ISBN (EPUB) 978-3-11-049143-2

Volume 45
Eugene Stefanovich
Elementary Particle Theory: Volume 1: Quantum Mechanics, 2018
ISBN 978-3-11-049088-6, e-ISBN (PDF) 978-3-11-049213-2,
e-ISBN (EPUB) 978-3-11-049103-6

Volume 44
Vladimir V. Kiselev
Collective Effects in Condensed Matter Physics, 2018
ISBN 978-3-11-058509-4, e-ISBN (PDF) 978-3-11-058618-3,
e-ISBN (EPUB) 978-3-11-058513-1

Volume 43
Robert F. Snider
Irreducible Cartesian Tensors, 2017
ISBN 978-3-11-056363-4, e-ISBN (PDF) 978-3-11-056486-0,
e-ISBN (EPUB) 978-3-11-056373-3

Volume 42
Javier Roa
Regularization in Orbital Mechanics: Theory and Practice, 2017
ISBN 978-3-11-055855-5, e-ISBN (PDF) 978-3-11-055912-5,
e-ISBN (EPUB) 978-3-11-055862-3

Bei Fragen zur Produktsicherheit wenden Sie sich bitte an:
If you have any questions regarding product safety,
please contact:

Walter de Gruyter GmbH
Genthiner Straße 13
10785 Berlin
productsafety@degruyterbrill.com